FOUNDATIONS OF EXPERIMENTAL DATA ANALYSIS

FOUNDATIONS OF EXPERIMENTAL DATA ANALYSIS

LUDMILA KUBÁČKOVÁ

GEOPHYSICAL INSTITUTE
OF THE SLOVAK ACADEMY
OF SCIENCES
BRATISLAVA
CZECHOSLOVAKIA

CRC

CRC PRESS
Boca Raton • Ann Arbor • London • Tokyo

Co-published by

Ister Science Press, Ltd. Bratislava
and
CRC Press, Inc., Boca Raton, Florida

Sole distributors for Czechoslovakia and Poland

Ister Science Press, Ltd.
Staromestská 6
811 06 Bratislava
Czechoslovakia

For the U.S.A. and Canada

CRC Press, Inc.
2000 Corporate Blvd., N.W.
Boca Raton, FL. 33431
U.S.A

CIP Data available from the Library of Congress

ISBN 0-8493-7736-6 (CRC Press)
ISBN 80-900486-6-8 (Ister Science Press)

Printed in Czechoslovakia

CONTENTS

Introduction . . . 1

LIST OF SYMBOLS . . . 7

1 PRELIMINARY DEFINITIONS FROM PROBABILITY THEORY AND MATHEMATICAL STATISTICS . . . 17

1.1 Random events, axiomatic definitions of probability, conditional probability . . . 17

1.2 The random variable . . . 25

1.2.1 The random variable, its distribution function . . . 25

1.2.2 Random vectors . . . 29

1.2.3 A marginal probability distribution . . . 35

1.2.4 Conditional probability distributions . . . 38

1.2.5 Independent random variables . . . 41

1.3 Parameters of the probability distribution . . . 42

1.3.1 One-dimensional random variable . . . 42

1.3.2 Random vectors . . . 49

1.3.3 Theoretical regression . . . 62

1.3.4 Theoretical correlation . . . 70

1.4 Characteristic functions . . . 73

1.5 Some probability distributions . . . 78

1.6 Random samples, the empirical distribution function, sample characteristics, their probability distributions . . . 91

1.6.1 Random samples, the empirical distribution function, fundamental theorems of mathematical statistics . . . 91

1.6.2 Sample random variables, empirical moments . . . 97

1.6.3 Probability distribution of some sample random variables . 102

2 ESTIMATION THEORY . . . 107

2.1 The aim of data processing; a theoretical (deterministic), stochastic and statistical model of the object being investigated . . . 107

2.2 Basic linear measurement models, linear estimators . . . 114

2.3 Method of least squares, theorem on equivalence of estimators . . 128

2.4 Linear unbiased and jointly efficient estimators of unknown parameters in basic linear models of measurement 129
2.5 Mixed linear models 145
2.5.1 A linear unbiased and jointly efficient estimator of the vector $(\Theta', \vartheta')'$ 148
2.5.2 Linear regression in estimation theory 163
2.6 The optimum elimination of systematic influences 174
2.7 Estimators of the unit dispersion 190
2.7.1 Estimators of the unit dispersion in the simple models of measurement 190
2.7.2 Estimators of the unit dispersion in mixed linear models . . 195
2.7.3 Estimators of the unit dispersion from data affected by a systematic deterministic influence 199
2.8 Estimators of the parameters of the covariance matrix $var(\zeta) = \sum_{i=1}^{p} \kappa_i V_i$; the estimation of an unknown covariance matrix of the observation vector in a replicated model of measurement 199
2.9 Normality of the observation vector 206
2.9.1 Properties of estimators of the first and second order parameters in the case of a normal observation vector . . . 206
2.9.2 The confidence regions of the first and second order parameters in the case of a normal observation vector . . . 213
2.9.3 Estimated covariance matrix of a normal observation vector in a replicated model, and its influence on the first order parameter estimators 226
2.10 The universal model 234
2.11 Correctness checks in computing the first and second order parameters 246
2.11.1 Correctness checks of the estimates of the first order parameters 246
2.11.2 Numerical correctness checks of the unit dispersion estimate 259
2.11.3 Numerical correctness checks of dispersion in the estimator of the function $p'\Delta\Theta$ and in the covariance matrix of the estimator of the vector function $(p_1, p_2, \ldots, p_n)\Delta\Theta$. 267
3 ANALYSIS OF A SET OF MEASURED DATA: HYPOTHESES TESTING 274
3.1 Preliminary definitions 274
3.1.1 Statistical hypotheses, statistical tests 274
3.1.2 The critical region for tests of the hypothesis H_0 against H_1 at a given significance level (the risk) 276
3.1.3 The power function of a test 285
3.1.4 The Neyman – Pearson lemma for a simple null and a simple alternative hypothesis 285

3.1.5 Uniformly most powerful tests of a simple hypothesis H_0 against a composite hypothesis H_1 290
3.2 Tests verifying the linear hypothesis on the first order parameters within linear models of estimation theory 292
3.3 Tests on stochastic relations 302
3.4 Tests on goodness of fit 312
3.4.1 The Pearson test of goodness of fit (known parameters of the probability distribution) 313
3.4.2 The Pearson test of goodness of fit (unknown parameters of the probability distribution) 315
3.4.3 The Pearson test of goodness of fit for verifying the normality with unknown m and σ^2 317
3.4.4 The Kolmogorov test of goodness of fit 319
3.4.5 The Rényi test of goodness of fit 321
3.5 Tests of normality 323
3.5.1 Empirical coefficient of asymmetry and empirical excess as test statistics for verifying normality 323
3.5.2 The Shapiro – Wilk test of normality 325
3.5.3 The D'Agostini test of normality 326
3.6 Tests of homogeneity 327
3.6.1 Tests for verifying the same probability distribution 327
3.6.1.1 The Smirnov test 327
3.6.1.2 The Wang test 331
3.6.1.3 The Wilcoxon test 332
3.6.1.4 The Kruskal – Wallis test of homogeneity 335
3.6.2 Tests verifying the identity of dispersions 336
3.6.2.1 The Fisher test 336
3.6.2.2 The Barlett test 338
3.6.2.3 The Cochran test 339
3.6.3 Tests verifying the identity of mean values 341
3.6.3.1 The Student's test 341
3.6.3.2 The Fisher – Behrens problem 342
3.7 Test for judging the significance of extremely deviated data 344
Tables 351
Conclusion 358
References 362
Subject index 365

INTRODUCTION

Probability theory and mathematical statistics have a unique position in helping us to learn about the surrounding world. Probabilistic or statistical disciplines interact with specific natural sciences and engineering by providing theoretical models to compare with the concrete objects being studied. A comparison between a theory on the objects and their reality is usually carried out within a suitably designed experiment which aims either at determining how well the theory reflects reality or at obtaining phenomenal data on the measured objects. The theoretical model expresses the relation between parameters which adequately characterize the measured objects and the directly observable parameters. Every theoretical model is intended to represent a more or less perfect approximation of the reality being investigated. The degree to which the approximation succeeds depends on the level of knowledge maintained by the theory about the measured objects. The specialist in a specific branch of natural science or engineering is indispensable in the development of the theoretical model that is consistent with the real facts discovered by research; the expert cannot be replaced by even the most perfect statistical apparatus used in processing the experimental data. However, statistical methods are able to disclose any discrepancy between the theoretical model used and reality.

Each unsuitable experimental design or badly founded heuristic procedure in data processing reduces the value of the effort and wastes the financial investment spent to perform the experiment. This, together with the fact that certain experiments are unique and nonreplicable, or that some tests are carried out on live organisms, are reasons why we develop and apply methods of data processing which enable us to obtain in the best possible way all the information we can on the objects involved in the research.

All this implies that mathematical statistics provides an experiment with direct support in three areas: optimum design, the testing of hypotheses, and estimation theory. This publication deals with the last two problem domains. As far as the first discipline is concerned, we can recommend the monograph by A. Pázman: Foundations of Optimum Experimental Design. D. Reidel, Dortrecht, Boston, Lancaster, Tokyo (1986). [26].

It is absolutely impossible within the framework of one publication to answer all questions connected with solving a given problem within a certain scientific discipline, the more so because the research relating to concrete objects reveals new problems many of which require solutions that have not yet been formulated by mathematical statistics. Meanwhile, mathematical statistics, which is a very diversified discipline, is undergoing very intensive development. New knowledge is being required that is of great importance from the viewpoint of application; however, accounts of this knowledge are often published in mathematical language not easily undersood by nonmathematicians. Therefore, a subsidiary aim of this publication is to acquaint the reader with the rigorous meaning of the basic mathematical notions and with the terminology used in mathematical statistics; this will simultaneously make it easier for the specialists in mathematical statistics and the users of its methods to understand one another, especially when formulating and solving complicated problems.

The first chapter presents a survey of the basic probabilistic notions necessary for stochastic modelling; this modelling, based on a theoretical (deterministic) model of the object under investigation, takes into account the fact that each measurement results in a real value, a realization of the random variable which models the measurement.

The most important task of the first chapter is to show the relationship between probability theory and mathematical statistics. This relationship has been validly characterized by Marek Fisz, as follows: "Hence the significance of probability theory for statistical problems in which we investigate certain regularities of random phenomena by observing their frequency is the same as the significance of geometry for geodesy" (Probability Theory and Mathematical Statistics. J.Wiley & Sons, Inc., New York, London (1963), p. 335).

Probability theory provides a foundation for the stochastic modelling of situations and processes which do not yield identical outcomes within an experiment. In this sense, the measurement process is analogous to the toss of a die, which can result in one of the numbers $1, 2, 3, 4, 5, 6$. This simple example demonstrates the essence of the problem.

The basic notion of probability theory assumes the set of all elementary events. In our case the set of all elementary events contains six elements, $\omega_1, \omega_2, \omega_3, \omega_4, \omega_5$ and ω_6; each element denotes an event consisting of casting for the number that equals the index subscripted. On this set of elementary events the σ-algebra is formed, i.e., a class of random events which are to be considered in the concrete case. A mapping of the set of all elementary events along the real line defines the random variable; the mapping used has to preserve the structure of the σ-algebra generated on the set of all elementary events (this structure reflects the way the random events are produced from the elementary events). The mapping mentioned assigns to each random variable modelling a concrete stochastic phenomenon a set of all of its possible realizations, the set of images of the elementary events. In our case of the mapping $\Omega \to \mathcal{R}^1 : \xi(\omega_i) = i,\ i = 1, 2, \ldots, 6$, the set of all possible realizations contains the elements 1 to 6. Furthermore, if individual random events are assigned the probability of their occurrence, then the same probability is assigned to their corresponding images obtained by the mapping $\Omega \to \mathcal{R}^1$ (the probability of inverse fibres and the corresponding images is the same). This is how the stochastic model of the toss of a die is derived. So far, we considered the toss of a die. By performing several tosses we participate in an experiment and thus engage mathematical statistics.

Let us now treat the basic notion of a random sample. If the die is thrown 1800 times (the Meissner experiment), we produce a 1800-dimensional random sample $(\xi_1, \xi_2, \ldots, \xi_{1800})'$ (in the case of a measurement this random sample is called the observation vector). Each of the sample's components can be realized by any value from the set of all realizations of the modelling random variable, in our case by any one of the numbers $1, 2, 3, 4, 5$, or 6. The 1800 outcomes of tossing a die comprise full information on the die obtained by the experiment (and obviously on the conditions under which the experiment was carried out). The methods of mathematical statistics enable us to determine the empirical (sample) characteristics describing the measured object, e.g., the empirical frequency, the empirical moments, the empirical quantiles, and to compare them, if necessary, with the corresponding theoretical values of the stochastic model (they can be determined without performing an experiment).

What situation does occur within a measurement? Consider a simple stochastic model whose randomness is attributable to the random measurement error; let it possess a normal probability distribution. When we set up a model of measurement under these conditions, we assume a concrete value for the mean value and the dispersion of the modelling random variable, i.e., concrete values for the two parameters that uniquely determine the probability density of the normal probability distribution of the error component in the measurement. The experiment, and specifically the measurement, serve to determine their true values or to verify, subject to a given risk, whether the error component possesses the probability distribution assumed by the stochastic model. What does this mean? It means that when we start out to solve a statistical problem, we begin to deal with a whole class of normal probability distributions in which the mean value may be any real number and the dispersion any positive number.

The measurement produces a realization of a finite-dimensional observation vector, each of whose components is a real number contained within a certain interval (theoretically within the interval $(-\infty, \infty)$) which represents the set of all realizations of the normal random variable modelling the measurement. Because mathematical statistics takes into account the whole class of probability distributions of the same type, which differ only according to the values of the parameters specifying them, certain notational problems arise. The reader can find in the literature symbols which emphasize this fact, e.g., $F_{\Theta,\sigma}(\cdot), f_{\Theta,\sigma}(\cdot), E_{\Theta,\sigma}(\cdot)$, where Θ and σ are the actual values of the parameters determining the concrete probability distribution, frequency, and mean value, respectively. For the sake of simplicity, this convention is not observed in the present publication; no misunderstanding can occur both in linear estimation theory and in hypothesis testing, where the null hypothesis and the alternative hypothesis precisely specify the models.

The essential chapters are Chapter 2 (Estimation theory) and Chapter 3 (The analysis of a set of measured data: hypothesis testing), based on Chapter 1 (Preliminary definitions from probability theory and mathematical statistics).

Chapter 2, dealing with the background of estimation theory, presents a survey of algorithms enabling us to obtain phenomenal data on the object being measured according to the fundamental six linear models of measurement; they describe all the known types of linear or linearized relations between directly observable parameters of the object under investigation and the unknown parameters characterizing it; the simple and the mixed versions of these models are discussed.

The mixed models, in contradistinction to the simple ones in which the error component is stochastic only, contain another random component connected with the measured object. As a typical example of such a measurement that is perturbed by another random component in addition to the error component, we can cite the measurement of settling large buildings, where the theoretical model gives a good picture of its settling in general, given certain assumptions concerning the structure of the subsoil and loading. The same theoretical model can apply to research in other localities with the same type of subsoil. The settling of an actual building is affected by the specific type of subsoil which, evidently, cannot be mapped in sufficient detail (if we permitted this as a possibility, we would obtain an unreasonably complicated and therefore inapplicable theoretical model). Analogous to the case of a die, where the randomness of the outcome is caused by a set of unknown factors affecting the toss, here, the unknown difference between the type of subsoil assumed in the model and the actual subsoil represents an unknown factor, which is the reason for the existence of an additional random component in the result of the measurement.

The need for algorithms to predict the useful component of the observation vector can be demonstrated by this example of the settling of large buildings.

From the viewpoint of the objectivity of our knowledge obtained by measuring, the opinion that the methods of optimum processing of measurement data become useless as a consequence of the increasing accuracy of measuring equipment, is very dangerous. Those who proclaim such views do not realize that there exist only few cases where the unknown parameter determined is directly observable. Moreover, they do not take into account the fact that the increasing accuracy of the measuring equipment requires the perturbing effects on the object to be fully detected (in the past perturbances were below the sensitivity threshold of the equipment used and therefore part of the measurement error). These effects are both deterministic and stochastic in nature. As far as the deterministic influences are concerned, two possibilities occur: Either the deterministic influence is so well investigated that it can be eliminated from the measurement findings in the form of a reduction known a priori (e.g., the tidal influence on gravimetric and levelling measurements) or this effect can be modelled, and the unknown parameters of the model can be estimated from the measurement results (the drift of the gravimeter used). The essential difference between these two procedures for eliminating the systematic effects is evident. In the former case the reduction is independent of the measured results, in the latter the unknown parameters of the systematic influence are determined from the measured data. It is necessary to pay great attention to eliminating the systematic influences affecting the measured results in order not to produce erroneous estimates of the first and second order parameters.

A natural, inseparable part of each calculation is to check the numerical stability and the correctness of the result obtained. The last part of Chapter 2 deals with this important problem that is often unreasonably neglected in the age of computers. The basic notions concerning the universal model of measurement have to be familiar before deriving the algorithms that are used in checking estimates of first and second order parameters. The fact that the universal model gives up its conditions of regularity has a useful consequence in the fact that it is not necessary to verify a priori whether the design matrix possesses a full rank in its columns and whether the covariance matrix of the observation vector is positive definite (from the numerical point of view this is a difficult problem especially in the case of large-dimensional matrices); on the other hand, within any universal model a problem arises concerning the estimability of the unknown vector parameter or a linear function of its components; from the numerical point of view some knowledge of matrix calculus with generalized matrix inverses is necessary here. Consequently, the last part of the chapter presents an introduction to the study of the universal model.

The basic principle applied in Chapter 2 in connection with estimating first order parameters is the method of least squares in its general form. The estimators based on it are unbiased linear and jointly efficient. The estimators of the second order parameters are unbiased and invariant with respect to the first order parameters and, if the observation vector possesses a normal probability distribution, even efficient.

Obviously, there exist other probability criteria for deriving optimum estimators of the unknown parameters. They are not dealt with in this publication.

The third chapter deals with the analysis of a set of measured data. From the logical point of view this chapter would precede the second chapter, as the analysis of the set of measured data represents the stage preceding the stage of their processing. The explanation of the procedures in the third chapter assumes some knowledge of the notions and results of estimation theory and for this reason this chapter follows the chapter concerning optimum estimation.

From the point of view of estimation theory the objectified approach to evaluate more or less deviated data and to verify the normality of the set which includes them is of great importance. If the hypothesis on the normality of the observation vector is not rejected, the possibility arises that we may obtain a greater amount of information on the estimators of the first order parameters, the residual vector, and the estimator of the unit dispersion than in the case where this hypothesis is rejected with the a priori given risk.

At the end of the Chapter 3 we introduce tables of those critical values of the test statistics that are not commonly accessible.

LIST OF SYMBOLS

ω	an elementary event
$\{\omega_1, \omega_2, \ldots, \omega_n\}$	a set consisting of elementary events $\omega_1, \omega_2, \ldots, \omega_n$
Ω	the set of elementary events
$A_i,\ i = 1, 2, \ldots$	sets, random events
A^c	a complement of the set A
$\{A_i\}_{i=1}^{\infty}$	a countable sequence of sets, random events
$A_1 \cup A_2$	the union of sets A_1 and A_2
$\bigcup_{i=1}^{\infty} A_i$	a countable union of sets $A_1, A_2, \ldots$
$\emptyset$	an empty set
$\mathcal{S}$	a probability field
$A_1 - A_2$	the difference of the sets A_1 and A_2
$A_1 \cap A_2$	the intersection of the sets A_1 and A_2
$\bigcap_{i=1}^{\infty} A_i$	a countable intersection of the sets $A_1, A_2, \ldots$
$P(\cdot),\ A \in \mathcal{S}$	a probability measure defined on $\mathcal{S}$
$P(A)$	the probability of the random event A
$(\Omega, \mathcal{S}, P)$	a probability space
$\mathcal{R}^i,\ i = 1, 2, \ldots, n$	an i-dimensional vector space
$P(A\|B)$	the probability of a random event A under the condition that the random event B has occurred (the conditional probability)

$f : D \to \mathcal{R}^i$	a function (a mapping) whose definition domain is D and the range $\mathcal{R}^i$
$\xi : \Omega \to \mathcal{R}^1$	a random variable (one-dimensional)
$\{\omega : \xi(\omega) < x\}$	the set of ω satisfying the condition $\xi(\omega) < x$
$F : \mathcal{R}^1 \to \mathcal{R}^1$, $F(\cdot)$, $x \in \mathcal{R}^1$, $F(x) = P(\{\omega : \xi(\omega) < x\})$	the distribution function of a random variable ξ
$p_i = P(\{\omega : \xi(\omega) = x_i\})$, $i = 1, 2, \ldots$	the probability function of a discrete random variable ξ
$f(\cdot)$, $x \in \mathcal{R}^1$	a probability density of a continuous random variable ξ
x', A'	the transposition of a vector x, a matrix A
$\xi : \Omega \to \mathcal{R}^n$	an n-dimensional random vector
$F : \mathcal{R}^n \to \mathcal{R}^1$, $F(\cdot)$, $x \in \mathcal{R}^n$	the distribution function of an n-dimensional random vector $\xi = (\xi_1, \xi_2, \ldots, \xi_n)'$
$p_{1k_1,1k_2,\ldots,nk_n} = P(\{\omega : \xi_1(\omega) = x_{1k_1}, \xi_2(\omega) = x_{2k_2}, \ldots, \xi(\omega) = x_{nk_n}\})$	a probability function of an n-dimensional discrete random variable $\xi = (\xi_1, \xi_2, \ldots, \xi_n)'$
$p_{1\cdot,2k_2}$, $p_{1k_1,2\cdot}$	marginal probabilities of a two-dimensional random vector $\xi = (\xi_1, \xi_2)'$
$f_1(\cdot)$, $x_1 \in \mathcal{R}^1$, $f_2(\cdot)$, $x_2 \in \mathcal{R}^1$	the marginal probability densities of a two-dimensional continuous random variable $\xi = (\xi_1, \xi_2)'$
$F_1(\cdot)$, $x_1 \in \mathcal{R}^1$, $F_2(\cdot)$, $x_2 \in \mathcal{R}^1$	the marginal probability distributions of a two-dimensional random variable $\xi = (\xi_1, \xi_2)'$
$F_1(\cdot\|x_2)$, $x_1 \in \mathcal{R}^1$, $F_2(\cdot\|x_1)$, $x_2 \in \mathcal{R}^1$	the conditional distribution functions of a two-dimensional random variable $\xi = (\xi_1, \xi_2)'$
$f_1(\cdot\|x_2)$, $x_1 \in \mathcal{R}^1$, $f_2(\cdot\|x_1)$, $x_2 \in \mathcal{R}^1$	the conditional probability densities of a two-dimensional continuous random variable $\xi = (\xi_1, \xi_2)'$

$E[\xi]$	the mean value of a random variable ξ
$D[\xi]$	the dispersion of a random variable ξ
γ_1	the asymmetry of a random variable
γ_2	the excess of a random variable
$F^{-1}(\cdot),\ \alpha \in (0,1)$	a quantile function of a random variable
$F^{-1}(\alpha)$	the α-quantile of a random variable
$F^{-1}(1/2) = x_{1/2}$	the median of a random variable with the distribution function $F(\cdot),\ x \in \mathcal{R}^1$
x_D	the mode of a random variable
$\xi \sim N(m, \sigma^2)$	the random variable ξ possesses the normal (Gauss) probability distribution with the mean value m and the dispersion σ^2
$F_{0,1}^{-1}(\alpha)$	the α-quantile of a random variable $\xi \sim N(0,1)$
$E[\xi]$	the mean value of the random vector $\xi = (\xi_1, \xi_2, \ldots, \xi_n)'$
$var(\xi)$	the covariance (variance) matrix of a random vector $\xi = (\xi_1, \xi_2, \ldots, \xi_n)'$
$cov(\xi_i, \xi_j)$	the covariance between the random variables ξ_i and ξ_j
$\rho_{i,j} = cor(\xi_i, \xi_j)$	the correlation coefficient between the random variables ξ_i and ξ_j
$E[\xi_1 \vert x_2],\ E[\xi_2 \vert x_1]$	conditional mean values
$\eta_{opt}(\xi)$	the optimum approximation of the random variable η by a random variable ξ
$\eta_{lopt}(\xi)$	the linear optimum approximation of the random variable η by a random variable ξ
$\rho_{\eta,\xi} = cor(\eta, \eta_{lopt}(\xi))$	the coefficient of a multiple correlation between a random variable η and a random vector ξ

$cor(\xi)$	the matrix of correlation coefficients between the components of a random vector ξ
$\rho_{\eta,\zeta\cdot\xi}$	the partial correlation coefficient between random variables η and ζ under the eliminated influence of the random variable ξ
$\varphi(t) = E[\exp(it\xi)],\ t \in \mathcal{R}^1$	the characteristic function of a random variable ξ
i	the imaginary unit
$\varphi(t) = E[\exp(it'\xi)],\ t \in \mathcal{R}^n$	the characteristic function of an n-dimensional random vector ξ
$\xi \sim N_n(m, \Sigma)$	the n-dimensional random vector ξ possesses the normal probability distribution characterized by the mean value m and the covariance matrix Σ
$n(x; m, \sigma^2)$	the probability density of a random variable $\xi \sim N(m, \sigma^2)$
$n(x; m, \Sigma)$	the probability density of an n-dimensional random vector $\xi \sim N_n(m, \Sigma)$
$\eta \sim \chi^2(n)$	a random variable η possesses the central chi-square probability distribution with n degrees of freedom
$\Gamma(\cdot),\ y > 0$	the gamma function
$\eta \sim \sigma^2\chi^2(n)$	the random variable η possesses the central σ^2chi-square probability distribution with n degrees of freedom
$\eta \sim \sigma^2\chi^2(n, \delta)$	the random variable η possesses the non-central σ^2chi-square probability distribution with n degrees of freedom and with the parameter of non-centrality δ
$\Xi \sim W_n(N, \Sigma)$	the random matrix Ξ of the type $n \times n$ has the Wishart probability distribution with N degrees of freedom
$x_1, x_2, \ldots, x_n, \ldots$	realizations of a random variable ξ

$S_n(\cdot),\ x \in \mathcal{R}^1$	a sample (empirical) probability distribution of a random variable ξ obtained from a random sample of size n
$\hat{\sigma}^2,\ \hat{\gamma}_1, \ldots$	an empirical (sample) dispersion, asymmetry, ...
$X = (X_1, X_2, \ldots, X_p)'$	the vector of the theoretical values of directly observable quantities of the measured object
$\Theta = (\Theta_1, \Theta_2, \ldots, \Theta_k)'$	the vector of the theoretical values of parameters of the measured object being determined
$f(\Theta) = (f_1(\Theta), f_2(\Theta), \ldots, f_p(\Theta))'$	a vector function expressing the relation between the theoretical values of the directly observable vector X and the vector Θ of the unknown parameters
$g(\Theta) = (g_1(\Theta), g_2(\Theta), \ldots, g_q(\Theta))'$	a vector function expressing the relation between the values $\Theta_1,\ \Theta_2, \ldots, \Theta_k$ of the unknown parameters
$\xi = (\xi_1, \xi_2, \ldots, \xi_n)'$	an observation vector
$\varepsilon = (\varepsilon_1, \varepsilon_2, \ldots, \varepsilon_n)'$	a vector of random measurement errors
$E[\xi]$	the mean value of the observation vector $\xi = (\xi_1, \xi_2, \ldots, \xi_n)'$
$var(\xi) = \Sigma$	the variance of the observation vector $\xi = (\xi_1, \xi_2, \ldots, \xi_n)'$
$i_{(n_i)}$	an n_i-dimensional vector consisting of unit elements
$0_{(n_i)}$	an n_i-dimensional zero vector
J	a $p \times n$-dimensional matrix reflecting the design of the experiment
$\hat{\Theta}(\xi),\ \tilde{\Theta}(\xi),\ \hat{\hat{\Theta}}(\xi)$	estimators of the vector parameter Θ
$\hat{\Theta}(x),\ \tilde{\Theta}(x),\ \hat{\hat{\Theta}}(x)$	estimates of the vector parameter Θ
$x_i = (x_{i1}, x_{i2}, \ldots, x_{in})'$	the ith realization of the observation vector $\xi = (\xi_1, \xi_2, \ldots, \xi_n)'$

$E[\hat{\Theta}(\xi)]$	the mean value of the estimator $\hat{\Theta}(\xi)$
$D[\hat{\Theta}(\xi)]$	the dispersion of a one-dimensional estimator
$var[\hat{\Theta}(\xi)]$	the variance of the vector estimator $\hat{\Theta}(\xi)$
$f_{lin}(\cdot)$, $g_{lin}(\cdot)$	the linearized form of the functions $f(\cdot)$ and $g(\cdot)$, respectively
$\partial f(\Theta_0)/\partial\Theta'$	the partial derivative of the vector function $f(\cdot)$ with respect to the row vector Θ' at the point Θ_0
$\Theta_0 = (\Theta_{10}, \Theta_{20}, \ldots, \Theta_{k0})'$	the k-dimensional vector of the approximate values of the unknown parameter Θ
$\Delta\Theta = \Theta - \Theta_0$	
$\mathcal{T} \subset \mathcal{R}^k$	the definition domain of the unknown vector parameter Θ
$A_{(p,k)} = \partial f_{(p)}(\Theta_{0(k)})/\partial\Theta'_{(k)}$	
JA	the $n \times k$-dimensional design matrix
$b = g(\Theta_0)$	the q-dimensional vector in a linearized system of conditions $b + B\Delta\Theta = 0$
$B = \partial g(\Theta_0)/\partial\Theta'$	the $q \times k$-dimensional matrix in a linearized system of conditions $b + B\Delta\Theta = 0$
Θ_1	the p-dimensional subvector of the k-dimensional vector of unknown parameters in Model 4 the l-dimensional subvector of the k-dimensional vector of unknown parameters in Model 6
Θ_2	the $k - p$-dimensional subvector of the k-dimensional vector of unknown parameters in Model 4 the $k - l$-dimensional subvector of the k-dimensional vector of unknown parameters in Model 6

C	the $q \times (k-p)$ and $q \times (k-l)$-dimensional matrix in the system of conditions $b+B\Delta\Theta+C\Delta\Theta_2=0$ in Model 4 and Model 6, respectively
$\eta = \xi - Jf(\Theta_0)$	a residual observation vector
$y = x - Jf(\Theta_0)$	a realization of the residual vector η
σ^2	a unit dispersion
H	a known positive definite matrix in the covariance matrix of the observation vector
P	a matrix of weights
λ	a scalar or vector Lagrange undetermined multiplier
φ	a random vector component in a mixed linear model
ζ	an observation vector in a mixed linear model
$s(\cdot), \vartheta \in \mathcal{R}^s$	an n-dimensional vector function of the mean value of the random vector φ in a mixed linear model
ϑ	an s-dimensional unknown parameter of the mean value of the random vector φ in a mixed linear model
$\varphi_0 = \varphi - s(\vartheta_0)$, $\vartheta = \vartheta_0 + \Delta\vartheta$	
$S = \partial s(\vartheta_0)/\partial\vartheta'$	
(JA, S)	an $n \times (k+s)$-dimensional design matrix in a mixed linear model
$\nu = \zeta - Jf(\Theta_0) - s(\vartheta_0)$	a residual observation vector in a mixed linear model
$\begin{pmatrix} \Theta \\ \vartheta \end{pmatrix} = \begin{pmatrix} \Theta_0 \\ \vartheta_0 \end{pmatrix} + \begin{pmatrix} \Delta\Theta \\ \Delta\vartheta \end{pmatrix}$	a $(k+s)$-dimensional vector of unknown parameters in a mixed linear model; Θ_0 and ϑ_0 are the approximate values of the vector parameters Θ and ϑ, respectively

$\begin{pmatrix} \widehat{\Delta\Theta}(\nu) \\ \widehat{\Delta\vartheta}(\nu) \end{pmatrix}$	an estimator of the hypervector of the unknown parameters in a mixed linear model
ζ_u	the useful part of the observation vector ζ (the useful signal)
ν_u	the useful part of the residual observation vector ν (the useful residual signal)
$\zeta_{up},\ \nu_{up}$	a predicted useful signal
$\hat{\zeta}_{up}(\zeta)$	the useful signal ζ_u predicted on the basis of the observation vector ζ
$\hat{E}[\nu]$	an estimator of the mean value of the random variable ν
$\hat{D}[\nu]$	an estimator of the dispersion of the random variable ν
$\nu^{(el)}$	the random vector ν after eliminating systematic influences
$\hat{\nu}^{(el)}(\nu)$	an optimum linear approximation of the random vector $\nu^{(el)}$ on the basis of the observation vector ν
$E,\ E_0$	elimination matrices, matrices eliminating systematic influences
$A^-_{m(H)}$	a minimum H seminorm generalized inverse (g-inverse) of the matrix A; H a positive definite matrix
v	an n-dimensional vector of residuals
Ψ_v	an $n^2 \times n^2$-dimensional matrix of the central fourth moments of the n-dimensinal random vector v
$A \otimes B$	the Kronecker product of matrices A and B
$Tr(V)$	the trace of a square matrix V
$\kappa_i,\ i = 1, 2, \ldots, p$	second order parameters ($var(\xi) = \sum_{i=1}^{p} \kappa_i V_i$, V_i, $i = 1, 2, \ldots, p$, known symmetric matrices)

$\sigma_{i,j}$	the (i,j)th element of a covariance matrix Σ
$e_i \in \mathcal{R}^n$	the ith unit vector of the n-dimensional linear space $\mathcal{R}^n$
$vec(A_{(n,m)})$	an nm-dimensional column vector consisting of ordered columns of the matrix A
$M * N$	the Hadamard multiplication of matrices M and N of the same type
$Diag(M)$	a square diagonal matrix of the type $n \times n$ whose diagonal elements equal the diagonal elements of the $n \times n$-dimensional matrix M
$diag(M)$	an n-dimensional vector whose elements equal the diagonal elements of the $n \times n$-dimensional matrix M
$1-\alpha,\ \alpha \in (0,1)$	a confidence level
$[\hat{\Theta}(\xi) - \Delta_{1\alpha}, \hat{\Theta}(\xi) + \Delta_{2\alpha}]$	a confidence interval on the confidence level $1-\alpha$
$t_{S_n}(1-\alpha/2)$	the $(1-\alpha)$-quantile of the Student random variable with n degrees of freedom
$\chi^2_n(1-\alpha)$	the $(1-\alpha)$-quantile of a random variable possessing a central chi-square probability distribution with n degrees of freedom
$F_{n,m}(1-\alpha)$	the $(1-\alpha)$-quantile of the Fisher – Snedecor probability distribution with n and m degrees of freedom
$\mathcal{E}$	a confidence ellipsoid
$\hat{\Sigma}(\eta)$	an estimator of the covariance matrix of the observation vector in a replicated model
A^-	a generalized inverse (g-inverse) of the matrix A
$\mathcal{A}^-$	a class of all g-inverses of the matrix A
$\mathcal{A}^-_{m(N)}$	a class of all minimum N seminorm g-inverses of the matrix A

$\widehat{\Delta\Theta}_N(\eta)$	an $N-LS$ estimator of the vector parameter $\Delta\Theta$ in a universal model
C_1, C_2, C_3, C_4	blocks of the Pandora's box matrix
$\mathcal{F}$	a class of probability densities
H_0, H_1	a zero and an alternative hypothesis, respectively
$A,\ W$	an acceptance and critical domain of a test statistic, respectively
$\tau,\ t$	a test statistic, its realization
$\alpha,\ \alpha \in (0,1)$	the level of the test significance
$t_{crit}(\alpha)$	the critical value of the test statistic at the significance level α
$\beta(\cdot; W)$	a power function of a test

Remark: The order of symbols corresponds to the order of their occurrence in the text.

Chapter 1

PRELIMINARY DEFINITIONS FROM PROBABILITY THEORY AND MATHEMATICAL STATISTICS

1.1 Random events, the axiomatic definition of probability, conditional probability

Data processing input resulting from various experiments is presented as an n-tuple of numbers; there can be differring n-tuples of results from multiple replications of the experiment performed under the same conditions. In the simplest case the differences between them are caused by measurement errors which are an organic part of every measurement process. For this reason, every measurement yields a realization of a certain random variable which models the measurement procedure used.

Before giving a definition of a random variable, we shall explain this notion by a simple example.

Let us assume we have an ideal die, each of its sides designated by one of the numbers $1, 2, 3, 4, 5, 6$. The result of a throw, i.e., of an experiment carried out by tossing the die, "casts" one number from the set $\{1, 2, 3, 4, 5, 6\}$. Naturally, before performing the experiment we are not able to say which number will be cast. Thus "casting" a certain number is a random event.

Consider an urn with white marbles only; there exists only one possible result of an experiment consisting in drawing marbles out of the urn — only a white marble may be drawn. This is an example of a sure event.

Let us summarize what has been indicated by these examples.

The possible results of an experiment, which is characterized by a complex of conditions γ, are random events. A random event may, but need not, be realized under the existing system of conditions. The extreme cases are represented by an impossible and a sure event. The sure event under the given system of conditions must always occur, the impossible one never occurs (the system of conditions in the last example consists of the existence of an urn which contains white marbles only and in drawing out one of them.

It is necessary to realize that a random event does not contradict the law of causality. A random variable may be associated with known and, moreover, completely unknown sets of conditions that influence the experimental results.

The following conditions occur for a die:

1. Known: We have a die with sides designated by the numbers 1, 2, 3, 4, 5 and 6;

2. Unknown: It is not known whether the die has the form of an ideal cube, if it is homogeneous; nor do we know its starting position, the direction of the toss, its speed, rotation, the physical properties of the surface on which the die falls, etc.

The events which may occur immediately as the outcome of an experiment are called elementary events, and given the notation ω_1, $\omega_2, \ldots,$ ω_n.

There exist 6 elementary events in the case of the die: ω_i is an elementary event consisting in the occurrence of the number i, $i = 1, 2, \ldots, 6$. The set $\{\omega_1, \omega_2, \ldots, \omega_6\}$ of all elementary events is denoted by Ω.

But we are not always interested only in elementary events. We may, for example, in connection with a die investigate such events as

a) the occurrence of an even number — denotated A_1;

b) the occurrence of an odd number — A_2;

c) the occurrence of 1 or 3 — A_3;

d) the occurrence of a number less than 4 — A_4.

Using the given notation, these events can be recorded in the following way: $A_1 = \{\omega_2, \omega_4, \omega_6\}$, $A_2 = \{\omega_1, \omega_3, \omega_5\}$ $A_3 = \{\omega_1, \omega_3\}$, $A_4 = \{\omega_1, \omega_2, \omega_3\}$. Question: What is the meaning of $\Omega = \{\omega_1, \omega_2, \omega_3, \omega_4, \omega_5, \omega_6\}$? Answer: The set Ω contains all elementary events, it is a sure event. One of the numbers from the set $\{1, 2, 3, 4, 5, 6\}$ must occur as a result of a single toss. The events A_1, A_2, A_3, and A_4 are subsets of the set Ω .

Summarizing these examples we can state:

The basic notion is a non-empty set of all elementary events Ω. The subject of our study is a system $\mathcal{S}$ consisting of subsets of Ω such that the conditions

(i) $\Omega \in \mathcal{S}$,
(ii) if $A \in \mathcal{S}$, then $A^c \in \mathcal{S}$ ($A^c = \Omega - A$),
(iii) if $\{A_i\}_{i=1}^{\infty}$ is a countable sequence of elements from $\mathcal{S}$, then $\bigcup_{i=1}^{\infty} A_i \in \mathcal{S}$ ($\bigcup_{i=1}^{\infty} A_i$ is a set of those elements of Ω which are contained in at least one set from the sequence $\{A_i\}_{i=1}^{\infty}$)

are satisfied.

Definition A non-empty system $\mathcal{S}$ composed of subsets of the set Ω, satisfying the above mentioned three conditions, is called a $\sigma-$algebra; in probability theory this σ-algebra is called a probability field.

The three fundamental properties of a σ-algebra imply further properties:

(iv) $\emptyset \in \mathcal{S}$ (a direct consequence of the property (ii); $\emptyset$ denotes an empty set);

(v) if $A \in \mathcal{S}$ and $B \in \mathcal{S}$, then $A \bigcup B \in \mathcal{S}$ (a consequence of the property (iii));

(vi) if $A \in \mathcal{S}$ and $B \in \mathcal{S}$, then $A - B \in \mathcal{S}$ ($A - B = [(A \bigcup B)^c \bigcup B]^c$);

(vii) if $A \in \mathcal{S}$ and $B \in \mathcal{S}$, then $A \bigcap B \in \mathcal{S}$ ($A \bigcap B = [B^c \bigcup (B - A)]^c$).

Definition Elements of the set Ω are called elementary events. Each element of the probability field is a (random) event.

Example Let us construct a σ-algebra of events corresponding to the example with a die.

1. Let $\mathcal{S}$ contain all one-element subsets of the set $\Omega = \{\omega_1, \omega_2, \omega_3, \omega_4, \omega_5, \omega_6\}$ of elementary events. A σ-algebra $\mathcal{S}$ fulfilling this requirement contains the $2^6 = 64$ following sets:
a) an empty set $\emptyset$,
b) one-element sets $\{\omega_1\}$, $\{\omega_2\}$, $\{\omega_3\}$, $\{\omega_4\}$, $\{\omega_5\}$, $\{\omega_6\}$,
c) two-element sets $\{\omega_1, \omega_2\}$, $\{\omega_1, \omega_3\}$, $\{\omega_1, \omega_4\}$, $\{\omega_1, \omega_5\}$, $\{\omega_1, \omega_6\}$, $\{\omega_2, \omega_3\}$, $\{\omega_2, \omega_4\}$, $\{\omega_2, \omega_5\}$, $\{\omega_2, \omega_6\}$, $\{\omega_3, \omega_4\}$, $\{\omega_3, \omega_5\}$, $\{\omega_3, \omega_6\}$, $\{\omega_4, \omega_5\}$, $\{\omega_4, \omega_6\}$, $\{\omega_5, \omega_6\}$,
d) three-element sets $\{\omega_1, \omega_2, \omega_3\}$, $\{\omega_1, \omega_2, \omega_4\}$, $\{\omega_1, \omega_2, \omega_5\}$, $\{\omega_1, \omega_2, \omega_6\}$, $\{\omega_1, \omega_3, \omega_4\}$, $\{\omega_1, \omega_3, \omega_5\}$, $\{\omega_1, \omega_3, \omega_6\}$, $\{\omega_1, \omega_4, \omega_5\}$, $\{\omega_1, \omega_4, \omega_6\}$, $\{\omega_1, \omega_5, \omega_6\}$, $\{\omega_2, \omega_3, \omega_4\}$, $\{\omega_2, \omega_3, \omega_5\}$, $\{\omega_2, \omega_3, \omega_6\}$, $\{\omega_2, \omega_4, \omega_5\}$, $\{\omega_2, \omega_4, \omega_6\}$, $\{\omega_2, \omega_5, \omega_6\}$, $\{\omega_3, \omega_4, \omega_5\}$, $\{\omega_3, \omega_4, \omega_6\}$, $\{\omega_3, \omega_5, \omega_6\}$, $\{\omega_4, \omega_5, \omega_6\}$,
e) four-element sets $\{\omega_1, \omega_2, \omega_3, \omega_4\}$, $\{\omega_1, \omega_2, \omega_3, \omega_5\}$, $\{\omega_1, \omega_2, \omega_3, \omega_6\}$, $\{\omega_1, \omega_2, \omega_4, \omega_5\}$, $\{\omega_1, \omega_2, \omega_4, \omega_6\}$, $\{\omega_1, \omega_2, \omega_5, \omega_6\}$, $\{\omega_1, \omega_3, \omega_4, \omega_5\}$, $\{\omega_1, \omega_3, \omega_4, \omega_6\}$, $\{\omega_1, \omega_3, \omega_5, \omega_6\}$, $\{\omega_1, \omega_4, \omega_5, \omega_6\}$, $\{\omega_2, \omega_3, \omega_4, \omega_5\}$, $\{\omega_2, \omega_3, \omega_4, \omega_6\}$, $\{\omega_2, \omega_3, \omega_5, \omega_6\}$, $\{\omega_2, \omega_4, \omega_5, \omega_6\}$, $\{\omega_3, \omega_4, \omega_5, \omega_6\}$,
f) five-element sets $\{\omega_1, \omega_2, \omega_3, \omega_4, \omega_5\}$, $\{\omega_1, \omega_2, \omega_3, \omega_4, \omega_6\}$, $\{\omega_1, \omega_2, \omega_3, \omega_5, \omega_6\}$, $\{\omega_1, \omega_2, \omega_4, \omega_5, \omega_6\}$, $\{\omega_1, \omega_3, \omega_4, \omega_5, \omega_6\}$, $\{\omega_2, \omega_3, \omega_4, \omega_5, \omega_6\}$,
g) the set $\Omega = \{\omega_1, \omega_2, \omega_3, \omega_4, \omega_5, \omega_6\}$.

The reader can easily convince himself, that the system $\mathcal{S}$ of all subsets of the given non-empty set Ω satisfies the requirements i – iii, p. 18, thus it is a σ-algebra of events.

2. Other σ-algebras may be constructed over the set of elementary events $\Omega = \{\omega_1, \omega_2, \omega_3, \omega_4, \omega_5, \omega_6\}$. For example, $\mathcal{S}_1 = \{\emptyset, \{\omega_2, \omega_4\}, \{\omega_1, \omega_3, \omega_5, \omega_6\}, \Omega\}$, or $\mathcal{S}_2 = \{\emptyset, \{\omega_2\}, \{\omega_1, \omega_3\}, \{\omega_1, \omega_2, \omega_3\}, \{\omega_4, \omega_5, \omega_6\}, \{\omega_2, \omega_4, \omega_5, \omega_6\}, \{\omega_1, \omega_3, \omega_4, \omega_5, \omega_6\}, \Omega\}$, $\mathcal{S}_3 = \{\emptyset, \{\omega_1\}, \{\omega_6\}, \{\omega_2, \omega_3\}, \{\omega_4, \omega_5\}, \{\omega_1, \omega_6\}, \{\omega_1, \omega_2, \omega_3\}, \{\omega_1, \omega_4, \omega_5\}, \{\omega_2, \omega_3, \omega_6\}, \{\omega_4, \omega_5, \omega_6\}, \{\omega_2, \omega_3, \omega_4, \omega_5\}, \{\omega_1, \omega_2, \omega_3, \omega_6\}, \{\omega_1, \omega_4, \omega_5, \omega_6\}, \{\omega_1, \omega_2, \omega_3, \omega_4, \omega_5\}, \{\omega_2, \omega_3, \omega_4, \omega_5, \omega_6\}, \Omega\}$.

Definition Two random events are disjoint if they have no common elementary event.

Example The random events A_1 and A_2 and the events A_1 and A_3 are disjoint. The random events A_1 and A_4, the events A_2 and A_3, the events A_2 and A_4 and the events A_3 and A_4 are not disjoint.

Definition An event containing all elements of the set Ω of all elementary events is a sure event. An event containing no element of the set Ω, i.e. an empty set, is an impossible event. The set difference $\Omega - A = A^c$ is a complementary event with respect to the event A.

Example The random events A_1 and A_2 are complementary events.

We have shown that the random variable depends on chance - a system of unknown causes influencing the results of an experiment, in consequence of which it is not possible to determine a priori its concrete outcome.

Let us take into account a large series of experiments. Many outstanding authors have performed extensive series of experiments, e.g., Pearson tossed a coin 24,000 times with the following result: heads came up 12,010 times. Meissner performed 1,800 throws of a die, number 1 occurred 299 times, number 2: 295 times, number 3: 303 times, number 4: 307 times, number 5: 289 times and number 6: 307 times.

Let us denote by n the number of all experiments performed and by m the number of experiments whose outcome is considered to be favourable. The ratio m/n determines the relative frequency of the investigated event.

The above mentioned examples of large series present the following regularities. In the case of Pearson, the relative frequency of the event $\{\omega_1\}$ is about 1/2 and the relative frequency of the complementary event is approximately the same. In the case of the die, if it satisfies with sufficient accuracy the requirement of the ideal cube form and of homogeneity, the relative frequency of each event $\{\omega_i\}$, $i = 1, 2, \ldots, 6$ is about 1/6.

The relative frequency of single events are non-negative numbers from the interval [0,1] (the relative frequency of the impossible event is 0, the relative frequency of the sure event is 1).

Thus we may conclude that, in some cases, it is possible to evaluate numerically how often the investigated event within an experiment replicated under the same conditions occurs. We are speaking of a probability that a random event will happen. The probability of a random variable expresses the measure of its realization.

In this connection, the notion probability measure often appears (strictly speaking, probability is the numerical value of the probability measure of the given random variable).

Before formulating the axiomatic definition of the probability measure it has to be pointed out that the foundations of the axiomatization of probability theory were laid by A.N. Kolmogorov in 1933 in his work Grundbegriffe der Wahrscheinlichkeitsrechnung. Up to that time, probability theory had been developing separately and, to some extent, intuitively.

Definition The probability measure is a set function defined on the probability field $\mathcal{S}$; its properties are defined by the following axioms:

Axiom I Every random event A is assigned a certain number $P(A)$ — its probability — such that $0 \leq P(A) \leq 1$.

Axiom II The probability of a sure event equals 1, $P(\Omega) = 1$.

Axiom III The probability of a sum of a finite or countable number of disjoint random events equals the sum of probabilities of these events (so called additivity and σ-additivity of the probability measure), $P\left(\bigcup_n A_n\right) = \sum_n P(A_n)$.

Theorem The sum of the probability of a random event A and the probability of the complementary random event equals 1: $P(A) + P(A^c) = 1$.

Proof. The random events A and A^c are disjoint and simultaneously $A \bigcup A^c = \Omega$; thus $P(A \bigcup A^c) = P(\Omega)$; using Axioms III and II we get $P(A) + P(A^c) = 1$.

Theorem The probability of an impossible event equals 0: $P(\emptyset) = 0$.

Proof. The assertion follows immediately from Axioms III and II, because $P(\Omega \bigcup \emptyset) = P(\Omega) + P(\emptyset)$ and $P(\Omega) = 1$.

Definition A triplet $(\Omega, \mathcal{S}, P)$, where Ω is the set of elementary events, $\mathcal{S}$ the probability field over Ω, and P the probability measure defined on $\mathcal{S}$, is called a probability space.

Example In the case of a die, let us consider the probability field $\mathcal{S}$ in the form of the σ-algebra of all subsets of the set of the elementary events $\Omega = \{\omega_1, \omega_2, \omega_3, \omega_4, \omega_5, \omega_6\}$ and the probability measure on $\mathcal{S}$ defined by the relation $P(\{\omega_i\}) = 1/6,\ i = 1, 2, \ldots, 6.$. As a consequence of the three axioms of probability theory, each element from $\mathcal{S}$ is assigned its probability, e.g., $P(A_1) = P(\{\omega_2, \omega_4, \omega_6\}) = P(\{\omega_2\}) + P(\{\omega_4\}) + P(\{\omega_6\}) = 1/2,\ P(A_3) = P(\{\omega_1, \omega_3\}) = P(\{\omega_1\}) + P(\{\omega_3\}) = 1/3$, etc.

Example Let the distance between the points A and B be measured. Every measurement is influenced by random errors, the set of elementary events corresponding to the experiment, which consists of direct measurement of a length, is given by a certain interval $[a, b]$, which contains the actual value of the measured distance; theoretically $\Omega = (-\infty, \infty) = \mathcal{R}^1$. In this case the set $\Omega = \mathcal{R}^1$ of elementary events $\omega = x$ has the power of the continuum. The probability field $\mathcal{S}$ is generated by events $\{(-\infty, x),\ x \in \mathcal{R}^1\}$. The probability measure P is defined on the elements generating the σ-algebra $\mathcal{S}$; the probability measure determines for each $x \in \mathcal{R}^1$ the probability that the measured value will be smaller than x. By applying the fundamental set operations (unions, differences, intersections and complements with respect to Ω) and the basic three axioms of the probability theory, we can specify the probabilities of random events that consist of the measurement results given in intervals of the form $(x, \infty),\ [x, \infty),\ (x_1, x_2)\ [x_1, x_2), (x_1, x_2]$, where $x_1 \leq x_2,\ x, x_1, x_2 \in \mathcal{R}^1$.

Before defining conditional probability, and, in connection with it, the notion of the independence of two random events, let us illustrate this notion with a simple example.

Example We have an urn containing 3 white and 2 black marbles. The experiment consists of two drawings performed one by one in such a way that the first marble drawn is not returned to the urn before the second is taken out. The set of elementary events contains the following elements:

– ω_{11} is the elementary event consisting of drawing out a white marble in both drawings,

– ω_{12} consists of drawing out a white marble in the first and a black one in the second drawing,

– ω_{21} consists of drawing out a black marble in the first and a white one in the second selection,

– ω_{22} consists of drawing out a black marble in both drawings (the index 1 corresponds to the white color, 2 to the black, the sequence of indices is identical with the sequence of drawings).

Thus $\Omega = \{\omega_{11}, \omega_{12}, \omega_{21}, \omega_{22}\}$, $\mathcal{S} = \{\emptyset, \{\omega_{11}\}, \{\omega_{12}\}, \{\omega_{21}\}, \{\omega_{22}\},$ $\{\omega_{11}, \omega_{12}\}, \{\omega_{11}, \omega_{21}\}, \{\omega_{11}, \omega_{22}\}, \{\omega_{12}, \omega_{21}\}, \{\omega_{12}, \omega_{22}\}, \{\omega_{21}, \omega_{22}\},$ $\{\omega_{11}, \omega_{12}, \omega_{21}\}, \{\omega_{11}, \omega_{12}, \omega_{22}\}, \{\omega_{11}, \omega_{21}, \omega_{22}\}, \{\omega_{12}, \omega_{21}, \omega_{22}\},$ $\{\omega_{11}, \omega_{12}, \omega_{21}, \omega_{22}\}\}$,
$P(\{\omega_{11}\}) = 3/10,\ P(\{\omega_{12}\}) = 3/10,\ P(\{\omega_{21}\}) = 3/10,\ P(\{\omega_{22}\}) = 1/10.$

Investigate the probability of drawing out a white marble in the second draw if we know that in the first a white one was drawn. Denote A as the event consisting in drawing out a white marble in the second draw ($A = \{\omega_{11}, \omega_{21}\}$), and B the event consisting in drawing out a white marble in the first draw ($B = \{\omega_{11}, \omega_{12}\}$). $A|B$ means an event consisting of drawing out a white marble in the second draw under the condition that in the first pick a white one was drawn out. In this case $B = \{\omega_{11}, \omega_{12}\}$ is the set of all elementary events (the elementary events $\{\omega_{21}\}$ and $\{\omega_{22}\}$ cannot set in as a consequence of the condition that a white marble was drawn in the first draw, thus the original set of elementary events $\{\omega_{11}, \omega_{12}, \omega_{21}, \omega_{22}\}$ must be reduced in the way we described). The random event $A|B$ investigated on the set $\{\omega_{11}, \omega_{12}, \omega_{21}, \omega_{22}\}$ of elementary events is identical to the random event $A \cap B = \{\omega_{11}\}$ studied in the set $\{\omega_{11}, \omega_{12}\} = B$ of the elementary events. As $P(B) = P(\{\omega_{11}\}) + P(\{\omega_{12}\}) = 6/10$ and $P(A \cap B)/P(B) = P(\{\omega_{11}\}) = 3/10$, we can easily convince ourselves, that $P(A|B) = P(A \cap B)/P(B) = 1/2$ (if a white marble was picked in the first draw from the urn containing 3 white and 2 black marbles at the beginning, the urn before the second draw contains 2 white and 2 black marbles; the probability of drawing a white one is now 2/4).

Definition The conditional probability of a random event A under the condition that B has occurred, i.e., $P(A|B)$, is defined in the following way

$$P(A|B) = P(A \cap B)|P(B),\ P(B) > 0. \tag{1.1.1}$$

In order to illustrate the independence of two random events, a notion very important in data processing, let us return to the previous example of the urn with 3 white and 2 black marbles. If the experiment is arranged in the way we mentioned - the marble drawn out is not returned to the urn before the following draw - the result of the first draw influences the result of the next one. The events A and B are dependent, $P(A|B) = 1/2 \neq 3/10 = P(A)$.

Let us modify the experiment in such a way that the marble drawn out in the first draw is returned to the urn before performing the following one. Evidently, the result of the first draw does not influence the result of the next. In this case, although the set of elementary events Ω, the probability field $\mathcal{S}$, and the events A and B are the same as in the non-modified experiment, the probability distribution has changed in the following way: $P(\{\omega_{11}\}) = 9/25$, $P(\{\omega_{12}\}) = 6/25$, $P(\{\omega_{21}\}) = 6/25$ and $P(\{\omega_{22}\}) = 4/25$; thus actually $P(A|B) = (9/25)/(15/25) = 9/25+6/25 = 15/25 = P(A)$.

It is to be pointed out that two disjoint events are not independent.

Definition Random events A and B are independent if $P(A \cap B) = P(A)P(B)$.

Theorem Let A and B be random events such that $P(A) > 0$ and $P(B) > 0$. The random events are independent if and only if $P(A|B) = P(A)$ or $P(B|A) = P(B)$.

Proof. a) The necessary condition: Let A and B be independent. Then, due to the definition of independence, $P(A \cap B) = P(A)P(B)$. According to the definition of the conditional probability $P(A|B) = P(A \cap B)/P(B)$. Thus, simultaneously, $P(A \cap B) = P(A)P(B)$ and $P(A \cap B) = P(A|B)P(B)$, which implies $P(A|B) = P(A)$. Analogously the relation $P(B|A) = P(B)$ can be proved.

b) The sufficient condition: Let $P(A|B) = P(A)$. Due to (1.1.1), $P(A|B) = P(A \cap B)/P(B)$, thus $P(A \cap B) = P(A)P(B)$, which means that the random events A and B are independent. Analogously it can be proved that $P(B|A) = P(B)$ implies the independence of the random events A and B.

The definition of independence of two random events may be extended to any finite number of random events.

Definition Random events $A_1, A_2, \ldots, A_n$ are independent if for each s-tuple $(s \leq n) 1 \leq k_1 < k_2 < \ldots < k_s \leq n$ of integers $P(A_{k_1} \cap A_{k_2} \cap \ldots \cap A_{k_s}) = P(A_{k_1})P(A_{k_2}) \ldots P(A_{k_s})$.

1.2 The random variable

1.2.1 The random variable and its distribution function

So far we have dealt with random events and some operations on the set of elementary events Ω and the σ-algebra of random events $\mathcal{S}$. The definitions which we have given can be illustrated by many examples from various fields of human activities, not only by the traditional examples of the die, an urn with vari-coloured marbles or by the toss of a coin.

In each possible example the random event represents something concrete, and different random events are studied for in various cases. Deducing from the homogeneity of concrete cases what is common to them and trying, simultaneously, to built up the simplest possible and most adequate general scheme for processing data, we come to the notion of a random variable. The main advantage of this notion consists in the fact that from a system whose elements are sets, we mowe, using a suitable mapping, to the region of real numbers. For example, in the case of the die, each elementary event $\omega_i \in \Omega$ is assigned the integer i, as its index. Each random variable is assigned the probability distribution corresponding to its inverse fibre.

Definition Let $(\Omega, \mathcal{S}, P)$ be a probability space. A finite real function ξ $(\xi : \Omega \to \mathcal{R}^1)$ defined on the set of elementary events such that for each real number x there exists a random event $A \in \mathcal{S}$ which is the inverse fibre of the interval $(-\infty, x)$ (in other words for any $x \in \mathcal{R}^1$, $\xi^{-1}((-\infty, x)) = \{\omega : \xi(\omega) < x\} = A \in \mathcal{S}$) is called a random variable. The mapping ξ with this property is said to be $\mathcal{S}$-measurable. $\xi(\omega)$ is a realization of the random variable ξ, i.e., the value of the function ξ at the point ω of its definition domain Ω.

As each event $\{\omega : \xi(\omega) < x\}$ in the definition of a random variable must be an element of $\mathcal{S}$, its probability is known. Thus each $x \in \mathcal{R}^1$ is assigned the probability $P(\{\omega : \xi(\omega) < x\})$.

Definition A real function $F(\cdot)$, $x \in \mathcal{R}^1$, such that

$$F(x) = P(\xi^{-1}((-\infty, x))) = P(\{\omega : \xi(\omega) < x\}),$$

is called a distribution function of the random variable $\xi(\cdot)$, $\omega \in \Omega$, on the probability space $(\Omega, \mathcal{S}, P)$.

The value of the distribution function at the point $x \in \mathcal{R}^1$ yields the probability that the random variable is realized into a value from the interval $(-\infty, x)$; this probability equals the probability of that random event which is the inverse fibre of the interval within the $\mathcal{S}$-measurable mapping ξ.

Example Consider a die and the probability space $(\Omega, \mathcal{S}, P)$, where $\Omega = \{\omega_1, \omega_2, \omega_3, \omega_4, \omega_5, \omega_6\}$, $\mathcal{S}$ is the σ-algebra consisting of the system of all subsets of the set Ω, see p. 19, $P(\{\omega_i\}) = 1/6$, $i = 1, 2, \ldots, 6$. The mapping $\xi : \Omega \to \mathcal{R}^1$ such that $\xi(\omega_i) = i$, $i = 1, 2, \ldots, 6$, is $\mathcal{S}$-measurable, thus ξ is a random variable with respect to the probability space $(\Omega, \mathcal{S}, P)$.

Let a_1, a_2, a_3, a_4, a_5, a_6, a_7 be real numbers such that $a_1 > 6$, $a_2 \in (5,6]$, $a_3 \in (4,5]$, $a_4 \in (3,4]$, $a_5 \in (2,3]$, $a_6 \in (1,2]$, $a_7 \leq 1$. Then $\xi^{-1}((-\infty, a_1)) = \{\omega_1, \omega_2, \omega_3, \omega_4, \omega_5, \omega_6\} = \Omega \in \mathcal{S}$, $\xi^{-1}((-\infty, a_2)) = \{\omega_1, \omega_2, \omega_3, \omega_4, \omega_5\} \in \mathcal{S}$, $\xi^{-1}((-\infty, a_3)) = \{\omega_1, \omega_2, \omega_3, \omega_4\} \in \mathcal{S}$, $\xi^{-1}((-\infty, a_4)) = \{\omega_1, \omega_2, \omega_3, \} \in \mathcal{S}$, $\xi^{-1}((-\infty, a_5)) = \{\omega_1, \omega_2\} \in \mathcal{S}$, $\xi^{-1}((-\infty, a_1)) = \{\omega_1\} \in \mathcal{S}, \xi^{-1}((-\infty, a_1)) = \{\emptyset\} \in \mathcal{S}$, thus the mapping ξ is $\mathcal{S}$-measurable (Fig 1.2.1).

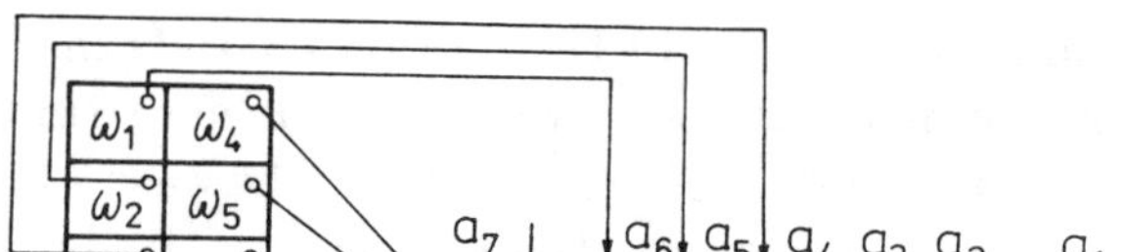

Fig. 1.2.1 The mapping $\xi : \{\omega_1, \omega_2, \omega_3, \omega_4, \omega_5, \omega_6\} \to \{1, 2, 3, 4, 5, 6\} \subset \mathcal{R}^1$ ($\xi(\omega_i) = i$, $i = 1, 2, \ldots, 6$) is $\mathcal{S}$-measurable; $\mathcal{S}$ is the system of all subsets of the set $\Omega = \{\omega_1, \omega_2, \omega_3, \omega_4, \omega_5, \omega_6\}$ (the example of the throw of a die).

What is the form of the distribution function of this random variable?

Due to the definition, $F(x) = P(\xi^{-1}((-\infty, x))) = P(\{\omega : \xi(\omega) < x\})$, therefore $F(x) = P(\xi^{-1}((-\infty, x))) = P(\{\omega : \xi(\omega) < 1\}) = P(\emptyset) = 0$ for any $x \in (-\infty, 1]$, $F(x) = P(\{\omega : \xi(\omega) < 2\}) = P(\{\omega_1\}) = 1/6$ for any $x \in (1, 2]$, $F(x) = P(\{\omega : \xi(\omega) < 3\}) = P(\{\omega_1\}) = 1/6 + P(\{\omega_2\}) = 2/6$ for any $x \in (2, 3]$, $F(x) = P(\{\omega : \xi(\omega) < 4\}) = P(\{\omega_1\}) + P(\{\omega_2\}) + P(\{\omega_3\}) = 3/6$ for any $x \in (3, 4]$, $F(x) = P(\{\omega : \xi(\omega) < 5\}) = P(\{\omega_1\}) + P(\{\omega_2\}) + P(\{\omega_3\}) + P(\{\omega_4\}) = 4/6$ for any $x \in (4, 5]$, $F(x) = P(\{\omega : \xi(\omega) < 6\}) = P(\{\omega_1\}) + P(\{\omega_2\}) + P(\{\omega_3\}) + P(\{\omega_4\}) + P(\{\omega_5\}) = 5/6$ for any $x \in (5, 6]$ and for any $x \in (6, \infty)$ $F(x) = P(\Omega) = 1$ (Fig.1.2.2).

Let us consider in the case of a die the probability space $(\Omega, \mathcal{S}_1, P_1)$, where $\Omega = \{\omega_1, \omega_2, \omega_3, \omega_4, \omega_5, \omega_6\}$, $\mathcal{S}_1 = \{\emptyset, \{\omega_1, \omega_3, \omega_5\}, \{\omega_2, \omega_4, \omega_6\}, \Omega\}$ (we are interested in the random event of obtaining either an odd or an even

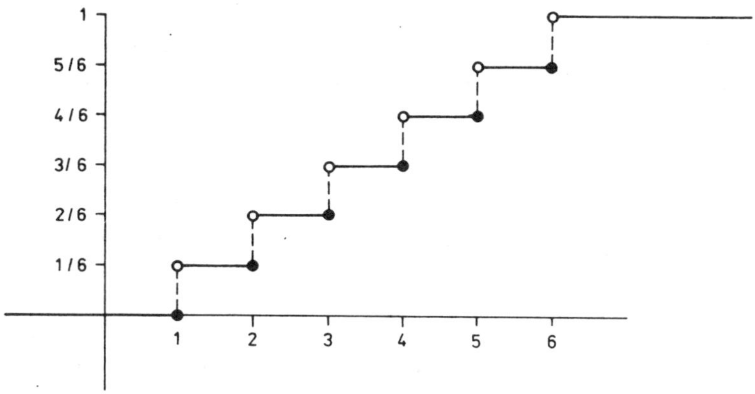

Fig. 1.2.2 The distribution function of the random variable $\xi : \{\omega_1, \omega_2, \omega_3, \omega_4, \omega_5, \omega_6\} \to \{1,2,3,4,5,6\} \subset \mathcal{R}^1$ on the probability space $(\Omega, \mathcal{S}, P)$, where $\Omega = \{\omega_1, \omega_2, \omega_3, \omega_4, \omega_5, \omega_6\}$, $\mathcal{S}$ is the system of all subsets of the set Ω, $\xi(\omega_i) = i$ and $P\{\omega : \xi(\omega_i) = i\} = p_i = 1/6$, $i = 1, 2, \ldots, 6$ (the example of the throw of a homogeneous die).

number), $P_1(\{\omega_1, \omega_3, \omega_5\}) = 1/2$, $P_1(\{\omega_2, \omega_4, \omega_6\}) = 1/2$, and the random variable ξ_1 defined in the following way: $\xi_1(\omega_{2k}) = 3$, $\xi_1(\omega_{2k-1}) = 1.7$, $k = 1, 2, 3$. This mapping defines a random variable, as $\xi_1^{-1}((-\infty, x)) = \emptyset \in \mathcal{S}_1$ for any $x \le 1.7$, for $1.7 < x \le 3$ $\xi_1^{-1}((-\infty, x)) = \{\omega_1, \omega_3, \omega_5\} \in \mathcal{S}_1$ and for $x > 3$ $\xi_1^{-1}((-\infty, x)) = \Omega \in \mathcal{S}_1$.

Simultaneously, it has to be realized that the mapping $\xi(\omega_k) = k$, $k = 1, 2, \ldots, 6$ is not a random variable with respect to the probability space $(\Omega, \mathcal{S}_1, P_1)$, because, e.g., $\xi^{-1}((-\infty, 1.5)) = \{\omega_1\} \notin \mathcal{S}_1$ (the mapping $\xi(\omega_k) = k$, $k = 1, 2, \ldots, 6$ being $\mathcal{S}$-measurable is not $\mathcal{S}_1$-measurable).

We can easily see that the distribution function of the random variable ξ_1 possesses a step form, being zero on the interval $(-\infty, 1.7]$, equalling $1/2$ on the interval $(1.7, 3]$ and 1 on the interval $(3, \infty)$.

Theorem The distribution function of a random variable ξ possesses the following properties:

(i) $\lim_{x \to -\infty} F(x) = 0$, $\lim_{x \to \infty} F(x) = 1$;

(ii) the distribution function is non-decreasing;

(iii) the distribution function is at least continuous from the left.

Proof. (i) $\lim_{x \to -\infty}(-\infty, x) = \emptyset$, $\xi^{-1}(\emptyset) = \emptyset$, $P(\emptyset) = 0$; $\lim_{x \to \infty}(-\infty, x) = \mathcal{R}^1$, $\lim_{x \to \infty} \xi^{-1}((-\infty, x)) = \Omega$, $P(\Omega) = 1$.

(ii) Let $x \le y$, then $F(y) = P(\{\omega : \xi(\omega) < y\}) = P(\{\omega : \xi(\omega) < x\}) + P(\{\omega : \xi(\omega) \in [x, y)\}) = F(x) + P(\{\omega : \xi(\omega) \in [x, y)\})$. By applying Axiom I in probability theory the proof is complete, because $F(y) -$

$F(x) = P(\{\omega : \xi(\omega) \in [x, y)\}) \geq 0$. Simultaneously the meaning of the difference between values of the distribution function at two points within its definition domain is explained. This difference expresses the probability that the random variable assumes any value from the interval $[x, y)$.

(iii) The problem is to prove that $\lim_{n\to\infty} F(x-1/n) = F(x)$ for an arbitrary $x \in \mathcal{R}^1$. Due to the definition, $F(x-1/n) = P(\{\omega : \xi(\omega) < x-1/n\})$, therefore the limit $\lim_{n\to\infty} P(\{\omega : \xi(\omega) < x - 1/n\})$ and its relation to the limit $P(\lim_{n\to\infty}\{\omega : \xi(\omega) < x - 1/n\}) = P(\{\omega : \xi(\omega) < x\})$ must be investigated. Let $A_n = \{\omega : \xi(\omega) < x - 1/n\}$ (apparently $A_n \in \mathcal{S}$, where $\mathcal{S}$ is a σ-algebra from the definition of the random variable $\xi(\cdot)$, $\omega \in \Omega$); the sequence $\{A_n\}_{n=1}^{\infty}$ of the random events is non-decreasing (each previous random event is involved in all following events), therefore $\lim_{n\to\infty} A_n = \cup_{n=1}^{\infty} A_n = \{\omega : \xi(\omega) < x\}$. As sets of the sequence $\{A_n\}_{n=1}^{\infty}$ are not disjoint, we have to modify it before applying σ-additivity of the probability measure. Let $B_1 = A_1$, $B_2 = A_2 - A_1$, $\ldots, B_n = A_n - A_{n-1}\ldots$, then the sequence $\{B_n\}_{n=1}^{\infty}$ is a sequence of disjoint sets from $\mathcal{S}$ (the difference between two sets belonging to $\mathcal{S}$ belongs to $\mathcal{S}$ as well) such that $\cup_{i=1}^{n} B_i = \cup_{i=1}^{n} A_i$ and $\cup_{i=1}^{\infty} B_i = \lim_{n\to\infty} A_n$ (but $\lim_{n\to\infty} B_n \neq \cup_{i=1}^{\infty} B_i$). Thus $P(\lim_{n\to\infty} A_n) = P(\cup_{i=1}^{\infty} A_i) = P(\cup_{i=1}^{\infty} B_i) = \sum_{i=1}^{\infty} P(B_i) = \lim_{n\to\infty} P(B_i) = \lim_{n\to\infty} P(\cup_{i=1}^{n} B_i) = \lim_{n\to\infty} P(\cup_{i=1}^{n} A_i)$. It has been proved that $P(\lim_{n\to\infty}\{\omega : \xi(\omega) < x - 1/n\}) = \lim_{n\to\infty} P(\cup_{i=1}^{n}\{\omega : \xi(\omega) < x - 1/n\})$. As $\lim_{n\to\infty}\{\omega : \xi(\omega) < x - 1/n\} = \{\omega : \xi(\omega) < x\}$ and $\cup_{i=1}^{n}\{\omega : \xi(\omega) < x - 1/n\} = \{\omega : \xi(\omega) < x - 1/n\}$ we have shown that $P(\{\omega : \xi(\omega) < x\}) = \lim_{n\to\infty} P(\{\omega : \xi(\omega) < x - 1/n\}$, which was the problem.

Definition A random variable is said to be discrete if it acquires a finite or countable number of values $x_1, x_2, \ldots, x_n, \ldots$.

The distribution function of a discrete random variable has a jump at each point x_i belonging to the set $\{x_1, x_2, \ldots, x_n, \ldots\}$ of its realizations. The height p_i of the jump at the point x_i is $p_i = P(\{\omega : \xi(\omega) = x_i\})$.

The random variable ξ that appears by mapping the set of elementary events in the example with a die can serve as an example of a discrete random variable.

Definition Let $\{x_1, x_2, \ldots, x_n, \ldots\}$ be the set of all realizations of a discrete random variable ξ. A function

$$p_i = P(\{\omega : \xi(\omega) = x_i\})$$

defined for each $x_i \in \{x_1, x_2, \ldots, x_n, \ldots\}$ is called a probability function (law of probability distribution) of the discrete random variable ξ.

$\sum_{i=1}^{n} p_i = 1$ if the number of all realizations of a discrete random variable is finite, $\sum_{i=1}^{\infty} p_i = 1$ if this set is countable.

The distribution function $F(\cdot)$ of a discrete random variable at a point $x \in \mathcal{R}^1$ is of the form $F(x) = \sum_{x_i < x} p_i$ (the sum includes all numbers that satisfy the inequality $x_i < x$).

Definition The random variable is said to be continuous if there exists a non-negative and piecewise continous function $f(\cdot)$, Riemann-integrable on $\mathcal{R}^1$, such that for all $x \in \mathcal{R}^1$

$$F(x) = \int_{-\infty}^{x} f(t)dt,$$

where $F(\cdot)$ is the distribution function of the random variable ξ. The function $f(\cdot)$, $x \in \mathcal{R}^1$, is the probability density of a continuous random variable ξ.

The properties of the distribution function imply the following properties of the probability density of a continuous random variable:

(i) $\int_{-\infty}^{\infty} f(t)dt = 1$;

(ii) let x and y be arbitrary real numbers such that $x < y$, then $P(\{\omega : \xi(\omega) \in [x, y)\}) = F(y) - F(x) = \int_x^y f(t)dt$;

(iii) at each point of continuity of the probability density $dF(x)/dx = \lim_{h \to 0} \frac{1}{h} \int_x^{x+h} f(t)dt = f(x)$.

1.2.2 Random vectors

Definition An n-dimensional random vector $\xi = (\xi_1, \xi_2, \ldots, \xi_n)'$ (in principle the column vectors are taken into account, $'$ denotes a transposition of a vector or a matrix) is an ordered n-tuple of random variables defined on the same probability space $(\Omega, \mathcal{S}, P)$.

Thus a finite vector mapping $\xi : \Omega \to \mathcal{R}^n$ is a random vector if for each point $x = (x_1, x_2, \ldots, x_n)'$ of the n-dimensional space $\mathcal{R}^n$ the set$\{\omega : \xi_1(\omega) < x_1 \ \& \ \xi_2(\omega) < x_2 \ \& \ \ldots \ \& \ \xi_n(\omega) < x_n, \omega \in \Omega\} = \cap_{k=1}^{n} \{\omega : \xi_k(\omega) < x_k, \omega \in \Omega\}$ is an element of the σ-algebra $\mathcal{S}$ of the probability space $(\Omega, \mathcal{S}, P)$ under consideration.

Definition A function $F : \mathcal{R}^n \to \mathcal{R}^1$

$$F(x_1, x_2, \ldots, x_n) = P(\{\omega : \xi_1(\omega) < x_1 \ \& \ \xi_2(\omega) < x_2 \ \& \ldots \ \& \ \xi_n(\omega) < x_n\}),$$

where P is the probability measure defined on the σ-algebra $\mathcal{S}$, is called a distribution function of the n-dimensional random vector ξ.

Thus the probability that the random vector ξ assumes a value from the interval $(-\infty, x_1) \times (-\infty, x_2) \times \ldots \times (-\infty, x_n)$ equals the probability of the random event which is the inverse fiber of this n-dimensional interval within the mapping $\xi \to \mathcal{R}^n$.

Example We are given a set of 177 mineral samples and want to study the relation between the hematite or magnetite content on the one hand and the value of the remanent magnetic polarization on the other. In accordance with problem as formulated, let us measure the remanent magnetic polarization and determine the amounts of hematite or magnetite by mineralogic analysis (the former property is quantitative, in nature, the latter qualitative). The set of the 177 samples is arranged in Table 1.1.

Table 1.1

The value of the remanent magnetic polarization	Hematite	Magnetite
$[10^{-11}, 10^{-10})$T	13	1
$[10^{-10}, 10^{-9})$T	28	6
$[10^{-9}, 10^{-8})$T	19	18
$[10^{-8}, 10^{-7})$T	8	35
$[10^{-7}, 10^{-6})$T	5	32
$[10^{-6}, 10^{-5}]$T	3	9

The set Ω of elementary events contains the following 12 elements (the meaning of the abbreviations is obvious):

ω_1=(he,$[10^{-11}, 10^{-10})$T), ω_7=(mag,$[10^{-11}, 10^{-10})$T),
ω_2=(he,$[10^{-10}, 10^{-9})$T), ω_8=(mag,$[10^{-10}, 10^{-9})$T),
ω_3=(he,$[10^{-9}, 10^{-8})$T), ω_9=(mag,$[10^{-9}, 10^{-8})$T),
ω_4=(he,$[10^{-8}, 10^{-7})$T), ω_{10}=(mag,$[10^{-8}, 10^{-7})$T),
ω_5=(he,$[10^{-7}, 10^{-6})$T), ω_{11}=(mag,$[10^{-7}, 10^{-6})$T),
ω_6=(he,$[10^{-6}, 10^{-5})$T] ω_{12}=(mag,$[10^{-6}, 10^{-5})$T].

Generate the σ-algebra $\mathcal{S}$ as a system of all subsets of the set Ω. It contains 2^{12}=4,096 elements; let the vector mapping $\xi : \Omega \to \mathcal{R}^2$ be defined by the relations:

$$\begin{aligned}
\xi(\omega_1) &= (\xi_1(\omega_1), \xi_2(\omega_1))' = (1,1)', & \xi(\omega_7) &= (\xi_1(\omega_7), \xi_2(\omega_7))' = (2,1)',\\
\xi(\omega_2) &= (\xi_1(\omega_2), \xi_2(\omega_2))' = (1,2)', & \xi(\omega_8) &= (\xi_1(\omega_8), \xi_2(\omega_8))' = (2,2)',\\
\xi(\omega_3) &= (\xi_1(\omega_3), \xi_2(\omega_3))' = (1,3)', & \xi(\omega_9) &= (\xi_1(\omega_9), \xi_2(\omega_9))' = (2,3)',\\
\xi(\omega_4) &= (\xi_1(\omega_4), \xi_2(\omega_4))' = (1,4)', & \xi(\omega_{10}) &= (\xi_1(\omega_{10}), \xi_2(\omega_{10}))' = (2,4)',\\
\xi(\omega_5) &= (\xi_1(\omega_5), \xi_2(\omega_5))' = (1,5)', & \xi(\omega_{11}) &= (\xi_1(\omega_{11}), \xi_2(\omega_{11}))' = (2,5)',\\
\xi(\omega_6) &= (\xi_1(\omega_6), \xi_2(\omega_6))' = (1,6)', & \xi(\omega_{12}) &= (\xi_1(\omega_{12}), \xi_2(\omega_{12}))' = (2,6)'.
\end{aligned}$$

This mapping actually defines a two-dimensional random variable:

Let $(x_1, x_2)' \in (-\infty, 1] \times (-\infty, 1]$, then $\{\omega : \xi_1(\omega) < x_1 \;\&\; \xi_2(\omega) < x_2\} = \emptyset \cap \emptyset = \emptyset \in \mathcal{S}$, let $(x_1, x_2)' \in (-\infty, 1] \times (-\infty, 2]$, then $\{\omega : \xi_1(\omega) < x_1 \;\&\; \xi_2(\omega) < x_2\} = \emptyset \cap \{\omega_1, \omega_7\} = \emptyset \in \mathcal{S}$, if $(x_1, x_2)' \in (-\infty, 2] \times (-\infty, 1]$, then $\{\omega : \xi_1(\omega) < x_1 \;\&\; \xi_2(\omega) < x_2\} = \{\omega_1, \omega_2, \omega_3, \omega_4, \omega_5, \omega_6\} \cap \emptyset = \emptyset \in \mathcal{S}$, if $(x_1, x_2)' \in (-\infty, 3] \times (-\infty, 2]$, then $\{\omega : \xi_1(\omega) < x_1 \;\&\; \xi_2(\omega) < x_2\} = \Omega \cap \{\omega_1, \omega_7\} = \{\omega_1, \omega_7\} \in \mathcal{S}$, etc., thus this mapping is $\mathcal{S}$-measurable (Fig.1.2.3).

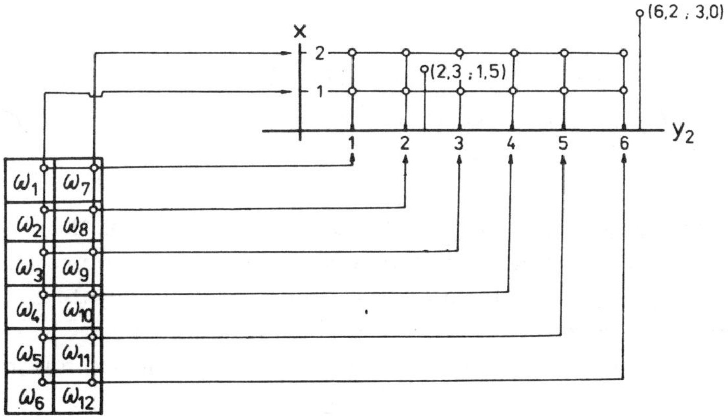

Fig. 1.2.3 The vector mapping $\xi : \{\omega_1, \omega_2, \ldots, \omega_{12}\} \to \{(1,1)', (1,2)', \ldots, (2,6)'\} \subset \mathcal{R}^2$ is $\mathcal{S}$-measurable (the example of the set of mineral samples containing hematite and magnetite).

The probability measure can be defined on Ω in the following way:

$$\begin{aligned}
&P(\{\omega : \xi_1(\omega) = 1 \;\&\; \xi_2(\omega) = 1\}) = 13/177,\\
&P(\{\omega : \xi_1(\omega) = 2 \;\&\; \xi_2(\omega) = 1\}) = 1/177,\\
&P(\{\omega : \xi_1(\omega) = 1 \;\&\; \xi_2(\omega) = 2\}) = 28/177,\\
&P(\{\omega : \xi_1(\omega) = 2 \;\&\; \xi_2(\omega) = 2\}) = 8/177,\\
&P(\{\omega : \xi_1(\omega) = 1 \;\&\; \xi_2(\omega) = 3\}) = 19/177,\\
&P(\{\omega : \xi_1(\omega) = 2 \;\&\; \xi_2(\omega) = 3\}) = 18/177,
\end{aligned}$$

$P(\{\omega : \xi_1(\omega) = 1 \ \& \ \xi_2(\omega) = 4\}) = 8/177,$
$P(\{\omega : \xi_1(\omega) = 2 \ \& \ \xi_2(\omega) = 4\}) = 35/177,$
$P(\{\omega : \xi_1(\omega) = 1 \ \& \ \xi_2(\omega) = 5\}) = 5/177,$
$P(\{\omega : \xi_1(\omega) = 2 \ \& \ \xi_2(\omega) = 5\}) = 32/177,$
$P(\{\omega : \xi_1(\omega) = 1 \ \& \ \xi_2(\omega) = 6\}) = 3/177,$
$P(\{\omega : \xi_1(\omega) = 2 \ \& \ \xi_2(\omega) = 6\}) = 9/177.$

For the distribution function, see Fig. 1.2.4.

Theorem The distribution function of an n-dimensional random vector possesses the following properties:

(i) $\lim_{x_k \to -\infty} F(x_1, x_2, \ldots, x_k) = 0, \ k = 1, 2, \ldots, n;$
$\lim_{x_1, x_2, \ldots, x_n \to \infty} F(x_1, x_2, \ldots, x_n) = 1;$

(ii) it is a non-decreasing function of each variable;

(iii) it is at least continuous from the left at each variable;

(iv) let $x_i \geq y_i, \ i = 1, 2, \ldots, n$, then

$$F(y_1, y_2, \ldots, y_n) - \sum_{j=1}^{n} F(y_1, y_2, \ldots, y_{j-1}, x_j, y_{j+1}, \ldots, y_n) + \\ + \sum_{j,k=1}^{n} F(y_1, y_2, \ldots, y_{j-1}, x_j, y_{j+1}, \ldots, y_{k-1}, x_k, y_{k+1}, \ldots, y_n) + \\ + \ldots + (-1)^n F(x_1, x_2, \ldots, x_n) \leq 0.$$

Proof. The way of proving the properties (i) - (iii) of the distribution function of an n-dimensional random vector is analogous to proving the corresponding properties of a (one-dimensional) random variable. The property (iv) will be proved in the simplest but very instructive case, namely for $n = 2$. The problem is to prove that for each pair of real numbers x_1, y_1 and x_2, y_2 such that $x_1 \leq y_1$ and $x_2 \leq y_2$

$$F(y_1, y_2) - F(x_1, y_2) - F(y_1, x_2) + F(x_1, x_2) \geq 0.$$

Let us modify the set argument of the probability measure in the relation $F(y_1, y_2) = P(\{\omega : \xi_1(\omega) < y_1, \xi_2(\omega) < y_1\})$ (which determines the value of the distribution function of a two-dimensional vector at the point $(y_1, y_2)'$) in the following way: $\{\omega : \xi_1(\omega) \in (-\infty, y_1), \xi_2(\omega) \in (-\infty, y_2)\} = \{\omega : \xi_1(\omega) \in (-\infty, x_1), \xi_2(\omega) \in (-\infty, y_2)\} \cup \{\omega : \xi_1(\omega) \in [x_1, y_1), \xi_2(\omega) \in (-\infty, y_2)\} = \{\omega : \xi_1(\omega) \in (-\infty, x_1), \xi_2(\omega) \in (-\infty, x_2)\} \cup \{\omega : \xi_1(\omega) \in (-\infty, x_1), \xi_2(\omega) \in [x_2, y_2)\} \cup \{\omega : \xi_1(\omega) \in [x_1, y_1), \xi_2(\omega) \in (-\infty, x_2)\} \cup \{\omega : \xi_1(\omega) \in [x_1, y_1), \xi_2(\omega) \in [x_2, y_2)\}$, which means that the input set (an element of the σ-algebra $\mathcal{S}$) is expressed in the form of 4 disjoint sets from $\mathcal{S}$, on which the probability measure is defined. Applying additivity of this

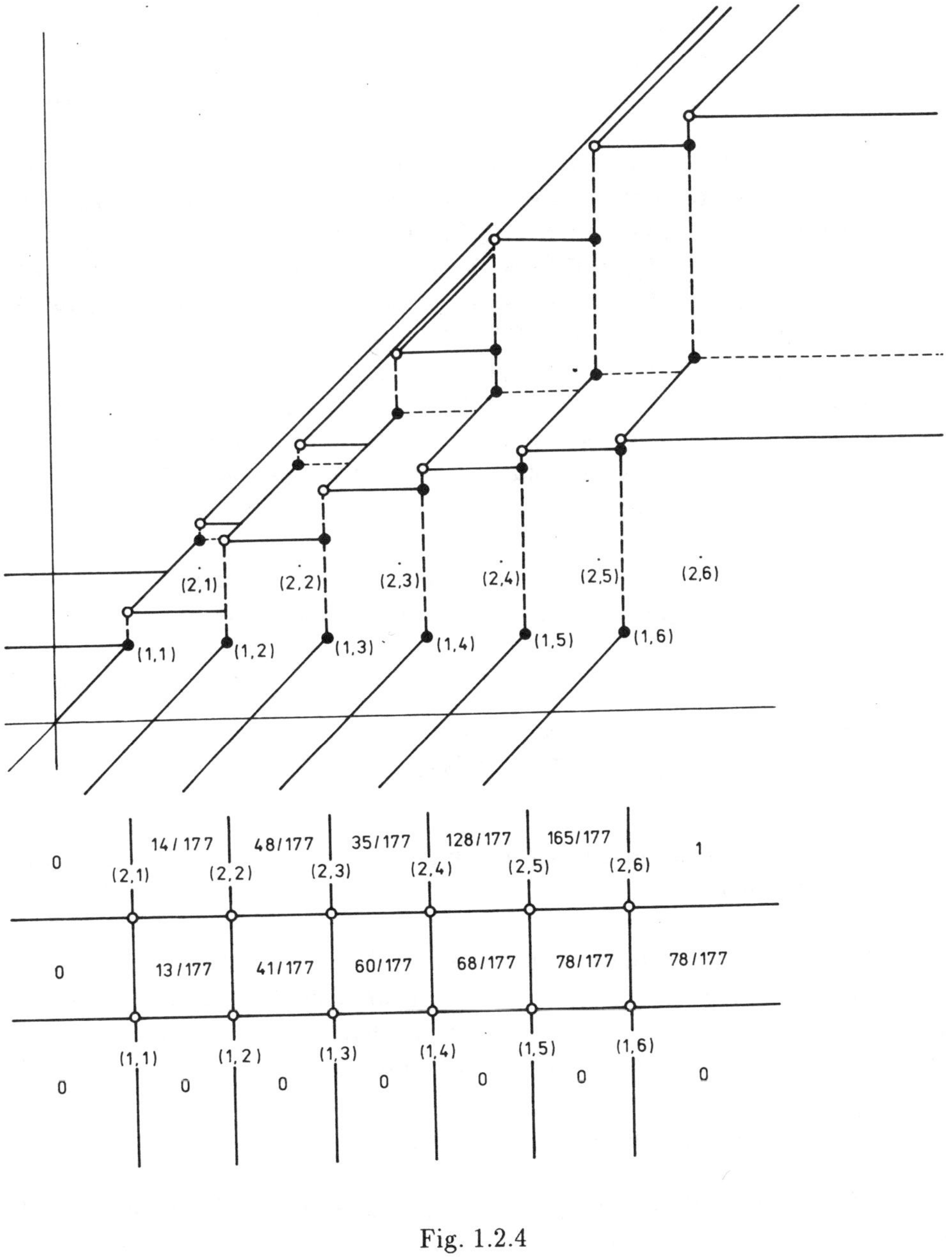
(2,1)
(2,2)
(2,3)
(2,4)
(2,5)
(2,6)
(1,1)
(1,2)
(1,3)
(1,4)
(1,5)
(1,6)
0
14/177
48/177
35/177
128/177
165/177
1
0
13/177
41/177
60/177
68/177
78/177
78/177
0
0
0
0
0
0
0

Fig. 1.2.4

measure (Axiom III), we obtain

$$\begin{aligned}F(y_1, y_2) = F(x_1, x_2) + P(\{\omega : \xi_1(\omega) \in (-\infty, x_1), \xi_2(\omega) \in [x_2, y_2)\} + \\ +P(\{\omega : \xi_1(\omega) \in [x_1, y_1), \xi_2(\omega) \in (-\infty, x_2)\} + \\ +P(\{\omega : \xi_1(\omega) \in [x_1, y_1), \xi_2(\omega) \in [x_2, y_2)\}.\end{aligned} \tag{1.2.1}$$

Next, we change the set arguments of the second and third function of the last relation. This relation is intendet to express probabilities by means of the distribution function. As $\{\omega : \xi_1(\omega) \in (-\infty, x_1), \xi_2(\omega) \in [x_2, y_2)\} = \{\omega : \xi_1(\omega) \in (-\infty, x_1), \xi_2(\omega) \in (-\infty, y_2)\} - \{\omega : \xi_1(\omega) \in (-\infty, x_1), \xi_2(\omega) \in (-\infty, x_2)\}$, we get $\{\omega : \xi_1(\omega) \in (-\infty, x_1), \xi_2(\omega) \in (-\infty, y_2)\} = \{\omega : \xi_1(\omega) \in (-\infty, x_1), \xi_2(\omega) \in [x_2, y_2)\} \cup \{\omega : \xi_1(\omega) \in (-\infty, x_1), \xi_2(\omega) \in (-\infty, x_2)\}$ and analogously $\{\omega : \xi_1(\omega) \in (-\infty, y_1), \xi_2(\omega) \in (-\infty, x_2)\} = \{\omega : \xi_1(\omega) \in (-\infty, x_1), \xi_2(\omega) \in (-\infty, x_2)\} \cup \{\omega : \xi_1(\omega) \in [x_1, y_1), \xi_2(\omega) \in (-\infty, x_2)\}$, thus $F(x_1, y_2) = F(x_1, x_2) + P(\{\omega : \xi_1(\omega) \in (-\infty, x_1), \xi_2(\omega) \in [x_2, y_2)\}$ and $F(y_1, x_2) = F(x_1, x_2) + P(\{\omega : \xi_1(\omega) \in [x_1, y_1), \xi_2(\omega) \in (-\infty, x_2)\}$. After insertion into (1.2.1) we get

$$\begin{aligned}F(y_1, y_2) = F(x_1, x_2) + F(x_1, y_2) - F(x_1, x_2) + F(y_1, x_2) - \\ -F(x_1, x_2) + P(\{\omega : \xi_1(\omega) \in [x_1, y_2), \xi_2(\omega) \in [x_2, y_2)\}),\end{aligned}$$

and finally

$$\begin{aligned}F(y_1, y_2) - F(x_1, y_2) - F(y_1, x_2) + F(x_1, x_2) = \\ = P(\{\omega : \xi_1(\omega) \in [x_1, y_1), \xi_2(\omega) \in [x_2, y_2)\}) \geq 0\end{aligned}$$

(the probability measure, due to Axiom I of probability theory, is a non-negative function).

Definition An n-dimensional random vector $\xi = (\xi_1, \xi_2, \ldots, \xi_n)'$ is said to be discrete if it can assume a finite or countable number of n-tuples of values $(x_{1k_1}, x_{2k_2}, \ldots, x_{nk_n})'$, $k_i = 1, 2, \ldots, K_i, \ldots$, $i = 1, 2, \ldots, n$, each characterized by the corresponding probability

$$P(\{\omega : \xi_1(\omega) = x_{1k_1}, \xi_2(\omega) = x_{2k_2}, \ldots, \xi_n(\omega) = x_{nk_n}\}) = p_{1k_1, 2k_2, \ldots, nk_n},$$

where simultaneously $\sum_{k_1, k_2, \ldots, k_n} p_{1k_1, 2k_2, \ldots, nk_n} = 1$ (the sum includes all the indices k_i, $i = 1, 2, \ldots, n$).

The distribution function $F(\cdot)$, $x \in \mathcal{R}^n$, of an n-dimensional random discrete vector is of the form

$$F(x_1, x_2, \ldots, x_n) = \sum_{x_{1k_1} < x_1, x_{2k_2} < x_2, \ldots, x_{nk_n} < x_n} p_{1k_1,2k_2,\ldots,nk_n}$$

(the sum includes all points x_{ik_i}, such that $x_{ik_i} < x_i$, $i = 1, 2, \ldots, n$).

The random variable discussed at the beginning of this section is an example of a two-dimensional discrete random vector; here $x_{11} = 1$, $x_{12} = 2\,(K_1 = 2)$ and $x_{21} = 1$, $x_{22} = 2$, $x_{23} = 3$, $x_{24} = 4$, $x_{25} = 5$, $x_{26} = 6\,(K_2 = 6)$; this random vector assumes 12 different values (the number of the set of all elementary events), namely $(x_{11}, x_{21})'$, $(x_{11}, x_{22})'$, $(x_{11}, x_{23})'$, $(x_{11}, x_{24})'$, $(x_{11}, x_{25})'$, $(x_{11}, x_{26})'$, $(x_{12}, x_{21})'$, $(x_{12}, x_{22})'$, $(x_{12}, x_{23})'$, $(x_{12}, x_{24})'$, $(x_{12}, x_{25})'$ and $(x_{12}, x_{26})'$ with the probabilities $p_{11,21} = P(\{\omega : \xi_1(\omega) = 1, \xi_2(\omega) = 1\}) = P(\{\omega_1)\}) = 13/177$, $p_{11,22} = 28/177, \ldots, p_{12,26} = 9/177$.

Definition An n-dimensional random vector ξ is said to be continuous if there exists a non-negative function $f(\cdot)$ Riemann integrable on $\mathcal{R}^n$ such that at any point $(x_1, x_2, \ldots, x_n)' \in \mathcal{R}^n$

$$F(x_1, x_2, \ldots, x_n) = \int_{-\infty}^{x_1} \int_{-\infty}^{x_2} \cdots \int_{-\infty}^{x_n} f(t_1, t_2, \ldots, t_n) dt_1 dt_2 \ldots dt_n,$$

where $F(\cdot)$, $x \in \mathcal{R}^n$, is the distribution function of the random vector ξ. The function $f(\cdot)$, $x \in \mathcal{R}^n$ is called a probability density.

The properties of the distribution function $F(\cdot)$, $x \in \mathcal{R}^n$, imply the following properties of the probability density of a continuous random vector:

(i) $\int_{\mathcal{R}^n} f(t_1, t_2, \ldots, t_n) dt_1 dt_2 \ldots dt_n = 1$;

(ii) at each point of continuity of the probability density of an n-dimensional

continuous random vector ξ

$$\partial^n F(x_1, x_2, \ldots, x_n)/\partial x_1 \partial x_2 \ldots \partial x_n = f(x_1, x_2, \ldots, x_n).$$

1.2.3 A marginal probability distribution

Let $(\xi_1, \xi_2)'$ be a discrete vector taking on the values $(x_{1k_1}, x_{2k_2})'$, $k_i = 1, 2, \ldots, K_i, \ldots$, $i = 1, 2$, with the probabilities $p_{1k_1,2k_2}$. What is the meaning of the expressions $\sum_{k_1} p_{1k_1,2k_2} = p_{1\cdot,2k_2}$ and $\sum_{k_2} p_{1k_1,2k_2} = p_{1k_1,2\cdot}$? According to the definition, $p_{1k_1,2k_2} = P(\{\omega : \xi_1(\omega) = x_{1k_1}, \xi_2(\omega) = x_{2k_2})$, which means that $\sum_{k_1} p_{1k_1,2k_2} = P(\{\omega : \xi_1(\omega) = x_{11}, \xi_2(\omega) = x_{2k_2}\}) + P(\{\omega : \xi_1(\omega) = x_{12}, \xi_2(\omega) = x_{2k_2}\}) + \ldots + P(\{\omega : \xi_1(\omega) = x_{1K_1}, \xi_2(\omega) = x_{2k_2}\}) + \cdots$. As the sets, being arguments of individual summands, are disjoint, the σ-additivity of the probability measure may be applied — $\sum_{k_1} p_{1k_1,2k_2} = P(\cup_{k_1}\{\omega : \xi_1(\omega) = x_{1k_1}, \xi_2(\omega) = x_{2k_2}\}) = P(\cup_{k_1}\{\omega : \xi_1(\omega) = x_{1k_1}\} \cap \{\omega : \xi_2(\omega) = x_{2k_2}\}) = P(\Omega \cap \{\omega : \xi_2(\omega) = x_{2k_2}\}) = P(\{\omega : \xi_2(\omega) = x_{2k_2}\})$. Analogously $\sum_{k_2} p_{1k_1,2k_2} = P(\{\omega : \xi_1(\omega) = x_{1k_1}\})$.

$p_{1\cdot,2k_2} = P(\{\omega : \xi_2(\omega) = x_{2k_2}\})$ expresses the probability that $\xi_2(\omega) = x_{2k_2}$ independently of the value the random variable ξ_1 assumes; the probability $p_{1k_1,2\cdot}$ has an analogous meaning. Obviously, $\sum_{k_2} p_{1\cdot,2k_2} = 1$, $\sum_{k_1} p_{1k_1,2\cdot} = 1$.

The probabilities $p_{1\cdot,2k_2}$ and $p_{1k_1,2\cdot}$ are called marginal probabilities.

A generalization of marginal probability for an n-dimensional random vector ($n > 2$) can easily be demonstrated using the example of the two-dimensional random vector (ξ_1, ξ_2) on the set of 177 samples examined at the beginning of this section:

ξ_1 \ ξ_2	1	2	3	4	5	6	Σ
1	13/177	28/177	19/177	8/177	5/177	3/177	76/177
2	1/177	6/177	18/177	35/177	32/177	9/177	101/177
Σ	14/177	34/177	37/177	43/177	37/177	12/177	177/177

Let us investigate the probability that samples containing hematite only will occur, regardless of the value of the remanent magnetic polarization; the problem is to determine the probability $p_{11,2\cdot}$. In accordance with the definition, $p_{11,2\cdot} = P(\{\omega : \xi_1(\omega) = 1, \xi_2(\omega) = 1\}) + P(\{\omega : \xi_1(\omega) = 1, \xi_2(\omega) = 2\}) + P(\{\omega : \xi_1(\omega) = 1, \xi_2(\omega) = 3\}) + P(\{\omega : \xi_1(\omega) = 1, \xi_2(\omega) = 4\}) + P(\{\omega : \xi_1(\omega) = 1, \xi_2(\omega) = 5\}) + P(\{\omega : \xi_1(\omega) = 1, \xi_2(\omega) = 6\})$.

The set arguments of the probability measures can be expressed in the form

$$\{\omega : \xi_1(\omega) = 1,\ \xi_2(\omega) = 1\} = \{\omega_1, \omega_2, \omega_3, \omega_4, \omega_5, \omega_6\} \cap \{\omega_1, \omega_7\} = \{\omega_1\},$$
$$\{\omega : \xi_1(\omega) = 1,\ \xi_2(\omega) = 2\} = \{\omega_1, \omega_2, \omega_3, \omega_4, \omega_5, \omega_6\} \cap \{\omega_2, \omega_8\} = \{\omega_2\},$$
$$\{\omega : \xi_1(\omega) = 1,\ \xi_2(\omega) = 3\} = \{\omega_1, \omega_2, \omega_3, \omega_4, \omega_5, \omega_6\} \cap \{\omega_3, \omega_9\} = \{\omega_3\},$$
$$\{\omega : \xi_1(\omega) = 1,\ \xi_2(\omega) = 4\} = \{\omega_1, \omega_2, \omega_3, \omega_4, \omega_5, \omega_6\} \cap \{\omega_4, \omega_{10}\} = \{\omega_4\},$$
$$\{\omega : \xi_1(\omega) = 1,\ \xi_2(\omega) = 5\} = \{\omega_1, \omega_2, \omega_3, \omega_4, \omega_5, \omega_6\} \cap \{\omega_5, \omega_{11}\} = \{\omega_5\},$$
$$\{\omega : \xi_1(\omega) = 1,\ \xi_2(\omega) = 6\} = \{\omega_1, \omega_2, \omega_3, \omega_4, \omega_5, \omega_6\} \cap \{\omega_6, \omega_{12}\} = \{\omega_6\}.$$

These sets are disjoint, which enables us (as a consequence of the σ-additivity of the probability measure) to write
$p_{11,2\cdot} = P\left(\cup_{k_2=1}^{6}\{\omega : \xi_1(\omega) = 1, \xi_2(\omega) = k_2\}\right)$ and because of $\cup_{k_2=1}^{6}\{\omega : \xi_1(\omega) = 1, \xi_2(\omega) = k_2\} = \{\omega_1,\omega_2,\omega_3,\omega_4,\omega_5,\omega_6\} \cap [\{\omega_1,\omega_7\} \cup \{\omega_2,\omega_8\} \cup \{\omega_3,\omega_9\} \cup \{\omega_4,\omega_{10}\} \cup \{\omega_5,\omega_{11}\} \cup \{\omega_6,\omega_{12}\}] = \{\omega_1,\omega_2,\omega_3,\omega_4,\omega_5,\omega_6\} \cap \Omega = \{\omega_1\}\cup\{\omega_2\}\cup\{\omega_3\}\cup\{\omega_4\}\cup\{\omega_5\}\cup\{\omega_6\}$; thus $p_{11,2\cdot} = 76/177$ is the probability of an occurrence of a sample containing hematite only, regardless of the value of the remanent magnetic polarization. Further, e.g., $p_{1\cdot,23} = P(\{\omega : \xi_2(\omega) = 3\}) = 37/177$ is the probability of an occurrence of a sample, whose remanent magnetic polarization lies in the interval $[10^{-9}, 10^{-8})$T, regardless whether it contains or does not contain hematite or magnetite. The boxed part of the table contains single probabilities $p_{1k_1,2k_2}$, the last column and row contain the marginal probabilities.

Let us consider a two-dimensional random vector $(\xi_1,\xi_2)'$ (discrete or continuous), whose distribution function is $F(\cdot)$, $(x_1,x_2)' \in \mathcal{R}^2$. We denote $F_1(\cdot)$, $x_1 \in \mathcal{R}^1$, the distribution function of the random variable ξ_1. It has the form

$$F_1(x_1) = \lim_{x_2\to\infty} F(x_1,x_2) = P(\{\omega : \xi_1(\omega) < x_1\}\cap\Omega) = P(\{\omega : \xi_1(\omega) < x_1\}).$$

Analogously the distribution function $F_2(\cdot)$, $x_2 \in \mathcal{R}^1$, of the random variable ξ_2 is

$$F_2(x_2) = \lim_{x_1\to\infty} F(x_1,x_2) = P(\{\omega : \xi_2(\omega) < x_2\}\cap\Omega) = P(\{\omega : \xi_2(\omega) < x_2\}).$$

The distribution functions of random variables ξ_1 and ξ_2 which are elements of a two-dimensional discrete random vector $(\xi_1,\xi_2)'$ are $F_1(x_1) = \sum_{x_{1k_1}<x_1} p_{1k_1,2\cdot}$ and $F_2(x_2) = \sum_{x_{2k_2}<x_2} p_{1\cdot,2k_2}$.

If $(\xi_1,\xi_2)'$ is a continuous random vector, the marginal distribution functions of its components are $F_1(x_1) = \lim_{x_2\to\infty} F(x_1,x_2) = \int_{-\infty}^{x_1}\int_{-\infty}^{\infty} f(t_1,t_2)dt_1dt_2 = \int_{-\infty}^{x_1} f_1(t_1)dt_1$ and $F_2(x_2) = \lim_{x_1\to\infty} F(x_1,x_2) = \int_{-\infty}^{\infty}\int_{-\infty}^{x_2} f(t_1,t_2)dt_1dt_2 = \int_{-\infty}^{x_2} f_2(t_2)dt_2$; here $f_1(\cdot)$ and $f_2(\cdot)$ are marginal probability densities defined by the relations $f_1(x_1) = \int_{\mathcal{R}^1} f(x_1,x_2)dx_2$ and $f_2(x_2) = \int_{\mathcal{R}^1} f(x_1,x_2)dx_1$, respectively, in which $f(\cdot)$, $x_1,x_2 \in \mathcal{R}^2$, is the probability density of the two-dimensional random vector $\xi = (\xi_1,\xi_2)'$.

Generalizing this notion for n-dimensional random vectors ($n > 2$) is obvious.

1.2.4 Conditional probability distributions

Let us consider a two-dimensional discrete random vector $(\xi_1, \xi_2)'$ that assumes the values $(x_{1k_1}, x_{2k_2})'$ with the probabilities

$$p_{1k_1,2k_2},\ k_i = 1, 2, \ldots, K_i, \ldots,\ i = 1, 2.$$

The probability $P(\{\omega : \xi_1(\omega) = x_{1k_1}\}|\{\omega : \xi_2(\omega) = x_{2k_2}\})$ is defined for each $x_{1k_1} \in \{x_{11}, x_{12}, \ldots, x_{1K_1} \ldots\}$ and each $x_{2k_2} \in \{x_{21}, x_{22}, \ldots, x_{2K_2} \ldots\}$. Due to the relation (1.1.1),

$$P(\{\omega : \xi_1(\omega) = x_{1k_1}\}|\{\omega : \xi_2(\omega) = x_{2k_2}\}) =$$

$$= \frac{P(\{\omega : \xi_1(\omega) = x_{1k_1}\} \cap \{\omega : \xi_2(\omega) = x_{2k_2}\})}{P(\{\omega : \xi_2(\omega) = x_{2k_2}\})} =$$

$$= \frac{P(\{\omega : \xi_1(\omega) = x_{1k_1}, \xi_2(\omega) = x_{2k_2}\})}{P(\{\omega : \xi_2(\omega) = x_{2k_2}\})} = p_{1k_1,2k_2}/p_{1\cdot,2k_2}$$

(if $p_{1\cdot,2k_2} \neq 0$).

Analogously, $P(\{\omega : \xi_2(\omega) = x_{2k_2}\}|\{\omega : \xi_1(\omega) = x_{1k_1}\}) = p_{1k_1,2k_2}/p_{1k_1,2\cdot}$, if $p_{1k_1,2\cdot} \neq 0$. The former relation expresses the probability that $\xi_1(\omega) = x_{1k_1}$ under the condition that the event $\xi_2(\omega) = x_{2k_2}$ has occurred, the latter the probability that $\xi_2(\omega) = x_{2k_2}$ takes place, subject to the condition that the event $\xi_1(\omega_1) = x_{1,k_1}$ has happened.

The distribution functions of the above mentioned conditioned events are

$$F_1(x_1|x_{2k_2}) = \sum_{x_{1k_1} < x_1} \frac{p_{1k_1,2k_2}}{p_{1\cdot,2k_2}},\ x_1 \in \mathcal{R}^1,$$

and

$$F_2(x_2|x_{1k_1}) = \sum_{x_{2k_2} < x_2} \frac{p_{1k_1,2k_2}}{p_{1k_1,2\cdot}},\ x_2 \in \mathcal{R}^1,$$

respectively, when obviously $\sum_{k_1} p_{1k_1,2k_2}/p_{1\cdot,2k_2} = p_{1\cdot,2k_2}/p_{1\cdot,2k_2} = 1$ and $\sum_{k_2} p_{1k_1,2k_2}/p_{1k_1,2\cdot} = 1$.

Example What is the probability that a sample will show up with remanent magnetic polarization having a value from the interval $[10^{-9}, 10^{-8})$T for all samples containing hematite? $P(\{\omega : \xi_2(\omega) = 3\}|\xi_1(\omega) = 1\}) = p_{11,23}/p_{11,2\cdot} = (19/177)/(76/177) = 19/76$.

What is the form of the distribution function of the random variable ξ_2 conditioned by the fact that $\xi_1(\omega) = x_1 = 1$? The following table contains values for the variable, see Fig. 1.2.4 as well.

x_2	$(-\infty,1]$	(1,2]	(2,3]	(3,4]	(4,5]	(5,6]	(6,∞]
$F_2(x_2\|1)$	0	13/76	41/76	60/76	68/76	73/76	1

Now let $(\xi_1, \xi_2)'$ be a continuous random vector characterized by the probability density $f(\cdot)$, $(x_1, x_2)' \in \mathcal{R}^2$. The problem is to investigate the probability of the random event $\{\omega : \xi_1(\omega) < x_1\}|\{\omega : \xi_2(\omega) = x_2\}$. As in the case of a continuous random variable $P(\{\omega : \xi_1(\omega) = x_1\}) = P(\emptyset) = 0$, the expression for the conditional probability applied in the previous case of a discrete random variable must be modified. The probability $P(\{\omega : \xi_1(\omega) < x_1\}|\{\omega : \xi_2 \leq \xi_2(\omega) < x_2 + h\})$ can be determined without any difficulty: $P(\{\omega : \xi_1(\omega) < x_1\}|\{\omega : x_2 \leq \xi_2(\omega) < x_2 + h\}) = P(\{\omega : \xi_1(\omega) < x_1\} \cap \{\omega : x_2 \leq \xi_2(\omega) < x_2 + h\})/P(\{\omega : x_2 \leq \xi_2(\omega) < x_2 + h\})$ if $\{\omega : x_2 \leq \xi_2(\omega) < x_2 + h\} \neq \emptyset$.

The probability sought for the random event that consists in the random variable assuming a number smaller than x_1, under the condition that the random variable ξ_2 assumed the number x_2, equals the limit of the probability mentioned for $h \to 0$ (if this limit exists):

$$P(\{\omega : \xi_1(\omega) < x_1\}|\{\omega : \xi_2(\omega) = x_2\}) = F_1(x_1|x_2) =$$

$$= \lim_{h \to 0} \frac{P(\{\omega : \xi_1(\omega) < x_1, x_2 \leq \xi_2(\omega) < x_2 + h\})}{P\{\omega : x_2 \leq \xi_2 < x_2 + h\})} =$$

$$= \lim_{h \to 0} \frac{\frac{1}{h}\int_{-\infty}^{x_1}\int_{x_2}^{x_2+h} f(u_1, u_2)du_1du_2}{\frac{1}{h}\int_{x_2}^{x_2+h} f_2(u_2)du_2}$$

under the condition that the probability density $f(\cdot)$, $(x_1, x_2)' \in \mathcal{R}^2$, is continuous at the point $(x_1, x_2)'$ and the marginal probability density is continuous and positive at the point x_2. Then

$$\lim_{h \to 0} \frac{1}{h}\int_{x_2}^{x_2+h} f_2(u_2)du_2 = \lim_{h \to 0} \frac{F_2(x_2 + h) - F_2(x_2)}{h} = f_2(x_2)$$

and

$$\lim_{h \to 0} \frac{1}{h}\int_{-\infty}^{x_1}\int_{x_2}^{x_2+h} f(u_1, u_2)du_1du_2 = \int_{-\infty}^{x_1} \lim_{h \to 0} \frac{1}{h}\int_{x_2}^{x_2+h} f(u_1, u_2)du_2du_1 =$$

$$= \int_{-\infty}^{x_1} f(u_1, x_2)du_1,$$

thus

$$F_1(x_1|x_2) = \frac{\int_{-\infty}^{x_1} f(u_1, x_2)du_1}{f_2(x_2)}$$

and analogously

$$F_2(x_2|x_1) = \frac{\int_{-\infty}^{x_2} f(x_1, u_2)du_2}{f_1(x_1)}$$

if the probability density $f(\cdot)$ is continuous at the point $(x_1, x_2)'$ and the marginal density $f_1(\cdot)$ is continuous and positive at the point x_1. Finally, the conditional probability densities are

$$f_1(x_1|x_2) = \frac{\partial F_1(x_1|x_2)}{\partial x_1} =$$

$$= \frac{\lim_{\Delta x_1 \to 0} \frac{1}{\Delta x_1} \int_{x_1}^{x_1+\Delta x_1} f(u_1, x_2)du_1}{f_2(x_2)} = \frac{f(x_1, x_2)}{f_2(x_2)}, \ f_2(x_2) > 0$$

and analogously

$$f_2(x_2|x_1) = \frac{f(x_1, x_2)}{f_1(x_1)}, \ f_1(x_1) > 0.$$

Example A certain seismological station records earthquakes beginning with a certain magnitude, according to its detection capability. The problem is to determine the probability that this station will detect and record an earthquake if the number of earthquakes whose magnitude is equal to or larger than m is given by the formula $N(m) = \exp(\alpha - \beta m)$, $\alpha, \beta \in \mathcal{R}^1$. The solution of this problem requires that we derive the conditional probability density.

Let ω denote the elementary event consisting of the occurrence of one earthquake; the function $\xi(\cdot)$, $\omega \in \Omega$, assigns to each earthquake its magnitude; the set $\{\omega : \xi(\omega) \geq m\}$ denotes the random event consisting of an occurrence of an earthquake whose magnitude is equal or larger than m. We denote by $N(m_{min}) = C < \infty$ the number of all recordable earthquakes, m_{min} is the minimum magnitude of the recordable earthquake. The probability that an earthquake sets in with a magnitude equal or larger than m is $P(\{\omega : \xi(\omega) \geq m\}) = \exp(\alpha - \beta m)/C$, which means that the distribution function of the random variable ξ is $F(m) = P(\{\omega : \xi(\omega) < m\}) = 1 - \exp(\alpha - \beta m)/C$, $m \in [m_{min}, m_{max}] \subset \mathcal{R}^1$. Our problem is to investigate the probability that an earthquake occurred with a magnitude less than m, subject to the condition that it could be detected and recorded by the station whose detection capability is characterized by the magnitude m_0. From the relation (1.1.1) $A = \{\omega : \xi(\omega) \geq m_0\}$, $m \geq m_0$ and by its successive modifications we get

$$P(\{\omega : \xi(\omega) < m\}|\{\omega : \xi(\omega) \geq m_0\}) =$$

$$= \frac{P(\{\omega : \xi(\omega) < m\} \cap \{\omega : \xi(\omega) \geq m_0\})}{P(\{\omega : \xi(\omega) \geq m_0\})} =$$

$$= \frac{P(\{\omega : \xi(\omega) \in [m_0, m)\})}{P(\{\omega : \xi(\omega) \geq m_0\})} = \frac{F(m) - F(m_0)}{\exp(\alpha - \beta m_0)/C} =$$

$$= \frac{1 - \exp(\alpha - \beta m)/C - 1 + \exp(\alpha - \beta m_0)/C}{\exp(\alpha - \beta m_0)/C} = 1 - \exp[-\beta(m - m_0)];$$

deriving this expression we obtain the conditional probability density sought in the form

$$f(m|\{\omega : \xi(\omega) \geq m_0\}) = \frac{d\{1 - \exp[-\beta(m - m_0)]\}}{dm} = \beta \exp[-\beta(m - m_0)].$$

The generalization of the notion of conditional probability density for n-dimensional random vectors ($n > 2$) is obvious.

1.2.5 Independent random variables

Definition Random variables ξ_1 and ξ_2 on the probability space $(\Omega, \mathcal{S}, P)$ are independent if, for any real numbers x_1 and x_2, the random events $\{\omega : \xi_1(\omega) < x_1\}$ and $\{\omega : \xi_2(\omega) < x_2\}$ are independent.

The next proposition is implied by the definition of the independence of two random events:

Two random variables ξ_1 and ξ_2 are independent if and only if

$$P(\{\omega : \xi_1(\omega) < x_1, \xi_2(\omega) < x_2\}) = P(\{\omega : \xi_1(\omega) < x_1\})P(\{\omega : \xi_2(\omega) < x_2\}),$$

i.e., if and only if

$$F(x_1, x_2) = F_1(x_1)F_2(x_2), \tag{1.2.2}$$

for any $x_1 \in \mathcal{R}^1$ and any $x_2 \in \mathcal{R}^1$, where $F(\cdot)$, $(x_1, x_2)' \in \mathcal{R}^2$, is the distribution function of the random vector $(\xi_1, \xi_2)'$ and $F_i(\cdot)$, $x_i \in \mathcal{R}^1$, $i = 1, 2$, are the marginal distribution functions.

Components of a two-dimensional continuous random vector $(\xi_1, \xi_2)'$ are stochastically independent if and only if at each point $(x_1, x_2)' \in \mathcal{R}^2$

$$f(x_1, x_2) = f_1(x_1)f_2(x_2),$$

where $f(\cdot)$, $(x_1, x_2)' \in \mathcal{R}^2$, is the probability density of the random vector $(\xi_1, \xi_2)'$ and $f_i(\cdot)$, $x_i \in \mathcal{R}^1$, $i = 1, 2$, are marginal probability densities (the assertion is a direct consequence of (1.2.2), since under the given assumptions $f(x_1, x_2) = \partial^2 F(x_1, x_2))/(\partial x_1 \partial x_2) = [\partial F_1(x_1)/\partial x_1][\partial F_2(x_2)/\partial x_2] = f_1(x_1) f_2(x_2)$, $(x_1, x_2)' \in \mathcal{R}^2$).

Two discrete random variables ξ_1 and ξ_2 are stochastically independent if for each $x_{1k_1} \in \{x_{11}, x_{12}, \ldots, x_{1n_1}\} \subset \mathcal{R}^1$ and each $x_{2k_2} \in \{x_{21}, x_{22}, \ldots, x_{2n_2}\} \subset \mathcal{R}^1$

$$P(\{\omega : \xi_1(\omega) = x_{1k_1}, \xi_2(\omega) = x_{2k_2}\}) =$$

$$= P(\{\omega : \xi_1(\omega) = x_{1k_1}\}) P(\{\omega : \xi_2(\omega) = x_{2k_2}\}),$$

i.e., if

$$p_{1k_1,2k_2} = p_{1\cdot,2k_2} p_{1k_1,2\cdot}.$$

Example The random variables in the table on page 36 are not independent, as, e.g., $p_{1\cdot,21} p_{11,2\cdot} = (14/177)(76/177) = 6.011/177 \neq 13/177 = p_{11,21}$.

The components of an n-dimensional random vector $\xi = (\xi_1, \xi_2, \ldots, \xi_n)'$ are independent if all pairs, triads,..., $(n-1)$-tuples and the n-tuple of its components are independent (it has to be pointed out here that the independence of all pairs of components of an n-dimensional random vector is not sufficient, in a telling way this fact is demonstrated by the Bernstein example ([6],p. 26) of a three-dimensional random vector, where all pairs of components are independent, but the triad is not independent).

1.3 Parameters of the probability distribution

1.3.1 One-dimensional random variable

Definition Let ξ be a discrete random variable on the probability space $(\Omega, \mathcal{S}, P)$ that assumes the real values $x_1, x_2, \ldots, x_k, \ldots$ with probabilities $P(\{\omega : \xi(\omega) = x_k\}) = p_k$; let c be an arbitrary real constant and l an arbitrary integer. Then the lth moment $E[(\xi - c)^l]$ of the random variable ξ at the point c is defined by the expression

$$E[(\xi - c)^l] = \sum_k p_k (x_k - c)^l \tag{1.3.1}$$

under the condition that the series $\sum_k p_k (x_k - c)^l$ converges absolutely.

Definition Let ξ be a continuous random variable on the probability space $(\Omega, \mathcal{S}, P)$ characterized by the probability density $f(\cdot)$, $x \in \mathcal{R}^1$; let c be an arbitrary real constant and l an arbitrary integer. Then the lth moment $E[(\xi - c)^l]$ of the random variable ξ at the point c is defined by the expression

$$E[(\xi - c)^l] = \int_{\mathcal{R}^1} (x - c)^l f(x) dx \tag{1.3.2}$$

under the condition that the subintegral function is absolutely integrable.

If $c = 0$, the moment is called initial, the notation used for it is usually m_l.

If $c = m_1 = E[\xi]$, the moment $E[(\xi - E[\xi])^l] = E[(\xi - m_1)^l]$ is called central; the notation used for it is usually μ_l.

Moments of a random variable ξ are real numbers.

Central moments of a random variable (discrete or continuous) can be expressed in terms of initial moments; before giving explicit formulas, it is necessary to show some fundamental properties of the moments:

(i) $E[(\xi_1 - c_1)^l + (\xi_2 - c_2)^k] = E[(\xi_1 - c_1)^l] + E[(\xi_2 - c_2)^l]$ (ξ_1 and ξ_2 are random variables, c_1 and c_2 arbitrary real numbers and l and k arbitrary integers);

(ii) $E[\{a(\xi - c)\}^l] = a^l E[(\xi - c)^l]$, $a, c \in \mathcal{R}^1$;

(iii) $E[a] = a$, $a \in \mathcal{R}^1$;

(iv) $E[a\xi + b] = aE[\xi] + b$, $a, b \in \mathcal{R}^1$.

These assertions are evident; they follow directly from the definition of moments. Applying them, we can write

$$\mu_1 = E[\xi - m_1] = m_1 - m_1 = 0,$$

$$\mu_2 = E[(\xi - m_1)^2] = E[(\xi^2 - 2m_1\xi + m_1^2)] = m_2 - m_1^2,$$

$$\mu_3 = E[(\xi - m_1)^3] = m_3 - 3m_1 m_2 + 2m_1^3.$$

Definition The first initial moment of a random variable is called its mean value.

Definition The second central moment of a random variable is called its dispersion, $D[\xi] = \mu_2 = E[(\xi - m_1)^2]$.

In accordance with the definition, the second moments of a random variable are real non-negative numbers.

From the set of all second moments the second central moment is the minimum

$$E[(\xi - c)^2] \geq E[(\xi - E[\xi])^2], \; c \in \mathcal{R}^1,$$

the equality occurs for $c = E[\xi]$ $(E[(\xi - c)^2] = E[\{(\xi - E[\xi]) + (E[\xi] - c)\}^2] = E[(\xi - E[\xi])^2] + 2E[(\xi - E[\xi])(E[\xi] - c)] + (E[\xi] - c)^2 = E[(\xi - E[\xi])^2 + (E[\xi] - c)^2$, where $(E[\xi] - c)^2 \geq 0$ and the equality holds for $c = E[\xi])$.

Definition The square root of the dispersion is called a standard deviation.

Definition A random variable whose mean value equals 0 is said to be centred; a random variable whose dispersion equals 1 is said to be normed.

Let η be a random variable such that $E[\eta] = m_1$ and $D[\eta] \neq 0$, then the random variable $\xi = \eta - m_1$ is centred and the random variable $\zeta = \eta / \sqrt{D[\eta]}$ is normed.

By means of the central moments of orders 2, 3 and 4 of a random variable ξ we can define its asymmetry γ_1

$$\gamma_1 = \mu_3 / \mu_2^{3/2}$$

and excess γ_2

$$\gamma_3 = (\mu_4 / \mu_2^2) - 3$$

(obviously under the assumption that all these moments are finite and the dispersion is non-zero). Further numerical characteristics of a random variable are its α-quantiles, median and mode.

Definition Let $F(\cdot)$, $x \in \mathcal{R}^1$, be the distribution function of a random variable ξ. The function $F^{-1}(\alpha) = \inf\{x : F(x) \geq \alpha\}$, $0 < \alpha < 1$, is called a quantile function of the random variable ξ, the number $F^{-1}(\alpha)$ is an α-quantile of the random variable ξ (remember that the quantile function is continuous from the left). The 0.5-quantile is called a median $x_{1/2}$ (Fig.1.3.1).

The reader can find also the following definition of a median: The median is each real number satisfying simultaneously the inequalities $F(x_{1/2}) \leq 1/2$ and $F(x_{1/2} + 0) \geq 1/2$; a median defined in this way is not determined uniquely (for example, in the case of the die, median is, in accordance to the latter definition, any number from the interval $[3, 4]$, as $F(3) = 1/3 < 1/2$ and simultaneously $F(3 + 0) = 1/2$ and $F(4) = 1/2$ and simultaneously

$F(4+0) = 2/3 > 1/2$; according to the former definition, the 0.5-quantile is determined uniquely: $F^{-1}(0.5) = \inf\{x : F(x) \geq 0.5\} = \inf\{x : x \in (3,\infty)\} = 3$).

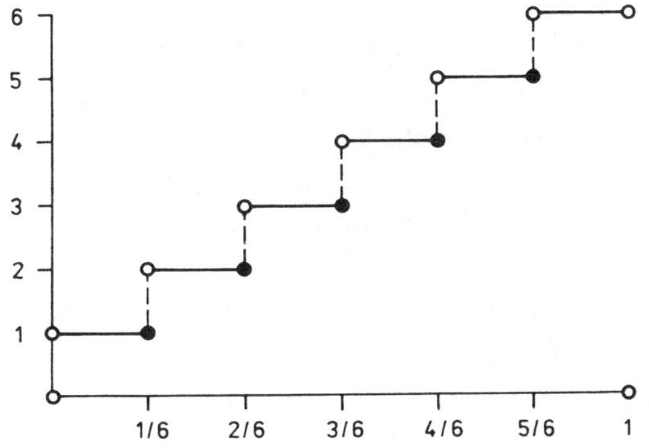

Fig. 1.3.1 The quantile function of the random variable $\xi : \{\omega_1, \omega_2, \omega_3, \omega_4, \omega_5, \omega_6\} \rightarrow \{1,2,3,4,5,6\} \subset \mathcal{R}^1$ on the probability space $(\Omega, \mathcal{S}, P)$, where $\Omega = \{\omega_1, \omega_2, \omega_3, \omega_4, \omega_5, \omega_6\}$, $\mathcal{S}$ is the system of all subsets of the set Ω, $\xi(\omega_i) = i$ and $P\{\omega : \xi(\omega_i) = i\} = p_i = 1/6$, $i = 1, 2, \ldots, 6$ (the example of the toss of a homogeneous die); for the distribution function which belongs to it, see Fig.1.2.2.

The notion of an α-quantile is very important in the domain of testing hypotheses.

Definition A mode (modal value) x_D of a discrete random variable ξ is its maximum probable value; a mode of a continuous random variable is the value at which the probability density becomes maximum.

Example 1. Consider a random variable that assumes the values $x_1 = 0$ and $x_2 = 1$ with the probabilities $P(\{\omega : \xi(\omega) = 0\}) = 1/5$ and $P(\{\omega : \xi(\omega) = 1\}) = 4/5$. The problem is to calculate all its numerical characteristics:

a) the mean value $E[\xi] = 4/5$,
b) the dispersion $D[\xi] = 4/25$,
c) the standard deviation $\sqrt{D[\xi]} = 2/5$,
d) the asymmetry $\gamma_1 = -(12/625)/(8/125) = -3/2$,
e) the excess $\gamma_2 = (52/625)/(16/625) - 3 = 13/4 - 3 = 1/4$,
f) the 0.1-quantile is 0, the 0.9-quantile equals 1,
g) the median $x_{1/2} = F^{-1}(0.5) = \inf\{x : F(x) \geq 0.5\} = \inf\{x : x \in (1,\infty)\} = 1$ (according to the second definition of a median $F(1) = 1/5 < 1/2$ and $F(1+0) = 4/5 > 1/2$),
h) the mode $x_D = 1$.

2. Consider a random variable ξ that assumes the values $x_k = k$ with the probabilities $P(\{\omega : \xi(\omega) = x_k\}) = 1/6,\ k = 1, 2, \ldots, 6$ (the case of a die). Compute in this case all numerical characteristics:

a) the mean value $E[\xi] = \sum_{k=1}^{6} k/6 = 21/6 = 3.5$,
b) the dispersion $D[\xi] = (1/6)\sum_{k=1}^{6}(k - 3.5)^2 = 2.916$,
c) the standard deviation $\sqrt{D[\xi]} = 1.708$,
d) the asymmetry $\gamma_1 = 0$ (because of $\mu_3 = (1/6)\sum_{k=1}^{6}(k - 3.5)^3 = 0$),
e) the excess $\gamma_2 = (14.7291\dot{6}/8.5069\dot{4}) - 3 = -1.26875$,
f) quantiles
$F^{-1}(0.25) = \inf\{x : F(x) \geq 0.25\} = \inf\{x : x \in (2, \infty)\} = 2$
$F^{-1}(0.75) = \inf\{x : F(x) \geq 0.75\} = \inf\{x : x \in (5, \infty)\} = 5$
$F^{-1}(0.10) = \inf\{x : F(x) \geq 0.10\} = \inf\{x : x \in (1, \infty)\} = 1$
$F^{-1}(0.90) = \inf\{x : F(x) \geq 0.90\} = \inf\{x : x \in (6, \infty)\} = 6$
$F^{-1}(0.3\dot{3}) = \inf\{x : F(x) \geq 0.3\dot{3}\} = \inf\{x : x \in (3, \infty)\} = 3$,

thus

1 is the α-quantile for each $\alpha \in (0, 1/6]$,
2 is the α-quantile for each $\alpha \in (1/6, 2/6]$,
3 is the α-quantile for each $\alpha \in (2/6, 3/6]$,
4 is the α-quantile for each $\alpha \in (3/6, 4/6]$,
5 is the α-quantile for each $\alpha \in (4/6, 5/6]$,
6 is the α-quantile for each $\alpha \in (5/6, 1)$,
g) if the median is defined as the 0.5-quantile, then $x_{1/2} = 3$; if the median is defined as a number fulfilling simultaneously the inequalities $F(x_{1/2}) \leq 1/2$ and $F(x_{1/2} + 0) \geq 1/2$, then the median is each number from the interval [4,3].
h) the mode is any number from the set $\{1, 2, 3, 4, 5, 6\}$.

3. Consider a continuous random variable whose probability density is of the form

$$f(x) = \frac{1}{\sigma\sqrt{2\pi}} \exp\left(-\frac{(x - m)^2}{2\sigma^2}\right), \quad x \in \mathcal{R}^1, \tag{1.3.3}$$

$m \in \mathcal{R}^1$ and $\sigma \in (0, \infty)$ are constants.

The probability distribution with the probability density (1.3.3) is called normal; the fact that a random variable ξ possesses a normal probability distribution is written as $\xi \sim N(m, \sigma^2)$ (Figs 1.3.2 and 1.3.3).

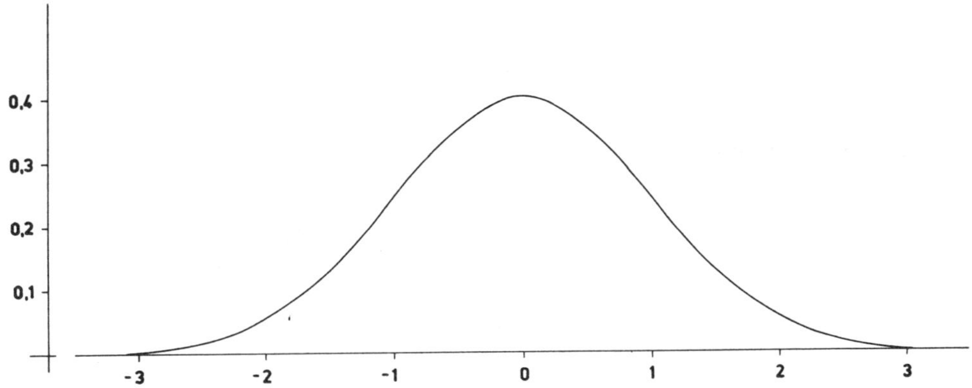

Fig. 1.3.2 The probability density of the random variable $\xi \sim N(0,1)$.

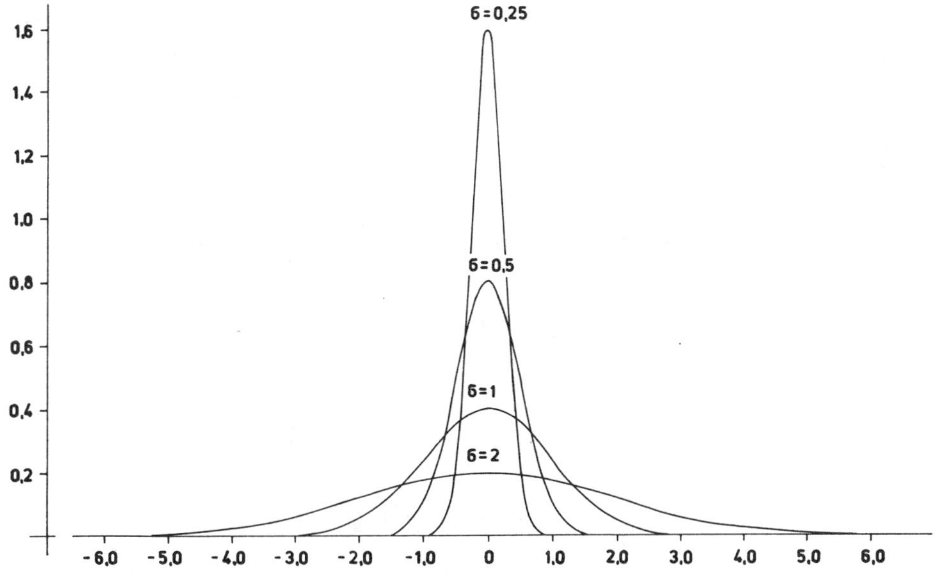

Fig. 1.3.3 The probability densities of the random variable $\xi \sim N(0,\sigma^2)$ for $\sigma = 0.25$, $\sigma = 0.5$, $\sigma = 1.0$ and $\sigma = 2.0$.

The problem is to compute the numerical characteristics of the normally distributed random variable:

a) the mean value

$$E[\xi] = \frac{1}{\sigma\sqrt{2\pi}} \int_{-\infty}^{\infty} x \exp\left(-\frac{(x-m)^2}{2\sigma^2}\right) dx =$$

$$= \frac{\sigma\sqrt{2}}{\sqrt{\pi}} \int_{-\infty}^{\infty} y \exp(-y^2) dy + \frac{m}{\sqrt{\pi}} \int_{-\infty}^{\infty} \exp(-y^2) dy = m$$

(for modifying the integral the substitution $(x-m)/(\sigma\sqrt{2}) = y$ and the relation 3.321 3 $\int_0^\infty \exp(-q^2x^2)dx = \sqrt{\pi}/(2q)$, $q > 0$ [8] were applied);

b) the dispersion

$$D[\xi] = \frac{1}{\sigma\sqrt{2\pi}}\int_{-\infty}^{\infty}(x-m)^2\exp\left(-\frac{(x-m)^2}{2\sigma^2}\right)dx =$$

$$= \frac{2\sigma^2}{\sqrt{\pi}}\int_{-\infty}^{\infty}y^2\exp(-y^2)dy = \sigma^2$$

(the same substitution and the relation 3.4614 $\int_{-\infty}^{\infty}(x+ai)^{2n}\exp(-x^2)dx = \frac{(2n-1)!!}{2^n}\sqrt{\pi}\sum_{k=0}^{n}(-1)^k\frac{(2a)^{2k}n!}{(2k)!(n-k)!}$ from [8] were used).

When calculating the mean value and the dispersion of a normally distributed random variable $\xi \sim N(m,\sigma^2)$ the meaning of the constants m and σ^2 in the relation (1.3.3) has been explained simultaneously;

c) central moments μ_3 and μ_4

$$\mu_3 = \frac{1}{\sigma\sqrt{2\pi}}\int_{-\infty}^{\infty}(x-m)^3\exp\left(-\frac{(x-m)^2}{2\sigma^2}\right)dx =$$

$$= \frac{2^{3/2}\sigma^3}{\sqrt{\pi}}\int_{-\infty}^{\infty}y^3\exp(-y^2)dy = 0,$$

$$\mu_4 = \frac{1}{\sigma\sqrt{2\pi}}\int_{-\infty}^{\infty}(x-m)^4\exp\left(-\frac{(x-m)^2}{2\sigma^2}\right)dx =$$

$$= \frac{2^2\sigma^4}{\sqrt{\pi}}\int_{-\infty}^{\infty}y^4\exp(-y^2)dy = 3\sigma^4,$$

thus $\gamma_1 = 0$ and $\gamma_2 = 0$. The reader can see that all odd central moments of the normal probability distribution are zero and its $2n$th central moment is $\mu_{2n} = (2n-1)!!\sigma^{2n} = 1\cdot 3\cdot 5\cdot\ldots\cdot(2n-1)\sigma^{2n}$, $n = 1,2,\ldots$.

d) the median $x_{1/2} = m$, as

$$F(m) = \frac{1}{\sigma\sqrt{2\pi}}\int_{-\infty}^{m}\exp\left(-\frac{(x-m)^2}{2\sigma^2}\right)dx = \frac{1}{\sqrt{\pi}}\int_{-\infty}^{0}\exp(-y^2)dy = \frac{1}{2};$$

e) the α-quantiles: In order to determine the numerical values of the quantile function $F^{-1}(\cdot)$, $\alpha \in (0,1)$ of a normally distributed random variable suitable statistical tables are to be used containing the values of the distribution function $F(x) = (2\pi)^{-1/2} \int_{-\infty}^{x} \exp(-u^2/2)du$ (it corresponds to the values $m = 0$ and $\sigma = 1$, i.e., to the values of a centred and normed random variable $\eta = (\xi - m)/\sigma$). In order to distinguish quantiles of a normal random variable corresponding to different values of m and σ, the symbol of the inverse distribution function is complemented by an index expressing their actual values.

$$
\begin{aligned}
&F_{0,1}^{-1}(0.005) = \inf\{x : F(x) \geq 0.005\} = -2.576, && F_{0,1}^{-1}(0.995) = 2.576,\\
&F_{0,1}^{-1}(0.01) = \inf\{x : F(x) \geq 0.01\} = -2.326, && F_{0,1}^{-1}(0.99) = 2.326,\\
&F_{0,1}^{-1}(0.025) = \inf\{x : F(x) \geq 0.025\} = -1.960, && F_{0,1}^{-1}(0.975) = 1.960,\\
&F_{0,1}^{-1}(0.05) = \inf\{x : F(x) \geq 0.05\} = -1.645, && F_{0,1}^{-1}(0.95) = 1.645,\\
&F_{0,1}^{-1}(0.1) = \inf\{x : F(x) \geq 0.1\} = -1.282, && F_{0,1}^{-1}(0.9) = 1.282,\\
&F_{0,1}^{-1}(0.2) = \inf\{x : F(x) \geq 0.2\} = -0.842, && F_{0,1}^{-1}(0.8) = 0.842,\\
&F_{0,1}^{-1}(0.5) = 0.
\end{aligned}
$$

If $m \neq 0$ and $\sigma \neq 1$, then $F_{m,\sigma}^{-1}(\alpha) = \sigma F_{0,1}^{-1}(\alpha) + m, \quad \alpha \in (0,1)$.

1.3.2 Random vectors

Definition Let $\xi = (\xi_1, \xi_2, \ldots, \xi_n)'$ be an n-dimensional discrete random vector that has the values $(x_1, x_2, \ldots, x_n)'$ with the probabilities $p_{1k_1,2k_2,\ldots,nk_n}$, $k_i = 1, 2, \ldots, K_i, \ldots,\ \ i = 1, 2, \ldots, n$. Let $c = (c_1, c_2, \ldots, c_n)'$ be an arbitrary n-dimensional real vector and $l_1, l_2, \ldots, l_n$ arbitrary integers; the $(l_1 + l_2 + \ldots + l_n)$th moment $E[(\xi_1 - c_1)^{l_1}(\xi_2 - c_2)^{l_2} \ldots (\xi_n - c_n)^{l_n}]$ of the random vector ξ at the point c is defined by the relation

$$
\begin{aligned}
&E[(\xi_1 - c_1)^{l_1}(\xi_2 - c_2)^{l_2} \ldots (\xi_n - c_n)^{l_n}] = \\
&= \sum_{k_1, k_2, \ldots, k_n} p_{1k_1,2k_2,\ldots,nk_n}(x_{1k_1} - c_1)^{l_1}(x_{2k_2} - c_2)^{l_2} \ldots (x_{nk_n} - c_n)^{l_n}
\end{aligned}
\tag{1.3.4}
$$

under the condition that the series converges absolutely.

Definition Let $\xi = (\xi_1, \xi_2, \ldots, \xi_n)'$ be an n-dimensional continuous random vector characterized by the probability density $f(\cdot)$, $x \in \mathcal{R}^n$.

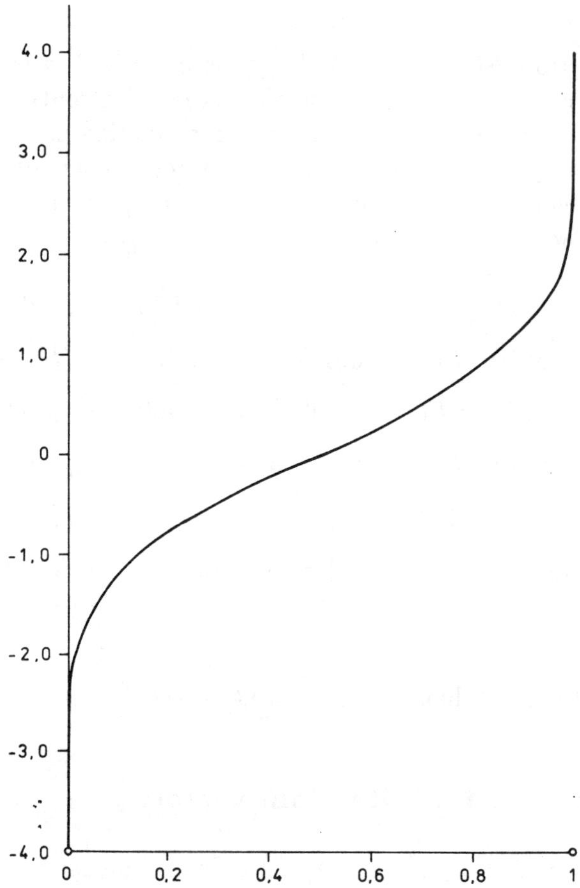

Fig. 1.3.4 The quantile function of the random variable $\xi \sim N(0,1)$.

Let $c = (c_1, c_2, \ldots, c_n)'$ be an arbitrary n-dimensional real vector and $l_1, l_2, \ldots, l_n$ arbitrary integers; the $(l_1 + l_2 + \ldots + l_n)$th moment $E[(\xi_1 - c_1)^{l_1}(\xi_2 - c_2)^{l_2} \ldots (\xi_n - c_n)^{l_n}]$ of the random vector ξ at the point c is defined by the relation

$$E[(\xi_1 - c_1)^{l_1}(\xi_2 - c_2)^{l_2} \ldots (\xi_n - c_n)^{l_n}] =$$
$$= \int_{\mathcal{R}^n} (x_1 - c_1)^{l_1}(x_2 - c_2)^{l_2} \ldots (x_n - c_n)^{l_n} f(x)dx \tag{1.3.5}$$

under the condition that

$$\int_{\mathcal{R}^n} |x_1 - c_1|^{l_1} |x_2 - c_2|^{l_2} \ldots |x_n - c_n|^{l_n} f(x)dx < \infty.$$

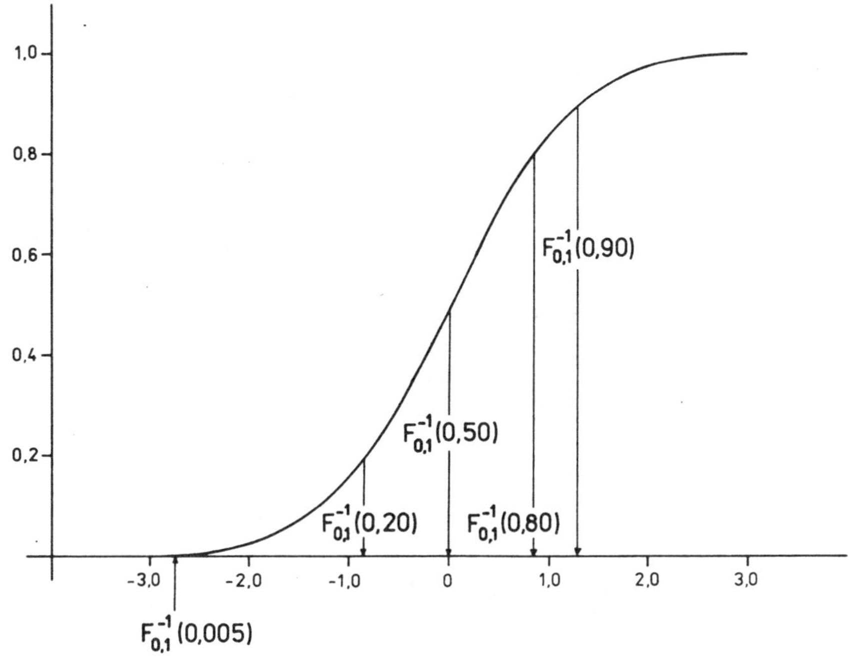

Fig. 1.3.5 The distribution function of the random variable $\xi \sim N(0,1)$ with α-quantiles for α=0.005, α=0.20, α=0.50, α=0.80 and α=0.90.

If $c = 0$, the relations (1.3.4) and (1.3.5) define the initial moments; if $c_i = E[\xi_i]$, $i = 1, 2, \ldots, n$, the moments are called central.

In order to explain precisely the notion of a moment of a random vector let us analyse the relations expressing the initial and the central moments of the first and the second order of a two-dimensional random vector $\xi = (\xi_1, \xi_2)'$.

$$E[\xi_1] = \sum_{k_1}\sum_{k_2} p_{1k_1,2k_2} x_{1k_1} = \sum_{k_1} x_{1k_1} \sum_{k_2} p_{1k_1,2k_2} = \sum_{k_1} x_{1k_1} p_{1k_1,2\cdot},$$

thus $E[\xi_1]$ is the mean value of the marginal probability distribution of the component ξ_1 of the vector $(\xi_1, \xi_2)'$. Analogously, the mean value of the second component ξ_2 of the vector $(\xi_1, \xi_2)'$ is

$$E[\xi_2] = \sum_{k_2} x_{2k_2} p_{1\cdot,2k_2}.$$

The second initial moments are

$$E[\xi_1^2] = \sum_{k_1}\sum_{k_2} x_{1k_1}^2 p_{1k_1,2k_2} = \sum_{k_1} x_{1k_1}^2 p_{1k_1,2\cdot},$$

$$E[\xi_2^2] = \sum_{k_1}\sum_{k_2} x_{2k_2}^2 p_{1k_1,2k_2} = \sum_{k_2} x_{2k_2}^2 p_{1\cdot,2k_2}$$

and

$$E[\xi_1\xi_2] = \sum_{k_1}\sum_{k_2} x_{1k_1} x_{2k_2} p_{1k_1,2k_2}.$$

As far as the central moments of the second order are concerned

$$E[(\xi_1 - E[\xi_1])^2] = \sum_{k_1}(x_{1k_1} - E[\xi_1])^2 p_{1k_1,2\cdot},$$

$$E[(\xi_2 - E[\xi_2])^2] = \sum_{k_2}(x_{2k_2} - E[\xi_2])^2 p_{1\cdot,2k_2}$$

and

$$E[(\xi_1 - E[\xi_1])(\xi_2 - E[\xi_2])] = cov(\xi_1, \xi_2) =$$

$$= \sum_{k_1}\sum_{k_2}(x_{1k_1} - E[\xi_1])(x_{2k_2} - E[\xi_2]) p_{1k_1,2k_2}.$$

The first of the mentioned relations expresses the dispersion of the first component of the random vector $(\xi_1, \xi_2)'$, the second the dispersion of the component ξ_2, and the third the covariance between the components ξ_1 and ξ_2 of the vector ξ.

Further, a survey of analogous relations for a two-dimensional continuous random vector ξ characterized by the probability density $f(\cdot)$, $x \in \mathcal{R}^2$ is given.

The mean values and the dispersions of the first and the second component of the vector ξ and their covariance are

$$E[\xi_i] = \int_{-\infty}^{\infty} x_i f_i(x_i) dx_i, \; i = 1, 2,$$

$$D[\xi_i] = \int_{-\infty}^{\infty} (x_i - E[\xi_i])^2 f_i(x_i) dx_i, \; i = 1, 2$$

and

$$cov(\xi_1, \xi_2) = \int_{-\infty}^{\infty}\int_{-\infty}^{\infty} (x_1 - E[\xi_1])(x_2 - E[\xi_2]) f(x_1, x_2) dx_1 dx_2,$$

respectively, where $f_i(\cdot)$, $x_i \in \mathcal{R}^1$, is the marginal probability density of the ith component $(i = 1, 2)$ of the random vector $\xi = (\xi_1, \xi_2)'$. Expressing now the second mixed central moment of a two-dimensional random vector ξ in terms of the corresponding initial moments, we get

$$cov(\xi_1, \xi_2) = E[(\xi_1 - E[\xi_1])(\xi_2 - E[\xi_2])] =$$
$$= E[\xi_1\xi_2 - E[\xi_1]\xi_2 - \xi_1 E[\xi_2] + E[\xi_1]E[\xi_2]] = E[\xi_1\xi_2] - E[\xi_1]E[\xi_2].$$

Analogously

$$D[\xi_i] = E[\xi_i] - (E[\xi_i])^2, \ i = 1, 2.$$

Thus each two-dimensional random vector $\xi = (\xi_1, \xi_2)'$ is characterized by a two-dimensional vector of mean values $E[\xi] = (E[\xi_1], E[\xi_2])'$ and a 2×2-dimensional covariance matrix

$$var(\xi) = E[(\xi - E[\xi])(\xi - E[\xi])'] = \begin{pmatrix} D[\xi_1], & cov(\xi_1, \xi_2) \\ cov(\xi_2, \xi_1), & D[\xi_2] \end{pmatrix};$$

each n-dimensional random vector $\xi = (\xi_1, \xi_2, \ldots, \xi_n)'$ is characterized by an n-dimensional vector of mean values $E[\xi] = (E[\xi_1], E[\xi_2], \ldots, E[\xi_n])'$ and an $n \times n$ covariance matrix $var(\xi)$

$$var(\xi) = E[(\xi - E[\xi])(\xi - E[\xi])'] =$$

$$= \begin{pmatrix} D[\xi_1], & cov(\xi_1, \xi_2), & \ldots, & cov(\xi_1, \xi_n) \\ cov(\xi_2, \xi_1), & D[\xi_2], & \ldots, & cov(\xi_2, \xi_n), \\ \ldots & \ldots & \ldots & \ldots \\ cov(\xi_n, \xi_1), & cov(\xi_n, \xi_2), & \ldots, & D[\xi_n] \end{pmatrix},$$

which has the following properties:

(i) its diagonal elements $D[\xi_k]$, $k = 1, 2, \ldots, n$, are non-negative;

(ii) it is symmetric;

(iii) $|cov(\xi_k, \xi_l)|^2 \le D[\xi_k]D[\xi_l]$, $k = 1, 2, \ldots, n$, $l = 1, 2, \ldots, n$;

(iv) it is positive definite, i.e., $\sum_{k=1}^{n}\sum_{l=1}^{n} a_k a_l cov(\xi_k, \xi_l) \ge 0$, for any n-tuple of real numbers $a_1, a_2, \ldots, a_n$.

The first property is evident; each even moment, by definition, is a non-negative number. The symmetry of the covariance matrix is a consequence of the fact that $cov(\xi_k, \xi_l) = cov(\xi_l, \xi_k)$; the third property is a direct consequence of the Schwarz inequality, due to which

$$\left|\sum_{k_i}\sum_{k_j} p_{ik_i, jk_j}(x_{ik_i} - E[\xi_i])(x_{jk_j} - E[\xi_j])\right|^2 \le$$

$$\le \sum_{k_i}\sum_{k_j} p_{ik_i, jk_j}(x_{ik_i} - E[\xi_i])^2(x_{jk_j} - E[\xi_j])^2, \ i = 1, 2, \ldots, n, \ j = 1, 2, \ldots, n,$$

if a discrete n-dimensional random vector ξ is taken into account and

$$|\int_{\mathcal{R}^2} f(x_i,x_j)(x_i - E[\xi_i])(x_j - E[\xi_j])dx_i dx_j|^2 \leq$$

$$\leq \int_{\mathcal{R}^2} f(x_i,x_j)(x_i - E[\xi_i])^2 dx_i dx_j \int_{\mathcal{R}^2} f(x_i,x_j)(x_j - E[\xi_j])^2 dx_i dx_j,$$

$i = 1,2,\ldots,n$ $j = 1,2,\ldots,n$, in the case of a continuous random vector ξ; the property (iv) is a direct consequence of the fact that the dispersion of any random variable, thus of the random variable of the form $\eta = \sum_{k=1}^{n} a_k\xi_k$ as well, is non-negative

$$D[\sum_{k=1}^{n} a_k\xi_k] = \sum_{k=1}^{n}\sum_{l=1}^{n} a_k a_l cov(\xi_k,\xi_l) \geq 0.$$

Definition Let $(\xi_1,\xi_2,\ldots,\xi_n)'$ be an n-dimensional random vector; if dispersions of its components exist and are non-zero, the expression

$$\varrho_{i,j} = \frac{cov(\xi_i,\xi_j)}{\sqrt{D[\xi_i]D[\xi_j]}} \tag{1.3.6}$$

defines the correlation coefficient between the random variables ξ_i and ξ_j, $i = 1,2,\ldots,n$, $j = 1,2,\ldots,n$.

The value of the correlation coefficient lies in the interval [−1,1]. This is a direct consequence of the proved inequality $|cov(\xi_i,\xi_j)|^2 \leq D[\xi_i]D[\xi_j]$.

Let us investigate now the value of the correlation coefficient belonging to two independent random variables ξ_i and ξ_j. If two discrete random variables that assume the values $(x_{ik_i}, x_{jk_j})'$ with probabilities p_{ik_i,jk_j} are considered, then (in the case of their independence) $p_{ik_i,jk_j} = p_{ik_i,j\cdot} p_{i\cdot,jk_j}$, $k_i = 1,2,\ldots,K_i,\ldots,$ $k_j = 1,2,\ldots,K_j,\ldots,$ and therefore

$$cov(\xi_i,\xi_j) = \sum_{k_i}\sum_{k_j} p_{ik_i,jk_j}(x_{ik_i} - E[\xi_i])(x_{jk_j} - E[\xi_j]) =$$

$$= \sum_{k_i} p_{ik_i,j\cdot}(x_{ik_i} - E[\xi_i]) \sum_{k_j} p_{i\cdot,jk_j}(x_{jk_j} - E[\xi_j]) = 0;$$

in the case of two continuous independent random variables with probability density $f(\cdot)$, $(x_i,x_j) \in \mathcal{R}^2$,

$$cov(\xi_i,\xi_j) = \int_{-\infty}^{\infty}\int_{-\infty}^{\infty} f(x_i,x_j)(x_i - E[\xi_i])(x_j - E[\xi_j])dx_i dx_j =$$

$$= \int_{-\infty}^{\infty} f_i(x_i)(x_i - E[\xi_i])dx_i \int_{-\infty}^{\infty} f(x_j)(x_j - E[\xi_j])dx_j = 0,$$

since as a consequence of the independence $f(x_i, x_j) = f_i(x_i)f_j(x_j)$, $x_i, x_j \in \mathcal{R}^1$. We have shown that the covariance and, consequently, the correlation coefficient of two independent random variables is zero; recall that the zero covariance and the correlation coefficient do not mean that the random variables are independent; the zero covariance and the zero correlation coefficient represent a necessary but not sufficient condition for the independence of two random variables; if the correlation coefficient of two random variables is zero, the random variables are said to be non-correlated.

What is the relation of two random variables whose correlation coefficient equals 1 or −1? The answer to this important and frequently occurring question is found in the following assertion.

The correlation coefficient $\varrho_{i,j}$ of the random variables ξ_i and ξ_j equals +1 if and only if with probability 1 $\xi_i = a + b\xi_j$, where a is a real and b a positive number; the correlation coefficient equals −1 if and only if with probability 1 $\xi_i = a + b\xi_j$, where $a \in \mathcal{R}^1$ and b is a negative number.

The assertion follows from the inequality $|cov(\xi_i, \xi_j)|^2 \leq D[\xi_i]D[\xi_j]$ rewritten in the form

$$|E[(\xi_i - E[\xi_i])(\xi_j - E[\xi_j])]| \leq \{E[(\xi_i - E[\xi_i])^2]E[(\xi_j - E[\xi_j])^2]\}^{1/2}.$$

The equality occurs if and only if with probability 1 $\xi_i = a + b\xi_j$, $a, b \in \mathcal{R}^1$, $b \neq 0$ (the assertion applies to random variables, therefore the probability with which it occurs must be stated). In this case $E[\xi_i] = a + bE[\xi_j]$, $\xi_i - E[\xi_i] = b(\xi_j - E[\xi_j])$, $E[(\xi_i - E[\xi_i])^2] = b^2E[(\xi_j - E[\xi_j])^2]$ and $E[(\xi_i - E[\xi_i])(\xi_j - E[\xi_j])] = bE[(\xi_j - E[\xi_j])^2]$, which ensures the equality (the trivial case of the equality for $\xi_i = E[\xi_i]$ or $\xi_j = E[\xi_j]$ cannot be taken into account because of the assumptions $D[\xi_i] > 0$ and $D[\xi_j] > 0$). As $\varrho_{i,j} = bE[(\xi_j - E[\xi_j])^2]/|b|E[(\xi_j - E[\xi_j])^2]$, actually $\varrho_{i,j} = 1$ if $b > 0$ and $\varrho_{i,j} = -1$ if $b < 0$.

The next problem is to investigate how a linear transformation of an n-dimensional random vector influences the fundamental moment characteristics.

Let ξ be an n-dimensional random vector characterized by its mean value $E[\xi]$ and the covariance matrix $var(\xi)$. Let T be a $m \times n$-dimensional matrix and t a m-dimensional vector. The mean value and the covariance matrix of the m-dimensional random vector $\eta = t + T\xi$ are

$$E[\eta] = TE[\xi] + t$$

and

$$var(\eta) = Tvar(\xi)T',$$

respectively.

Both assertions follow immediately from the definitions of the mean value and the covariance matrix of a random vector η : $E[\eta] = E[T\xi + t] = TE[\xi] + t$; $var(\eta) = E[(\eta - E[\eta])(\eta - E[\eta])'] = E[T(\xi - E[\xi])(\xi - E[\xi])'T'] = TE[(\xi - E[\xi])(\xi - E[\xi])']T' = Tvar(\xi)T'$.

If $T = i'$ (an n-dimensional row vector whose all elements are 1), then $\eta = i'\xi$ is a one-dimensional random variable such that $E[\eta] = i'E[\xi]$ (the mean value of the sum of random variables equals the sum of their mean values) and

$$D[\eta] = i'var(\xi)i = i'\begin{pmatrix} D[\xi_1], & cov(\xi_1,\xi_2), & \dots, & cov(\xi_1,\xi_n) \\ cov(\xi_2,\xi_1), & D[\xi_2], & \dots, & cov(\xi_2,\xi_n) \\ \dots & \dots & \dots & \dots \\ cov(\xi_n,\xi_1), & cov(\xi_n,\xi_2), & \dots, & D[\xi_n] \end{pmatrix} i,$$

and $D[\eta] = \sum_{i=1}^{n} D[\xi_i]$ if and only if $cov(\xi_k, \xi_l) = 0$ for all possible pairs of the random variables ξ_k, ξ_l, $k = 1, 2, \dots, n$, $l = 1, 2, \dots, n$, $k \neq l$.

Analogous results may be obtained for any random variable which is a linear combination of components of a random vector $\xi = (\xi_1, \xi_2, \dots, \xi_n)'$; especially, $E[a_i\xi_i \pm a_j\xi_j] = a_iE[\xi_i] \pm a_jE[\xi_j]$ and $D[a_i\xi_i \pm a_j\xi_j] = a_i^2D[\xi_i] + a_j^2D[\xi_j] \pm 2a_ia_jcov(\xi_i, \xi_j)$.

Definition Let $(\xi_1, \xi_2)'$ be a two-dimensional discrete random vector that assumes the values $(x_{1k_1}, x_{2k_2})'$ with probabilities

$$p_{1k_1,2k_2},\ k_1 = 1, 2, \dots, K_1, \dots,\ k_2 = 1, 2, \dots, K_2, \dots,;$$

a random variable $E[\xi_1\xi_2]$ defined for each realization x_{2k_2} of the random variable ξ_2 by the relation

$$E[\xi_1|x_{2k_2}] = \sum_{k_1} x_{1k_1}\frac{p_{1k_1,2k_2}}{p_{1\cdot,2k_2}},\ x_{2k_2} \in \{x_{21}, x_{22}, \dots, x_{2K_2}, \dots\},$$

is called a conditional mean value of the random variable ξ_1.

The discrete (finite or countable) random variable $E[\xi_1|\xi_2]$ assumes the values

$$E[\xi_1|x_{21}] = \sum_{k_1} x_{1k_1}\frac{p_{1k_1,21}}{p_{1\cdot,21}}, E[\xi_1|x_{22}] = \sum_{k_1} x_{1k_1}\frac{p_{1k_1,22}}{p_{1\cdot,22}}, \dots, E[\xi_1|x_{2K_2}] =$$

$$= \sum_{k_1} x_{1k_1} \frac{p_{1k_1,2K_2}}{p_{1\cdot,2K_2}}, \ldots$$

with probabilities $p_{1\cdot,21}, p_{1\cdot,22}, \ldots, p_{1\cdot,2K_2}, \ldots$.

The mean value and the dispersion of the random variable $E[\xi_1|\xi_2]$ are

$$E[E[\xi_1|\xi_2]] = \sum_{k_2} p_{1\cdot,2k_2} E[\xi_1|x_{2k_2}] = \sum_{k_2} p_{1\cdot,2k_2} \sum_{k_1} x_{1k_1} \frac{p_{1k_1,2k_2}}{p_{1\cdot,2k_2}} =$$

$$= \sum_{k_1} x_{1k_1} \sum_{k_2} p_{1k_1,2k_2} = \sum_{k_1} x_{1k_1} p_{1k_1,2\cdot} = E[\xi_1],$$

$$D[E[\xi_1|\xi_2]] = \sum_{k_2} p_{1\cdot,k_2} (E[\xi_1|\xi_2])^2 - (E[\xi_1])^2.$$

Definition Let $(\xi_1, \xi_2)'$ be a continuous random vector characterized by the probability density $f(\cdot)$, $(x_1, x_2)' \in \mathcal{R}^2$; a random variable $E[\xi_1|\xi_2]$ defined for each realization x_2 of the random variable ξ_2 by the relation

$$E[\xi_1|x_2] = \int_{\mathcal{R}^1} x_1 \frac{f(x_1, x_2)}{f_2(x_2)} dx_1, \ x_1 \in \mathcal{R}^1,$$

is called a conditional mean value of the random variable ξ_1.

Thus $E[\xi_1|\xi_2]$ is a one-dimensional random variable, its probability density equals the marginal probability density $f_2(\cdot)$, $x_2 \in \mathcal{R}^1$.

The mean value and the dispersion of the random variable $E[\xi_1|\xi_2]$ are

$$E[E[\xi_1|\xi_2]] = \int_{\mathcal{R}^1} E[\xi_1|x_2] f_2(x_2) dx_2 = \int_{\mathcal{R}^1} x_1 \int_{\mathcal{R}^1} f(x_1, x_2) dx_2 dx_1 = E[\xi_1]$$

and

$$D[E[\xi_1|\xi_2]] = \int_{\mathcal{R}^1} f_2(x_2)(E[\xi_1|x_2])^2 dx_2 - (E[\xi_1])^2,$$

respectively.

If the components of the two-dimensional random vector $(\xi_1, \xi_2)'$ are independent, $E[\xi_1|\xi_2]] = E[\xi_1]$; e.g., if a continuous random vector $(\xi_1, \xi_2)'$ characterized by the probability density $f(\cdot)$, $(x_1, x_2) \in \mathcal{R}^2$, is considered,

$$E[\xi_1|\xi_2] = \int_{\mathcal{R}^1} x_1 \frac{f(x_1, x_2)}{f_2(x_2)} dx_1 = \int_{\mathcal{R}^1} x_1 \frac{f_1(x_1) f_2(x_2)}{f_2(x_2)} dx_1 = E[\xi_1].$$

An analogical relation can be obtained for a random variable $E[\xi_2|\xi_1]$. The same method can be used for defining random variables $D[\xi_1|\xi_2]$ and $D[\xi_2|\xi_1]$ and their mean values and dispersions.

Example. The two-dimensional discrete random vector $(\xi_1, \xi_2)'$ whose probability distribution is described by the table on page 36 has the following numerical characteristics:

$E[\xi_1] = 278/177 = 1.57062,$ $E[\xi_2] = 622/177 = 3.51412,$
$E[\xi_1^2] = 480/177 = 2.71186,$ $E[\xi_2^2] = 2578/177 = 14.28249,$
$D[\xi_1] = 43.36677/177 = 0.24501,$ $D[\xi_2] = 342.21534/177 = 1.93342,$
$E[\xi_1\xi_2] = 1043/177 = 5.89266,$ $cov(\xi_1, \xi_2) = 66.07410/177 = 0.373300,$
$\varrho_{1,2} = 0.54238.$

Before determining the values that are possible realizations of the random variables $E[\xi_1|\xi_2]$ and $E[\xi_2|\xi_1]$ and their mean values and dispersions, the conditional probabilities $P(\{\omega : \xi_1(\omega) = x_{1k_1}\}|\{\omega : \xi_2(\omega) = x_{2k_2}\})$ and $P(\{\omega : \xi_2(\omega) = x_{2k_2}\}|\{\omega : \xi_1(\omega) = x_{1k_1}\})$, $k_1 = 1, 2,$ $k_2 = 1, 2, \ldots, 6$, have to be determined. For the numerical data see Tables 1.2 and 1.3.

Table 1.2

Conditional probabilities
$P(\{\omega : \xi_1(\omega) = x_{1k_1}\}|\{\omega : \xi_2(\omega = x_{2k_2}\}) = p_{1k_1,2k_2}/p_{1\cdot,2k_2}$

$x_{2k_2} \setminus x_{1k_1}$	1	2
1	13/14	1/14
2	28/34	6/34
3	19/37	18/37
4	8/43	35/43
5	5/37	32/37
6	3/12	9/12

Table 1.3

The random variables $E[\xi_1|\xi_2]$ and $D[\xi_1|\xi_2]$ assume the values $E[\xi_1|x_{2k_2}]$ and $D[\xi_1|x_{2k_2}]$ with probabilities $p_{1\cdot,2k_2}$

x_{2k_2}	$E[\xi_1\|x_{2k_2}]$	$D[\xi_1\|x_{2k_2}]$	$p_{1\cdot,2k_2}$
1	15/14=1.071 43	$17/14-(15/14)^2$=0.0663	14/177
2	40/43=1.176 47	$52/34-(40/34)^2$=0.1453	34/177
3	55/37=1.486 49	$91/37-(55/37)^2$=0.2498	37/177
4	78/43=1.813 95	$148/43-(78/43)^2$=0.1514	43/177
5	69/37=1.864 86	$133/37-(69/37)^2$=0.1169	37/177
6	21/12=1.750 00	$39/12-(21/12)^2$=0.1875	12/177

$E[E[\xi_1|\xi_2]] = 278/177 = 1.57062,$

$E[(E[\xi_1|\xi_2])^2] = 451.80106/177 = 2.55255,$

$D[E[\xi_1|\xi_2]] = 0.08570$
and analogously (cf. Tables 1.4. and 1.5)

Table 1.4

Conditional probabilities
$P(\{\omega : \xi_2(\omega) = x_{2k_2}\}|\{\omega : \xi_1(\omega) = x_{1k_1}\}) = p_{1k_1,2k_2}/p_{1k_1,2\cdot}$

$x_{1k_1} \setminus x_{2k_2}$	1	2	3	4	5	6
1	13/76	28/76	19/76	8/76	5/76	3/76
2	1/101	6/101	18/101	35/101	32/101	9/101

Table 1.5

The random variables $E[\xi_2|\xi_1]$ and $D[\xi_2|\xi_1]$ assume the values $E[\xi_2|x_{1k_1}]$ and $D[\xi_2|x_{1k_1}]$ with probabilities $p_{1k_1,2\cdot}$

x_{1k_1}	$E[\xi_2\|x_{1k_1}]$	$D[\xi_2\|x_{1k_1}]$	$p_{1k_1,2\cdot}$
1	201/76=2.644 74	657/76−(201/76)^2=1.6501	76/177
2	421/101=4.168 32	1 871/101−(421/101)^2=1.1500	101/177

$E[E[\xi_2|\xi_1]] = 622/177 = 3.51412,$
$E[(E[\xi_2|\xi_1])^2] = 2286.45349/177 = 12.91782,$
$D[E[\xi_2|\xi_1]] = 0.56878.$

2. Let us consider a two-dimensional random variable with the probability density

$$f(x_1,x_2) = \frac{1}{2\pi\sigma_1\sigma_2\sqrt{1-\varrho_{1,2}^2}} \exp\left(-\frac{(x_1-m_1)^2}{2\sigma_1^2(1-\varrho_{1,2}^2)} + \right.$$

$$\left. + \frac{\varrho_{1,2}(x_1-m_1)(x_2-m_2)}{\sigma_1\sigma_2(1-\varrho_{1,2}^2)} - \frac{(x_2-m_2)^2}{2\sigma_2^2(1-\varrho_{1,2}^2)}\right),$$

$$(x_1,x_2) \in \mathcal{R}^2,\ m_1,m_2 \in \mathcal{R}^1,\ \sigma_1,\sigma_2 \in (0,\infty),\ \varrho \in (-1,1) \tag{1.3.7}$$

(the probability density of a normal probability distribution). The problem is to determine the fundamental numerical characteristics of the random vector $(\xi_1,\xi_2)'$.

First the marginal probability densities will be determined:

$$f_1(x_1) = \int_{-\infty}^{\infty} f(x_1, x_2)dx_2 = \frac{1}{2\pi\sigma_1\sigma_2\sqrt{1-\varrho_{1,2}^2}} \exp\left(-\frac{(x_1-m_1)^2}{2\sigma_1^2(1-\varrho_{1,2}^2)}\right) \cdot$$

$$\cdot \int_{-\infty}^{\infty} \exp\left(-\frac{(x_2-m_2)}{2\sigma_2^2(1-\varrho_{1,2}^2)} + \frac{\varrho_{1,2}(x_1-m_1)(x_2-m_2)}{\sigma_1\sigma_2(1-\varrho_{1,2}^2)}\right) dx_2 =$$

$$= \frac{1}{\sqrt{2\pi}\sigma_1} \exp\left(-\frac{(x_1-m_1)^2}{2\sigma_1^2(1-\varrho_{1,2}^2)} + \frac{\varrho_{1,2}^2(x_1-m_1)^2}{2\sigma_1^2(1-\varrho_{1,2}^2)}\right) =$$

$$\frac{1}{\sqrt{2\pi}\sigma_1} \exp\left(-\frac{(x_1-m_1)^2}{2\sigma_1^2}\right)$$

(the substitution $x_2 - m_2 = y_2$ and the relation 3 322 2

$$\int_{-\infty}^{\infty} \exp(-p^2x^2 \pm qx)dx = \exp(q^2/4p^2)\sqrt{\pi}/p$$

from [8] have been used).

Analogously,

$$f_2(x_2) = \frac{1}{\sqrt{2\pi}\sigma_2} \exp\left(-\frac{(x_2-m_2)^2}{2\sigma_2^2}\right).$$

If we compare the expressions obtained for marginal probability densities $f_1(\cdot)$ and $f_2(\cdot)$ with the relation (1.3.3), we can state that each of the components of a normal two-dimensional random vector $(\xi_1, \xi_2)'$ is also normal and simultaneously $E[\xi_1] = m_1$, $E[\xi_2] = m_2$, $D[\xi_1] = \sigma_1^2$ and $D[\xi_2] = \sigma_2^2$. Further, let us determine the covariance between the components of the random vector $(\xi_1, \xi_2)'$

$$cov(\xi_1, \xi_2) =$$

$$= \frac{1}{2\pi\sigma_1\sigma_2\sqrt{1-\varrho_{1,2}^2}} \int_{-\infty}^{\infty}\int_{-\infty}^{\infty} (x_1-m_1)(x_2-m_2) \exp\left(-\frac{(x_1-m_1)^2}{2\sigma_1^2(1-\varrho_{1,2}^2)} + \right.$$

$$\left. + \frac{\varrho_{1,2}(x_1-m_1)(x_2-m_2)}{\sigma_1\sigma_2(1-\varrho_{1,2}^2)} - \frac{(x_2-m_2)^2}{2\sigma_2^2(1-\varrho_{1,2}^2)'}\right) dx_1dx_2 =$$

$$= \frac{1}{2\pi\sigma_1\sigma_2\sqrt{1-\varrho_{1,2}^2}} \cdot \int_{-\infty}^{\infty} (x_1-m_1) \exp\left(-\frac{(x_1-m_1)^2}{2\sigma_1^2(1-\varrho_{1,2}^2)}\right) \cdot$$

$$\cdot \int_{-\infty}^{\infty} (x_2 - m_2) \exp\left(\frac{\varrho_{1,2}(x_1 - m_1)(x_2 - m_2)}{\sigma_1 \sigma_2 (1 - \varrho_{1,2})} - \frac{(x_2 - m_2)^2}{2\sigma_2^2 (1 - \varrho_{1,2}^1)} \right) dx_1 dx_2 =$$

$$= \frac{\sigma_2 \varrho_{1,2}}{2\pi\sigma_1^2 \sqrt{1 - \varrho_{1,2}^2}} \int_{-\infty}^{\infty} (x_1 - m_1)^2 \exp\left(\exp - \frac{(x_1 - m_1)^2}{2\sigma_1^2} \right) dx_1 =$$

$$= \sigma_1 \sigma_2 \varrho_{1,2}$$

(the substitutions $x_i - m_i = y_i$, $i = 1, 2$, the relations $\int_{-\infty}^{\infty} x \exp(-px^2 + 2qx)dx = \frac{q}{p}\sqrt{\frac{\pi}{p}} \exp(q^2/p)$, $p > 0$, see [8], the relation 3 462 6 and $\int_{-\infty}^{\infty} x^2 \exp(-\mu x^2)dx = \frac{1}{2\mu}\sqrt{\frac{\pi}{\mu}}$, see [8], the relation 3 462 8, have been applied). Thus

$$cov(\xi_1, \xi_2) = \sigma_1 \sigma_2 \varrho_{1,2}$$

and

$$var((\xi_1, \xi_2)') = \begin{pmatrix} \sigma_1^2, & \sigma_1 \sigma_2 \varrho_{1,2} \\ \sigma_1 \sigma_2 \varrho_{1,2} & \sigma_2^2 \end{pmatrix}.$$

Analyse the conditional mean values $E[\xi_1|\xi_2]$ and $E[\xi_2|\xi_1]$. For this purpose the conditional probability densities $f(\cdot|x_2)$, $x_1 \in \mathcal{R}^1$, and $f(\cdot|x_1)$, $x_2 \in \mathcal{R}^1$ have to be determined:

$$f(x_1|x_2) = f(x_1, x_2)/f_2(x_2) =$$

$$= \frac{1}{2\sigma_1 \sqrt{1 - \varrho_{1,2}^2}} \exp\left(- \frac{[x_1 - m_1 - \varrho_{1,2}\sigma_1(x_2 - m_2)/\sigma_2]^2}{2\sigma_1^2 (1 - \varrho_{1,2}^2)} \right).$$

Realizing the meaning of the numerical coefficients in the relation for the probability density of a normally distributed random variable, taking simultaneously into account that the given relation is true for an arbitrary but fixed $x_2 \in \mathcal{R}^1$, we can state that$E[\xi_1|x_2] = m_1 + \varrho_{1,2}\sigma_1(x_2 - m_2)/\sigma_2$ is a realization of the random variable

$$E[\xi_1|\xi_2] = m_1 + \varrho_{1,2}\sigma_1(\xi_2 - m_2)/\sigma_2$$

whose mean value equals $[E[\xi_1|\xi_2]] = m_1$ and the dispersion is $D[E[\xi_1|\xi_2]] = \varrho_{1,2}^2 \sigma_1^2$.

Analogously

$$E[\xi_2|\xi_1] = m_2 + \varrho_{1,2}\sigma_2(\xi_1 - m_1)/\sigma_1$$

and $E[E[\xi_2|\xi_1]] = m_2$, $D[E[\xi_2|\xi_1]] = \varrho_{1,2}^2 \sigma_2^2$.

For the series of realizations of the conditional mean values $E[\xi_1|x_2]$ and $E[\xi_2|x_1]$ of a normal random vector $(\xi_1, \xi_2)'$ such that $E[\xi_1] = 0$, $E[\xi_2] = 0$, $D[\xi_1] = 1$, $D[\xi_2] = 1$ and $cov(\xi_1, \xi_2) = 1/2$, see Fig. 1.3.6; the dash lines indicate the value of the standard deviation.

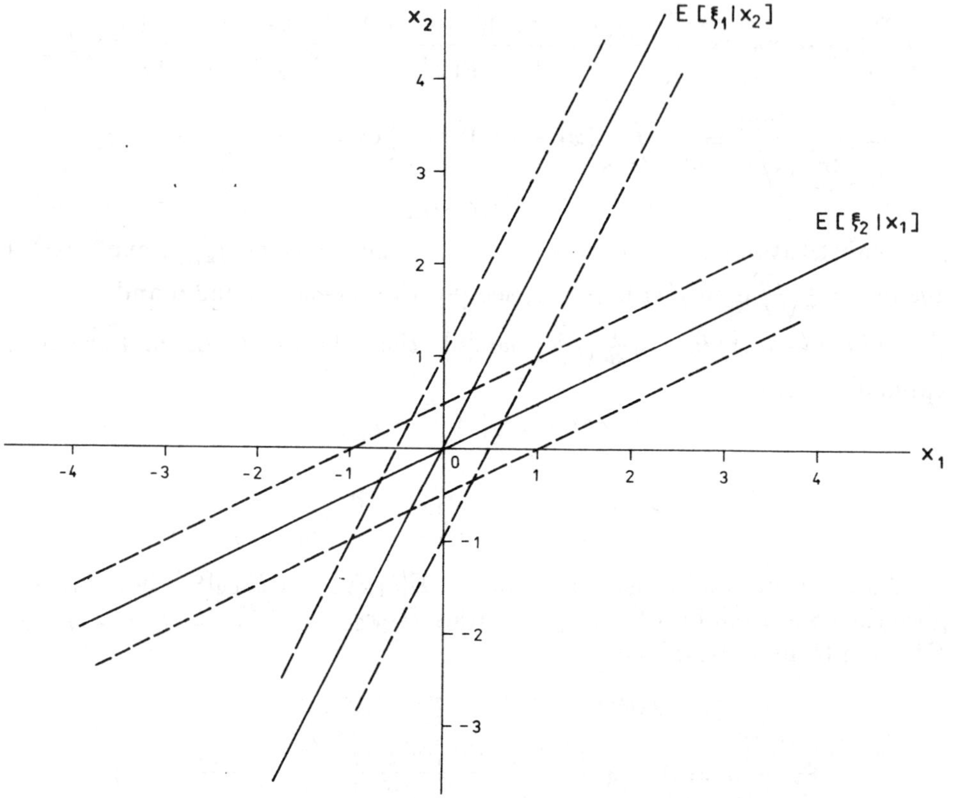

Fig. 1.3.6

1.3.3 Theoretical regression

The value of the covariance between two random variables ξ and η characterizes the strength of the stochastic coupling between them. Naturally, the question arises, how to utilize this information for approximating one random variable by means of the other. The approximation $\eta_{opt}(\xi) = u(\xi)$ fulfilling the condition

$$E[(\eta - \eta_{opt}(\xi))^2] = min \tag{1.3.8}$$

is understood to be the optimum approximation of the random variable η by means of the random variable ξ.

Let $(\xi,\eta)'$ be a two-dimensional discrete random vector characterized by the probability distribution $p_{k,l} = P(\{\omega : \xi(\omega) = x_k,\ \eta(\omega) = y_l\}), k = 1,2,\dots,K,\dots,\ l = 1,2,\dots,L,\dots$ Then

$$E[(\eta - u(\xi))^2] = \sum_k \sum_l (y_l - u(x_k))^2 p_{k,l} =$$

$$= \sum_k \sum_l (y_l - u(x_k))^2 P(\{\omega : \eta(\omega) = y_l\}|\{\omega : \xi(\omega) = x_k\}) p_{k,\cdot} =$$

$$= \sum_k p_{k,\cdot} \sum_l (y_l - u(x_k))^2 P(\{\omega : \eta(\omega) = y_l\}|\{\omega : \xi(\omega) = x_k\})$$

and this expression acquires its minimum if and only if $u(x_k) = E[\eta|x_k]$, since $\sum_l (y_l - u(x_k))^2 P(\{\omega : \eta(\omega) = y_l\}|\{\omega : \xi(\omega) = x_k\})$ is the conditional second moment of the random variable η at the point $u(x_k) \in \mathcal{R}^1$ and it acquires its minimum at the point $E[\eta|x_k] \in \mathcal{R}^1$ (from the set of all second moments the central one is minimum - cf. the properties of the second moments, p. 44).

Let $(\xi,\eta)'$ be a continuous random vector characterized by the probability density $f(\cdot)$, $(x,y)' \in \mathcal{R}^2$. Then analogously

$$E[(\eta - u(\xi))^2] = \int_{-\infty}^{\infty}\int_{-\infty}^{\infty} (y - u(x))^2 f(x,y)\,dx\,dy =$$

$$= \int_{-\infty}^{\infty}\int_{-\infty}^{\infty} (y - u(x))^2 f(y|x) f_1(x)\,dx\,dy =$$

$$= \int_{-\infty}^{\infty} f_1(x) \int_{-\infty}^{\infty} (y - u(x))^2 f(y|x)\,dy\,dx$$

and this expression is minimum at the point $E[\eta|x]$, $x \in \mathcal{R}^1$. It has been proved that the random variable $E[\eta|\xi]$ is, in the sense of the criterion $E[(\eta - \eta_{opt}(\xi))^2]$, the optimum approximation of the random variable η.

In the same way it can be shown that the random variable $E[\xi|\eta]$ is in the sense of the criterion $E[(\xi - \xi_{opt}(\eta))^2]$ the optimum approximation of the random variable ξ. In the previous section the expressions for the mean value and the dispersion of the random variable $E[\eta|\xi]$ have been derived. According to them $E[\eta_{opt}(\xi)] = E[\eta]$ and $D[\eta_{opt}(\xi)] = E[\eta_{opt}^2(\xi)] - (E[\eta])^2$. The covariance $cov(\eta,\eta_{opt}(\xi))$, due to its definition, is

$$cov(\eta,\eta_{opt}(\xi)) = E[(\eta - E[\eta])(\eta_{opt}(\xi) - E[\eta_{opt}(\xi)])] =$$

$$= E[\eta\eta_{opt}(\xi)] - (E[\eta])^2$$

(the fact that $E[\eta_{opt}(\xi)] = E[\eta]$ was applied), therefore the next problem is to determine the initial second moment $E[\eta\eta_{opt}(\xi)]$.

First, consider a two-dimensional random vector $(\xi, \eta)'$ characterized by a probability distribution $p_{k,l} = P(\{\omega : \xi(\omega) = x_k, \eta(\omega) = y_l\})$, $k = 1, 2, \ldots, K \ldots$, $l = 1, 2, \ldots, L, \ldots$. According to the definition

$$E[\eta\eta_{opt}(\xi)] = E[\eta E[\eta|\xi]] = \sum_k \sum_l E[\eta|x_k] y_l p_{k,l} =$$

$$\sum_k E[\eta|x_k] p_{k,\cdot} \sum_l y_l \frac{p_{k,l}}{p_{k,\cdot}} =$$

$$= \sum_k (E[\eta|x_k])^2 p_{k,\cdot} = E[(E[\eta|\xi])^2].$$

Taking into account a continuous random vector $(\xi, \eta)'$ characterized by the probability density $f(\cdot)$, $(x, y)' \in \mathcal{R}^2$, we obtain the same result:

$$E[\eta\eta_{opt}(\xi)] = E[\eta E[\eta|\xi]] = \int_{-\infty}^{\infty} \int_{-\infty}^{\infty} E[\eta|x] y f(x, y) dx dy =$$

$$= \int_{-\infty}^{\infty} E[\eta|x] f_1(x) \int_{-\infty}^{\infty} y \frac{f(x, y)}{f_1(x)} dx dy =$$

$$= \int_{-\infty}^{\infty} (E[\eta|x])^2 f_1(x) dx = E[(E[\eta|\xi])^2],$$

thus we have proved that

$$cov(\eta, \eta_{opt}(\xi)) = D[\eta_{opt}(\xi)],$$

which means that the dispersion of the residuum $\eta - \eta_{opt}(\xi)$ is

$$D[\eta - \eta_{opt}(\xi)] = D[\eta] - D[\eta_{opt}(\xi)].$$

Definition Let $(\xi, \eta)'$ be a two-dimensional discrete random vector that assumes the values $(x_k, y_l)' \in \mathcal{R}^2$ with the probabilities $p_{k,l}$, $k = 1, 2, \ldots, K, \ldots$, $l = 1, 2, \ldots, L \ldots$. A dotted line consisting of the points $(x_k, E[\eta|x_k] \equiv \eta_{opt}(x_k))'$ is called a regression line of the first kind with respect to the random variable ξ; a dotted curve built by the points $(E[\xi|y_l] \equiv \xi_{opt}(y_l),\ y_l)'$ is a regression line of the first kind with respect to the random variable η.

Let $(\xi, \eta)'$ be a two-dimensional continuous random variable characterized by the probability density $f(\cdot)$, $(x, y)' \in \mathcal{R}^2$. The curve consisting of points $(x, E[\eta|x] \equiv \eta_{opt}(x))'$ is called a regression line of the first kind with respect to the random variable ξ; the curve consisting of points $(y, E[\xi|y] \equiv \xi_{opt}(y))'$ is called a regression line of the first kind with respect to the random variable η.

Finally, consider two independent random variables ξ and η. The reader can convince himself that in this case $E[\eta|\xi] = E[\eta]$ and $E[\xi|\eta] = E[\xi]$), which means that $\eta_{opt}(\xi) = E[\eta]$ and $\xi_{opt}(\eta) = E[\xi]$, respectively, thus the regression line of the first kind with respect to the random variable ξ is a straight line parallel with the axis x at the distance $E[\eta]$ and the regression line of the first kind with respect to the random variable η is a straight line parallel with the axis y at the distance $E[\xi]$.

Let us analyse these notions using the example of the discrete random vector $(\xi_1, \xi_2)'$ possessing the probability distribution given on p. 36. Taking into account the numerical data from Table 1.4, we can see that the regression line of the first kind with respect to the random variable ξ_1 is given by two points, namely $(1, 2.64)'$ and $(2, 4.17)'$, represented in Fig. 1.3.7 by solid circles; it means that the optimum (in the sense of the criterion $E[(\xi_2 - u(\xi_1))^2] = min$) approximation of the random variable ξ_2 is realized by the value 2.64 under the condition that the random variable ξ_1 has been realized by the value 1; if ξ_1 has been realized by the value 2, the optimum approximation of the random variable ξ_2 is realized by the value $\xi_{opt}(2) = E[\xi_2|2] = 4.17$ (full triangles in Fig. 1.3.7 represent the value of the standard deviations of the given approximations).

Similarly, using data from Table 1.2, we can draw the regression line of the first kind with respect to the random variable ξ_2, Fig. 1.3.8.

Consider a normal two-dimensional continuous random vector $(\xi_1, \xi_2)'$. At the end of the previous section it has been shown that $E[\xi_1|\xi_2] = m_1 + \varrho_{1,2}^2\sigma_1(\xi_2 - m_2)/\sigma_2$, which means that the regression line of the first kind with respect to the random variable ξ_2 is the straight line of the form

$$E[\xi_1|x_2] = \xi_{1opt}(x_2) = m_1 + \varrho_{1,2}\sigma_1(x_2 - m_2)/\sigma_2, \ x_2 \in \mathcal{R}^1,$$

(see Fig. 1.3.6 representing the regression line with respect to the random variable ξ_1 of the form $\xi_{2opt}(x_1) = E[\xi_2|x_1] = m_2 + \varrho_{1,2}\sigma_2(x_1 - m_1)/\sigma_1$, $x_1 \in \mathcal{R}^1$, for $m_1 = m_2 = 0$, $\sigma_1 = \sigma_2 = 1$ and $\varrho_{1,2} = 1/2$).

The optimum approximation of the random variable η by the random variable ξ (and the optimum approximation of the random variable ξ by η) in the case of a normally distributed two-dimensional random vector $(\xi, \eta)'$ is linear. This fact enables us to seek the optimum approximation of one

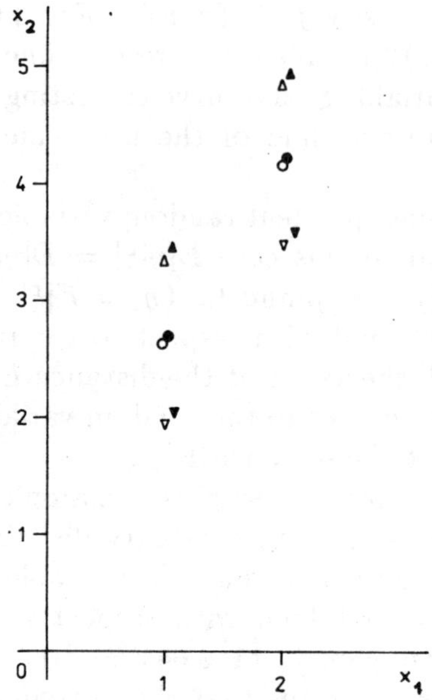

Fig. 1.3.7

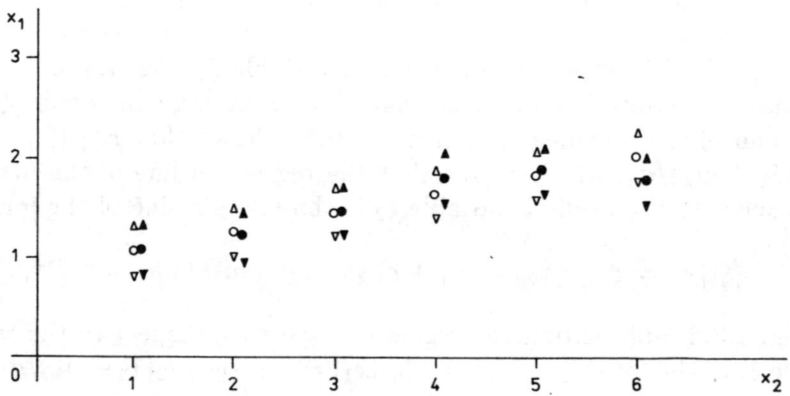

Fig. 1.3.8

random variable by the other in the class of linear approximations of the form $\eta_{opt}(\xi) = a + b\xi$ only. Applying the optimality criterion (1.3.8) we get relations determining the coefficients a and b uniquely: $E[(\eta - a - b\xi)^2] = E[\{(\eta - E[\eta]) - b(\xi - E[\xi]) + (E[\eta] - bE[\xi] - a)\}^2] = D[\eta] + b^2 D[\xi] + (E[\eta] - bE[\xi] - a)^2 - 2bcov(\xi, \eta)$, which shows that the requirement to minimize the expression $E[(\eta - a - b\xi)^2]$ is equivalent to the problem of finding the points of the minimum of the function $\Phi(a, b) = D[\eta] + b^2 D[\xi] + (E[\eta] - bE[\xi] - a)^2 - 2bcov(\xi, \eta)$. Its extremum point is determined by the solution of the system of equations

$(\partial\Phi(a, b)/\partial a =)\ - 2(E[\eta] - bE[\xi] - a) = 0$

$(\partial\Phi(a, b)/\partial b =)\ 2bD[\xi] - 2(E[\eta] - bE[\xi] - a)E[\xi] - 2cov(\xi, \eta) = 0.$

They imply

$$a = E[\eta] - D^{-1}[\xi]cov(\xi, \eta)E[\xi],\ b = D^{-1}[\xi]cov(\xi, \eta);$$

because of

$$[\partial^2\Phi(a, b)/\partial a^2][\partial^2\Phi(a, b)/\partial b^2] - [\partial^2\Phi(a, b)/\partial a\partial b]^2 =$$

$$4D[\xi] + 4(E[\xi])^2 - 4(E[\xi])^2 > 0$$

and

$$\partial^2\Phi(a, b)/\partial a^2 = 2 > 0,$$

this extremum point is the point of the minimum of the function Φ and the optimum linear approximation of the random variable η by the random variable ξ is

$$\eta_{1opt}(\xi) = E[\eta] + D^{-1}[\xi]cov(\xi, \eta)(\xi - E[\xi]);$$

its mean value and dispersion are

$$E[\eta_{1opt}(\xi)] = E[\eta] \quad \text{and} \quad D[\eta_{1opt}(\xi)] = D^{-1}(\xi)cov^2(\xi, \eta).$$

The linear function $\eta_{1opt}(x) = a + bx$, $x \in \mathcal{R}^1$ is said to be a regression straight line with respect to ξ (regression of the second kind), a and b are regression coefficients, $E[(\eta - \eta_{1opt}(\xi))^2] = D[\eta - \eta_{1opt}(\xi)]$ expresses the dispersion of the residuum:

$$E[(\eta - \eta_{1opt}(\xi))^2] = D[\eta - \eta_{1opt}(\xi)] =$$

$$D[\eta] - D[\eta_{1opt}(\xi)] = D[\eta] - D^{-1}[\xi]cov^2(\xi, \eta),$$

because of

$$cov(\eta, \eta_{1opt}(\xi)) = E[(\eta - E[\eta])D^{-1}[\xi]cov(\xi, \eta)(\xi - E[\xi])] =$$

$$= D^{-1}[\xi]cov^2(\xi, \eta).$$

The substance of the linear regression has been explained for the simplest case of a two-dimensional random vector $(\xi, \eta)'$. A generalization for an $(n+1)$-dimensional random vector $(\xi_1, \xi_2, \ldots, \xi_n, \eta)' = (\xi', \eta)'$ is relatively easy.

The problem is to determine the values of coefficients a, b_1, $b_2, \ldots,$ b_n that minimize the expression $E[(\eta - a - b'\xi)^2]$. Analogous to the previous case it is necessary to modify this expression in the following way

$$E[(\eta - a - b'\xi)^2] = E[\{\eta - E[\eta]) - b'(\xi - E[\xi]) + (E[\eta] - b'E[\xi] - a)\}^2] =$$

$$= D[\eta] + b'var(\xi)b + (E[\eta] - b'E[\xi] - a)^2 - 2b'cov(\xi, \eta).$$

The coordinates a, b_1, $b_2, \ldots,$ b_n of the stationary point of the function $\Phi(a, b) = D[\eta] + b'var(\xi)b + (E[\eta] - b'E[\xi] - a)^2 - 2b'cov(\xi, \eta)$ are given by a solution of the system of $n+1$ equations of the form

$$(\partial\Phi(a,b)/\partial a =) \; - 2(E[\eta] - b'E[\xi] - a) = 0,$$

$$(\partial\Phi(a,b)/\partial b =) \; 2var(\xi)b + 2(E[\eta] - b'E[\xi] - a)E[\xi] - 2cov(\xi, \eta) = 0$$

(in order to obtain the partial derivatives of the function $\Phi(a, b)$ the following rules were applied: if $f(x) = a'x$, where $a \in \mathcal{R}^n$ and $x \in \mathcal{R}^n$, then $\partial f(x)/\partial x = (\partial f(x)/\partial x_1, \partial f(x)/\partial x_2, \ldots, \partial f(x)/\partial x_n)' = a, \partial f(x)/\partial x' = (\partial f(x)/\partial x_1, \partial f(x)/\partial x_2, \ldots, \partial f(x)/\partial x_n) = a'$; if M is a symmetric matrix of the type n and $g(x) = x'Mx$, $x \in \mathcal{R}^n$, then $\partial g(x)/\partial x = 2Mx$, $\partial g(x)/\partial x' = 2x'M$ and $\partial^2 g(x)/\partial x \partial x' = 2M$). The sought optimum (in the sense of the criterion (1.3.8)) linear approximation of the random variable η by the random vector ξ (the linear regression with respect to ξ) is

$$\eta_{lopt}(\xi) = E[\eta] + cov(\eta, \xi)var^{-1}(\xi)(\xi - E[\xi]), \tag{1.3.9}$$

and

$$E[\eta_{lopt}(\xi)] = E[\eta], \; D[\eta_{lopt}(\xi)] = cov(\eta, \xi)var^{-1}(\xi)cov(\xi, \eta),$$

$$cov(\eta_{lopt}(\xi), \eta) = cov(\eta, \xi)var^{-1}(\xi)cov(\xi, \eta)$$

and

$$D[\eta - \eta_{lopt}(\xi)] = E[(\eta - \eta_{lopt}(\xi))^2] = D[\eta] - cov(\eta, \xi)var^{-1}(\xi)cov(\xi, \eta).$$

Example Let the approximation be chosen in the form $\eta_{lopt}(\xi) = \gamma'\xi$ under the condition $E[\eta] = E[\eta_{lopt}(\xi)]$ (an approximation satisfying this condition is said to be unbiased). The problem is to prove that in this case the optimality criterion $E[(\eta - \eta_{lopt}(\xi))^2] = min$ yields the optimum approximation of the form

$$\eta_{lopt}(\xi) = E[\eta]\{E[\xi']var^{-1}(\xi)E[\xi]\}^{-1}E[\xi']var^{-1}(\xi)\xi+$$

$$+cov(\eta, \xi)var^{-1}(\xi)\{\xi - E[\xi](E[\xi']var^{-1}(\xi)E[\xi])^{-1}E[\xi']var^{-1}(\xi)\xi\};$$

in the case as formulated the problem consists in determining the minimum of a function under a subsidiary condition; the auxiliary Lagrange function has the form

$$\Phi(\gamma) = D[\eta] + \gamma' var(\xi)\gamma - 2\gamma' cov(\xi, \eta) - 2\lambda(E[\eta] - \gamma' E[\xi]),$$

the coefficient γ and the Lagrange multiplier λ are solutions of the system of equations

$$\gamma' var(\xi) - cov(\eta, \xi) - \lambda E[\xi'] = 0$$

$$\gamma' E[\xi] - E[\eta] = 0.$$

The next problem is to prove that

a) in the preceding case without the subsidiary condition of unbiasedness the optimality criterion $E[(\eta - \eta_{lopt}(\xi))^2] = min$ yields the optimum approximation in the form

$$\eta_{lopt}(\xi) = \delta'\xi = \{cov(\eta, \xi) + E[\eta]E[\xi]\}\{var(\xi) + E[\xi]E[\xi']\}^{-1}\xi;$$

δ minimizes the expression $\delta' var(\xi)\delta - 2\delta' cov(\xi, \eta) + D[\eta] + \delta' E[\xi]E[\xi']\delta + 2\delta' E[\xi]E[\eta] + (E[\eta])^2$;

b) if the approximation function is chosen in the form $\eta_{lopt}(\xi) = \kappa'\xi$, the subsidiary condition of unbiasedness is abandoned, and the criterion of optimality is of the form $D[\eta - \eta_{lopt}(\xi)] = min$, then

$$\eta_{lopt}(\xi) = \kappa'\xi = cov(\eta, \xi)var^{-1}(\xi)\xi;$$

the function $\Phi(\kappa) = \kappa' var(\xi)\kappa - 2\kappa' cov(\xi, \eta) + D[\eta]$ is minimized.

Realize that the optimality criterion $E[(\eta - \eta_{lopt}(\xi))^2] = min$ under the condition of unbiasedness $E[\eta_{lopt}(\xi)] = E[\eta]$ is identical with the criterion $D[\eta - \eta_{lopt}(\xi)] = min$; if the subsidiary condition need not be fulfilled, the criteria mentioned yield different approximations.

The optimum (in the sense of the criterion (1.3.8)) linear approximation $\xi_{1lopt}(\xi_2)$ in the case of the random vector $(\xi_1, \xi_2)'$ (its probability distribution is given on the p. 36) has the form $\xi_{1lopt}(\xi_2) = 1.57 + 0.19(\xi_2 - 3.51)$, $D^{1/2}[\xi_{1lopt}(\xi_2)] = 0.27$, $D[\xi_1 - \xi_{1lopt}(\xi_2)] = 0.173$; its realization at the points $x_{21} = 1$, $x_{22} = 2, \ldots, x_{26} = 6$ is represented in Fig. 1.3.8 by empty circles; the empty triangles represent the value of the standard deviation. The situation depicted by this figure enables us to compare the difference between the regression of the second kind and the linear regression with respect to the random variable ξ_2. Analogously, $\xi_{2lopt}(\xi_1) = 3.51 + 1.52(\xi_1 - 1.57)$, $D^{1/2}[\xi_{2lopt}(\xi_1)] = 0.75$, $D[\xi_2 - \xi_{2lopt}(\xi_1)] = 1.365$; its realization at the points $x_{11} = 1$ and $x_{12} = 2$ is represented in Fig. 1.3.7. The random variable ξ_1 is realized by two points only, thus the regression curve of the second kind is identical with the regression line with respect to the random variable ξ_1.

1.3.4 Theoretical correlation

Let us consider a two-dimensional random vector $(\xi, \eta)'$. The correlation coefficient $\varrho_{\xi,\eta} = cor(\xi, \eta)$, defined by the relation (1.3.6)

$$\varrho_{\xi,\eta} = \frac{cov(\xi, \eta)}{\sqrt{D[\xi]D[\eta]}}, \quad D[\xi] > 0, \ D[\eta] > 0,$$

expresses the measure of the stochastic dependence between the random variables ξ and η.

We now take into account an $(n+1)$-dimensional random vector $(\xi_1, \xi_2, \ldots, \xi_n, \eta)'$. The measure of the stochastic dependence between the random variable η and the random vector $\xi = (\xi_1, \xi_2, \ldots, \xi_n)'$ is presented by the coefficient of a multiple correlation $\varrho_{\eta,\xi}$ defined by the relation

$$\varrho_{\eta,\xi} = cor(\eta, \eta_{lopt}(\xi)),$$

where $\eta_{lopt}(\xi) = a + b'\xi$ defined by the relation (1.3.9) is the optimum (in the sense of the criterion $E[(\eta - \eta_{lopt}(\xi))^2] = min$) linear approximation of the random variable η by the random vector ξ. Applying the results obtained in the preceding section, we can get an explicit form of the coefficient of the multiple correlation

$$\varrho_{\eta,\xi} = \left(\frac{cov(\eta,\xi)var^{-1}(\xi)cov(\xi,\eta)}{D[\eta]} \right)^{1/2}, \tag{1.3.10}$$

which implies that the coefficient of the multiple correlation is always non-negative; the zero value is taken on if $cov(\xi,\eta) = (cov(\xi_1,\eta), cov(\xi_2,\eta), \ldots, cov(\xi_n,\eta))' = 0$.

The ways of writing relations that occur in correlation theory can be simplified substantially by introducing a new operator $cor(\xi,\eta)$, after determining its relation to the known operator $cov(\xi,\eta)$, which appears in various types of correlation coefficients.

Consider an $(n+1)$-dimensional random vector $(\xi_1,\xi_2,\ldots,\xi_n,\eta)'$; then

$$cor(\xi,\eta) = (cor(\xi_1,\eta), cor(\xi_2,\eta), \ldots, cor(\xi_n,\eta))',$$

where $cor(\xi_i,\eta)$ is the correlation coefficient between the random variables ξ_i and η. Thus

$$cor(\xi,\eta) = (cov(\xi_1,\eta)D^{-1/2}[\xi_1]D^{-1/2}[\eta], \ldots, cov(\xi_n,\eta)D^{-1/2}[\xi_n]D^{-1/2}[\eta])'.$$

If $D_\xi = Diag(D[\xi_1], D[\xi_2], \ldots, D[\xi_n])$ denotes a diagonal matrix with zero elements beside the diagonal and with the diagonal built by dispersions of components of the vector $\xi = (\xi_1,\xi_2,\ldots,\xi_n)'$, then, as a consequence of the assumption

$$D[\xi_i] > 0, \; i = 1,2,\ldots,n,$$

$$D_\xi^{-1/2} = Diag(D^{-1/2}[\xi_1], D^{-1/2}[\xi_2], \ldots, D^{-1/2}[\xi_n])$$

and

$$cor(\xi,\eta) = D_\xi^{-1/2} cov(\xi,\eta) D_\xi^{-1/2}[\eta].$$

Denoting further

$$cor(\xi) = \begin{pmatrix} 1, & cor(\xi_1,\xi_2), & \ldots, & cor(\xi_1,\xi_n) \\ cor(\xi_2,\xi_1), & 1, & \ldots, & cor(\xi_2,\xi_n) \\ \ldots & \ldots & \ldots & \ldots \\ cor(\xi_n,\xi_1), & cor(\xi_n,\xi_2), & \ldots, & 1 \end{pmatrix},$$

we can write $cor(\xi) = D_\xi^{-1/2} var(\xi) D_\xi^{-1/2}$ and

$$\varrho^2_{\eta,\xi} = cor(\eta,\xi)cor^{-1}(\xi)cor(\xi,\eta). \tag{1.3.11}$$

Example Let $\xi = (\xi_1, \xi_2)'$. By applying the relation (1.3.11) we obtain the coefficient of the multiple correlation in the form

$$\varrho^2_{\eta,(\xi_1,\xi_2)'} = (\varrho_{\eta,\xi_1}, \varrho_{\eta,\xi_2}) \begin{pmatrix} 1, & \varrho_{\xi_1,\xi_2} \\ \varrho_{\xi_1,\xi_2}, & 1 \end{pmatrix}^{-1} \begin{pmatrix} \varrho_{\eta,\xi_1} \\ \varrho_{\eta,\xi_2} \end{pmatrix} =$$

$$= \frac{\varrho^2_{\eta,\xi_1} + \varrho^2_{\eta,\xi_2} - 2\varrho_{\eta,\xi_1}\varrho_{\eta,\xi_2}\varrho_{\xi_1,\xi_2}}{1 - \varrho^2_{\xi_1,\xi_2}}.$$

In what follows, an $(n+2)$-dimensional random vector $(\xi', \eta, \zeta)'$ is taken into account. Let the coefficients of the multiple correlation $\varrho_{\eta,\xi} = cor(\eta, \eta_{lopt}(\xi))$ and $\varrho_{\zeta,\xi} = cor(\zeta, \zeta_{lopt}(\xi))$ be non-zero. The correlation coefficient of the random residua $\eta - \eta_{lopt}(\xi)$ and $\zeta - \zeta_{lopt}(\xi)$ (the notation $\varrho_{\eta,\zeta\cdot\xi}$ is used for it) is said to be a partial correlation coefficient of the random variables η and ζ under the eliminated influence of the random vector ξ. It is of the form

$$\varrho_{\eta,\zeta\cdot\xi} = cor(\{\eta - E[\eta] - cov(\eta,\xi)var^{-1}(\xi)(\xi - E[\xi])\},$$

$$\{\zeta - E[\zeta] - cov(\zeta,\xi)var^{-1}(\xi)(\xi - E[\xi])\}).$$

Applying the expression for the dispersion of a residuum (see Section 1.3.3) and the fact that $cov(\eta - \eta_{lopt}(\xi), \zeta - \zeta_{lopt}(\xi)) = cov(\eta - E[\eta] - cov(\eta,\xi)var^{-1}(\xi)(\xi - E[\xi]), \zeta - E[\zeta] - cov(\zeta,\xi)var^{-1}(\xi)(\xi - E[\xi])) = cov(\eta,\zeta) - cov(\eta,\xi)var^{-1}cov(\xi,\zeta)$, we obtain an explicit form of the partial correlation coefficient

$$\varrho_{\eta,\zeta\cdot\xi} =$$

$$= \frac{cov(\eta,\zeta) - cov(\eta,\xi)var^{-1}(\xi)cov(\xi,\zeta)}{([D[\eta] - cov(\eta,\xi)var^{-1}(\xi)cov(\xi,\eta)][D[\zeta] - cov(\zeta,\xi)var^{-1}(\xi)cov(\xi,\zeta)])^{1/2}}$$

$$= \frac{cor(\eta,\zeta) - cor(\eta,\xi)cor^{-1}(\xi)cor(\xi,\zeta)}{([1 - cor(\eta,\xi)cor^{-1}(\xi)cor(\xi,\eta)][1 - cor(\zeta,\xi)cor^{-1}(\xi)cor(\xi,\zeta)])^{1/2}}.$$

Example Consider a random vector $(\xi, \eta, \zeta)'$. The partial correlation coefficient of the random variables η and ζ under the eliminated influence of the random variable ξ is

$$\varrho_{\eta,\zeta\cdot\xi} = \frac{\varrho_{\eta,\zeta} - \varrho_{\eta,\xi}\varrho_{\zeta,\xi}}{[(1-\varrho_{\eta,\xi}^2)(1-\varrho_{\zeta,\xi}^2)]^{1/2}}, \tag{1.3.12}$$

because $cor(\xi) = 1$.

1.4 Characteristic functions

Definition The function

$$\varphi(t) = E[\exp(it\xi)], \ t \in \mathcal{R}^1, \tag{1.4.1}$$

where i is the imaginary unit, is called a characteristic function of a one-dimensional random variable ξ.

The characteristic function of a discrete random variable ξ that assumes the values x_k with probabilities p_k, $k = 1, 2, \ldots, K, \ldots$, has the form

$$\varphi(t) = \sum_k p_k \exp(itx_k), \ t \in \mathcal{R}^1.$$

The characteristic function of a continuous random variable ξ characterized by the probability density $f(\cdot)$, $x \in \mathcal{R}^1$, is

$$\varphi(t) = \int_{-\infty}^{\infty} \exp(itx) f(x) dx.$$

The definition of a characteristic function by Eq. (1.4.1) implies that the characteristic function is a continuous function of the argument t on $\mathcal{R}^1$ with the following properties:

(i) $\varphi(0) = 1$ $(\varphi(0) = E[1] = 1)$;

(ii) $|\varphi(t)| \leq \varphi(0)$ $(|\varphi(t)| = |E[\exp(it\xi)]| \leq E[|\exp(it\xi)|] = E[1] = \varphi(0))$;

(iii) $\varphi(-t) = \overline{\varphi(t)}$ $(\varphi(-t) = E[\exp(-it\xi)] = \overline{\varphi(t)}$, the bar denotes a complex conjugate function).

If ξ is a random variable with the characteristic function $\varphi(\cdot)$, $t \in \mathcal{R}^1$, then the characteristic function of the random variable $\eta = a\xi + b$, $a, b \in \mathcal{R}^1$, is

$$\varphi_\eta(t) = \exp(itb)\varphi_\xi(at), \ t \in \mathcal{R}^1. \tag{1.4.2}$$

Eq. (1.4.2) follows immediately from Eq. (1.4.1) that defines the characteristic function. As a consequence of Eq. (1.4.2)

$$\varphi_\eta(t) = E[\exp(it\eta)] = E[\exp(it(a\xi + b))] =$$

$$= \exp(itb)E[\exp(ita\xi)] = \exp(itb)\varphi_\xi(at).$$

Example

a) Determine the characteristic function of a random variable $\xi \sim N(0,1)$.

Following Eq. (1.4.1),

$$\varphi_\xi(t) = \frac{1}{\sqrt{2\pi}}\int_{-\infty}^{\infty} \exp(itx)\exp\left(-\frac{x^2}{2}\right)dx =$$

$$= \exp\left(-\frac{t^2}{2}\right)\frac{1}{\sqrt{2\pi}}\int_{-\infty}^{\infty} \exp\left(-\frac{(x-it)^2}{2}\right)dx = \exp\left(-\frac{t^2}{2}\right), \tag{1.4.3}$$

because $\frac{1}{\sqrt{2\pi}}\int_{-\infty}^{\infty}\exp[-(x-it)^2/2]dx = 1$ (see p. 48).

b) The function defined by Eq. (1.4.3) represents the characteristic function of a centred and normed random variable $\xi \sim N(0,1)$. The problem is to determine the characteristic function of a random variable $\eta \sim N(m, \sigma^2)$ which uses it.

Due to the definition of a centred and normed random variable $\xi = \frac{\eta - m}{\sigma}$, i.e., $\eta = \sigma\xi + m$ and in accordance with Eq. (1.4.3)

$$\varphi_\eta(t) = \exp(imt)\varphi_\xi(\sigma t) = \exp(imt)\exp\left(-\frac{\sigma^2t^2}{2}\right) = \exp\left(imt - \frac{\sigma^2t^2}{2}\right).$$

Definition A function

$$\varphi(t) = E[\exp(it'\xi)], \ t \in \mathcal{R}^n, \tag{1.4.4}$$

where i is the imaginary unit, is called a characteristic function of an n-dimensional random vector $\xi = (\xi_1, \xi_2, \ldots, \xi_n)'$.

If ξ is an n-dimensional discrete random vector that assumes the values $(x_{1k_1}, x_{2k_2}, \ldots, x_{nk_n})'$ with probabilities $p_{1k_1,2k_2,\ldots,nk_n}$, $k_i = 1, 2, \ldots, K_i, \ldots$, $i = 1, 2, \ldots, n$, its characteristic function is of the form

$$\varphi(t) = \sum_{k_1,k_2,\ldots,k_n} p_{1k_1,2k_2,\ldots,nk_n} \exp\left(i\sum_{j=1}^{n} t_j x_{jk_j}\right), \quad t \in \mathcal{R}^n;$$

in the case of an n-dimensional continuous random variable ξ with the probability density $f(\cdot)$, $t \in \mathcal{R}^n$, the characteristic function is of the form

$$\varphi(t) = \int_{-\infty}^{\infty} \exp(it'x) f(x) dx, \quad t \in \mathcal{R}^n.$$

The characteristic function of an n-dimensional random vector is continuous on $\mathcal{R}^n$; it has the following properties:

(i) $\varphi(0) = 1$ $(\varphi(0) = E[\exp(0)] = 1)$;

(ii) $|\varphi(t)| \leq \varphi(0)$ $(|\varphi(t)| = |E[\exp(it'\xi)]| \leq E[|\exp(it\xi)|] = E[1] = = \varphi(0))$;

(iii) $\varphi(-t) = \overline{\varphi(t)}$ $(\varphi(-t) = E[\exp(-it'\xi)] = \overline{\varphi(t)})$.

Let ξ be an n-dimensional random vector, A a $p \times p$-dimensional matrix, and b a p-dimensional vector. The characteristic function of a p-dimensional random variable $\eta = A\xi + b$ is

$$\varphi_\eta(t) = \exp(it'b)\varphi_\xi(A't), \quad t \in \mathcal{R}^p. \tag{1.4.5}$$

Similarly, as in the case of a one-dimensional random variable, the assertion follows immediately from the definition of the characteristic function. Consequently,

$$\varphi_\eta(t) = E[\exp(it'\eta)] = E[\exp(it'(A\xi + b))] = \exp(it'b)E[\exp(it'A\xi)] =$$

$$= \exp(it'b)\varphi_\xi(A't).$$

Finally, the characteristic function of the sum of independent random variables will be derived.

Let $\xi_1, \xi_2, \ldots, \xi_n$ be stochastically independent random variables with characteristic functions $\varphi_{\xi_1}(\cdot), \varphi_{\xi_2}(\cdot), \ldots, \varphi_{\xi_n}(\cdot)$, $t \in \mathcal{R}^1$, respectively. The characteristic function of the random variable $\eta = \sum_{j=1}^n \xi_i$ is

$$\varphi_\eta(t) = \sqcap_{j=1}^n \varphi_{\xi_j}(t), \; t \in \mathcal{R}^1. \tag{1.4.6}$$

We shall prove this assertion for an an n-dimensional continuous random vector. The procedure for proving it in the case of a discrete random vector is analogous.

Let $(\xi_1, \xi_2, \ldots, \xi_n)' = \xi$ be a continuous random vector characterized by the probability density $f(\cdot)$, $x \in \mathcal{R}^n$. In accordance with the assumption, its components are stochastically independent, therefore at any point in the definition domain $\mathcal{R}^n$

$$f(x) = f_1(x_1) f_2(x_2) \ldots f_n(x_n),$$

where

$$f_i(x_i) = \int_{\mathcal{R}^{n-1}} f(x_1, \ldots, x_{i-1}, x_i, x_{i+1}, \ldots, x_n) dx_1 \ldots dx_{i-1} dx_{i+1} \ldots dx_n,$$

$i = 1, 2, \ldots, n$, are marginal probability densities of individual components. Then

$$\varphi_\eta(t) = E[\exp(it\eta)] = E\left[\exp\left(it\sum_{j=1}^n \xi_j\right)\right] =$$

$$= \int_{\mathcal{R}^n} \exp\left(it\sum_{j=1}^n x_j\right) f(x_1, x_2, \ldots, x_n) dx_1 dx_2 \ldots, dx_n =$$

$$= \int_{\mathcal{R}^n} \exp\left(it\sum_{j=1}^n x_j\right) f_1(x_1) f_2(x_2) \ldots f_n(x_n) dx_1 dx_2 \ldots, dx_n =$$

$$= \sqcap_{j=1}^n \int_{-\infty}^{\infty} \exp(itx_j) f_j(x_j) dx_j = \sqcap_{j=1}^n \varphi_{\xi_j}(t), \; t \in \mathcal{R}^1.$$

Example The problem is to determine the characteristic function of an arithmetic mean of independent random variables $\xi_j \sim N(m, \sigma^2)$, $j = 1, 2, \ldots, n$.

The characteristic function of a random variable $\xi \sim N(m, \sigma^2)$ is

$$\varphi_\xi(t) = \exp\left(itm - \frac{\sigma^2 t^2}{2}\right), \quad t \in \mathcal{R}^1,$$

see the previous example. Due to Eq. (1.4.6), the characteristic function of a sum $\eta = \sum_{j=1}^n \xi_j$ of independent random variables with the same probability distribution $N(m, \sigma^2)$ is

$$\varphi_\eta(t) = \sqcap_{j=1}^n \varphi_{\xi_j}(t) = \exp\left(intm - \frac{n\sigma^2 t^2}{2}\right), \quad t \in \mathcal{R}^1.$$

Applying Eq. (1.4.2) we can easily derive the characteristic function sought for the arithmetic mean $\overline{\xi} = \eta/n$ of independent random variables $\xi_j \sim N(m, \sigma^2)$ in the form

$$\varphi_{\overline{\xi}}(t) = \exp\left(itm - \frac{\sigma^2 t^2}{2n}\right), \quad t \in \mathcal{R}^1.$$

Remark Expressions defining the characteristic function of a random variable include its probability distribution. This means that the characteristic function is uniquely assigned to the given probability distribution. A very important and stronger assertion can be proved, namely that this relation is one-to-one (cf., e.g., [20], Theorem 1.4.4.1).

The proof of the Lindeberg — Lévy central limit theorem (cf. p. 105) is based on the following assertion:

If there exist initial moments $E[\xi^k]$, $k = 1, 2, \ldots, n$, of a random variable ξ, then there exist derivatives $d\varphi(\cdot)/dt$, $d^2\varphi(\cdot)/dt^2, \ldots, d^n\varphi(\cdot)/dt^n$ of its characteristic function and $d^k\varphi(0)/dt^k = i^k E[\xi^k]$; then the characteristic function may be expressed in terms of the Taylor series

$$\varphi(t) = \sum_{k=0}^{n} \frac{(it)^k E[\xi^k]}{k!} + \mathcal{O}(t^n), \tag{1.4.7}$$

where $\mathcal{O}(t^n)$ is its remainder.

First, consider a discrete random variable whose characteristic function is $\varphi(t) = \sum_j p_j \exp(itx_j)$. Its kth derivative is

$$\frac{d^k\varphi(t)}{dt^k} = \sum_j p_j i^k x_j^k \exp(itx_j) = E[i^k \xi^k \exp(it\xi)] \Rightarrow \frac{d^k\varphi(0)}{dt^k} = i^k E[\xi^k];$$

analogously, if a continuous random variable characterized by the probability density $f(\cdot)$, $x \in \mathcal{R}^1$, is taken into account,

$$\frac{d^k\varphi(t)}{dt^k} = \int_{-\infty}^{\infty} f(x)i^k x^k \exp(itx)dx = E[i^k\xi^k \exp(it\xi)] \Rightarrow \frac{d^k\varphi(0)}{dt^k} = i^k E[\xi^k];$$

as $\varphi(\cdot)$ is a continous function of the argument t on $\mathcal{R}^1$ and all the necessary derivatives exist, it may be developed into the Taylor series.

1.5 Some probability distributions

In this part we deal only with the fundamental probability distributions that occur most frequently in data processing both, in their analysis and in the optimum estimation of unknown parameters.

Definition A continuous random variable ξ whose probability density is

$$n(x; m, \sigma^2) = \frac{1}{\sigma\sqrt{2\pi}} \exp\left(-\frac{(x-m)^2}{2\sigma^2}\right), \; x \in \mathcal{R}^1, \tag{1.5.1}$$

where $m \in \mathcal{R}^1$ and $\sigma \in (0, \infty)$, possesses a normal (Gauss) probability distribution ($\xi \sim N(m, \sigma^2)$).

The mean value, dispersion, asymmetry, and excess of the random variable $\xi \sim N(m, \sigma^2)$ are

$$E[\xi] = m, \; D[\xi] = \sigma^2, \; \gamma_1 = 0, \; \gamma_2 = 0,$$

respectively (see pp. 46-49). The characteristic function is

$$\varphi(t) = \exp\left(imt - \frac{\sigma^2 t^2}{2}\right), \; t \in \mathcal{R}^1, \tag{1.5.2}$$

(see p. 74).

If $\xi \sim N(m, \sigma^2)$, the random variable $\eta = a\xi + b$, where a, $b \in \mathcal{R}^1$, has the normal probability distribution $N(am + b, a^2\sigma^2)$, since due to Eq. (1.4.2), $\varphi_\eta(t) = \exp(itb)\varphi_\xi(at)$ and due to Eq. (1.5.2), $\varphi_\xi(at) = \exp(iatm - \sigma^2 a^2 t^2/2)$, by which (because of the one-to one relation between the characteristic and the distribution function) the assertion is proved.

Definition An n-dimensional continuous random vector ξ with the probability density of the form

$$n(x;m,\Sigma) = \frac{1}{(2\pi)^{n/2}\sqrt{det(\Sigma)}} \exp\left(-\frac{(x-m)'\Sigma^{-1}(x-m)}{2}\right), \ x \in \mathcal{R}^n, \tag{1.5.3}$$

where $m \in \mathcal{R}^n$ and Σ is a positive definite $n \times n$ matrix, possesses a normal (Gauss) probability distribution ($\xi \sim N_n(m,\Sigma)$).

The mean value and the covariance matrix of the random vector $\xi \sim N(m,\Sigma)$ are

$$E[\xi] = m \quad \text{and} \quad var(\xi) = \Sigma,$$

respectively (for $n = 2$ the assertion is proved on p. 60; the idea of proving it for $n > 2$ is the same). The characteristic function of an n-dimensional normal random vector is

$$\varphi(t) = \exp(it'm - t'\Sigma t/2), \ t \in \mathcal{R}^n. \tag{1.5.4}$$

If $\xi \sim N_n(m,\Sigma)$, the random variable $\eta = A\xi + b$, where A is a $p \times p$-dimensional matrix and b a p-dimensional vector, has a p-dimensional normal probability distribution $N_p(Am+b, A\Sigma A')$, as, due to (1.4.5), $\varphi_\eta(t) = \exp(it'b)\varphi_\xi(A't)$, due to (1.5.4), $\varphi_\xi(t)(A't) = \exp(it'Am - t'A\Sigma A't/2)$ and, therefore, $\varphi_\eta(t) = \exp[it'(b+Am) - t'A\Sigma A't/2]$, which (as a consequence of the one-to-one relation between the characteristic and the distribution function) proves the assertion.

The normal probability distribution is natural within the measurement processing. The measurement errors influencing each result even, under very general conditions, have just a normal probability distribution.[1] If the resulting random error ε may be assumed to be a sum of n independent random elementary errors $\eta_1, \eta_2, \ldots, \eta_n$ having the same probability

[1] This does not imply that each set of measured data must possess a normal probability distribution. The very general conditions mentioned are not satisfied in all situations. This fact does not mean any restriction on applications of the methods of linear estimation theory; they are derived without any assumptions concerning the probability distribution of the set of processed data.

distribution, the probability distribution of the random error $\varepsilon_n = \sum_{i=1}^n \eta_1$ for $n \to \infty$ converges to normal. Under the formulated assumptions

$$E[\eta_1] = E[\eta_2] = \ldots = E[\eta_n] = 0 \Rightarrow E[\varepsilon_n] = E\left[\sum_{i=1}^n \eta_i\right] = 0$$

(since elementary measurement errors are taken into account, the additional assumption $E[\eta_i] = 0,\ i = 1, 2 \ldots, n$, is natural),

$$D[\eta_1] = D[\eta_2] = \ldots = D[\eta_n] = \sigma^2/n \Rightarrow D[\varepsilon_n] = \sigma^2 < \infty$$

(this form of the dependence of the dispersions of individual elementary errors on n ensures that the requirement $D[\varepsilon] = \sigma^2 < \infty$ is fulfilled for any n) and

$$\varphi_{\eta_1}(\cdot) = \varphi_{\eta_2}(\cdot) = \ldots = \varphi_{\eta_n}(\cdot) = \varphi_\eta(\cdot) \Rightarrow \varphi_{\varepsilon_n}(\cdot) = [\varphi_\eta(\cdot)]^n,\ t \in \mathcal{R}^n.$$

The characteristic function $\varphi_{\varepsilon_n}(\cdot)$, due to Eq. (1.4.7), may be expressed for any n in the form

$$\varphi_{\varepsilon_n}(t) = \left[1 - \frac{t^2\sigma^2}{2n} + \mathcal{O}\left(\frac{t^2}{n}\right)\right]^n;$$

since $\lim_{n\to\infty} n\mathcal{O}\left(\frac{t^2}{n}\right) = 0$ and $\lim_{n\to\infty}\left(1 - \frac{t^2\sigma^2}{2n}\right)^n = \exp(-t^2\sigma^2/2)$, for a sufficiently large n

$$\varphi_{\varepsilon_n}(t) = \left[1 - \frac{t^2\sigma^2}{2n} + \mathcal{O}\left(\frac{t^2}{n}\right)\right]^n \doteq \exp(-t^2\sigma^2/2),\ t \in \mathcal{R}^1;$$

$\exp(-t^2\sigma^2/2),\ t \in \mathcal{R}^1$, represents the characteristic function of a random variable $\xi \sim N(0,1)$; the relation between the characteristic and the distribution function of a random variable is one-to-one (see Remark on p. 77), therefore the measurement error $\varepsilon_n = \sum_{i=1}^n \eta_i$, under the given assumptions, has an approximately normal probability distribution $N(0, \sigma^2)$. If all elementary errors have the same normal probability distribution, the measurement error ε_n has an exactly normal probability distribution $N(0, \sigma^2)$.

This simple example of the genesis of a normal probability distribution of the set of measurement errors is based on the simplest of the central limit theorems, namely on the Lindeberg—Lévy central limit theorem (which will be dealt with later). However, it can be shown that the measurement errors possess a normal probability distribution even under more general conditions concerning the properties of the elementary errors, e.g., if they are assumed to be stochastically independent, each of them will have a probability distribution characterized by a distribution function $F_i(\cdot)$, $x \in \mathcal{R}^1$, $i = 1, 2, \ldots, n$, $F_i(\cdot) \neq F_j(\cdot)$, $i \neq j$ and, simultaneously, the dispersions of the elementary errors are required to be homogeneous in the sense that none of them exceeds the contribution of all the others. This example of the genesis of a normal probability distribution is based on the Lindeberg—Feller central limit theorem (for more detail see [6], p. 181, [15], p. 45).

Definition A one-dimensional continuous random variable η_n with the probability density

$$g_n(y) = \begin{cases} \frac{1}{2^{n/2}\Gamma(n/2)} y^{(n/2)-1} \exp(-y/2) & , \quad y > 0,^{2} \\ 0 & , \quad y \leq 0, \end{cases} \tag{1.5.5}$$

is said to possess a central chi-square probability distribution with n degrees of freedom.

The mean value, the dispersion and the characteristic function of the random variable η_n are

$$E[\eta_n] = n, \; D[\eta_n] = 2n \quad \text{and} \quad \varphi_n(t) = (1 - 2it)^{-n/2},$$

respectively.

Let $\xi_1, \xi_2, \ldots, \xi_n$ be independent random variables, $\xi_i \sim N(0,1)$, $i = 1, 2, \ldots, n$. The random variable $\eta_n = \sum_{i=1}^{n} \xi_i^2$ has a central chi-square probability distribution with n degrees of freedom ($\eta \sim \chi^2(n)$).

In order to prove this assertion, let us examine the probability distribution of the random variable $\eta = \xi^2$ if $\xi \sim N(0,1)$. The distribution function $G(\cdot)$, $y \in \mathcal{R}^1$, of the random variable η is zero on the interval $(-\infty, 0]$, as the random variable can assume positive values only with a non-zero probability. If $y > 0$, then

$$G(y) = P(\{\omega : \eta(\omega) < y\}) = P(\{\omega : \xi^2(\omega) < y\}) =$$

---

[2] A gamma function is defined by the relation

$$\Gamma(p) = \int_0^\infty x^{p-1} \exp(-x) dx, \; p > 0.$$

The recurrent relation $p\Gamma(p) = \Gamma(p+1)$ is true for any $p > 0$ (an integration *per partes* of the relations defining the gamma function yields $\int_0^\infty x^{p-1} \exp(-x) dx = (1/p) \int_0^\infty x^p \exp(-x) dx$).

Since $\Gamma(1) = 1$, $\Gamma(n+1) = n!$ for any integer n. Furthemore, $\Gamma(1/2) = \sqrt{\pi}$.

$$= P(\{\omega : \xi(\omega) \in (-\sqrt{y}, \sqrt{y})\}) = F(\sqrt{y}) - F(-\sqrt{y}),$$

where $F(\cdot)$, $x \in \mathcal{R}^1$, is the distribution function of the normal probability distribution $N(0,1)$. The fact that the probability density $f(x) = df(x)/dx$ of a normal probability distribution exists implies the existence of a probability density $g(y) = dG(y)/dy$ and the relation $G(y) = F(\sqrt{y}) - F(-\sqrt{y})$ yields

$$g(y) = \frac{1}{2\sqrt{y}} f(\sqrt{y}) + \frac{1}{2\sqrt{y}} f(-\sqrt{y}) = \frac{1}{\sqrt{2\pi}} y^{-1/2} \exp(-y/2) =$$

$$= \frac{1}{2^{1/2}\Gamma(1/2)} y^{-1/2} \exp(-y/2),$$

which means that the random variable $\eta = \xi^2$ possesses a chi-square probability distribution with one degree of freedom.

The method of complete induction will be applied as follows: Let $\eta_m = \sum_{i=1}^m \xi_i^2$ and $\eta_{m+1} = \eta_m + \xi_{m+1}^2$, where $\xi_1, \xi_2, \ldots, \xi_m, \xi_{m+1}$, be independent random variables such that $\xi_i \sim N(0,1)$, $i = 1, 2, \ldots, n$. Let the random variable η_m be assumed to have a central chi-square probability distribution with m degrees of freedom, i.e.,

$$g_m(y) = \begin{cases} \frac{1}{2^{m/2}\Gamma(m/2)} y^{(m/2)-1} \exp(-y/2) & , \quad y > 0, \\ 0 & , \quad y \le 0. \end{cases}$$

The random variable ξ_{m+1}^2 has been proved to possess a central chi-square probability distribution with one degree of freedom. The probability density of the random variable $\eta_{m+1} = \eta_m + \xi_m^2$, based on the theorem of the probability density of the sum of two random variables [3], is

$$g_{m+1}(y) = \int_0^y g_m(y-z) g(z) dz =$$

$$= \frac{\exp(-y/2)}{2^{(m+1)/2}\Gamma(m/2)\Gamma(1/2)} \int_0^y (y-z)^{(m/2)-1} z^{-1/2} dz, \ y > 0,$$

(the form of the lower limit of intergration is determined by the fact that both subintegral functions are non-zero for non-positive values of the argument only). Substituting $z = yu$, applying the definition of a beta function

[3] **Theorem of the probability density of the sum of two random variables** Let $f(\cdot)$, $(x_1, x_2)' \in \mathcal{R}^2$, be the probability density of a random variable $(\xi_1, \xi_2)'$. The probability density $g(\cdot)$, $z \in \mathcal{R}^1$, of a random variable $\zeta = \xi_1 + \xi_2$ is $g(z) = \int_{-\infty}^{\infty} f(x_1, z - x_1) dx_1$.

and the relation between a beta and a gamma function [4] we obtain

$$\int_0^1 (y-z)^{(m/2)-1} z^{-1/2} dz = y^{(m/2)-1} \int_0^1 (1-u)^{(m/2)-1} u^{-1/2} du =$$

$$= y^{(m-1)/2} B(m/2, 1/2) = y^{(m-1)/2} \frac{\Gamma(m/2)\Gamma(1/2)}{\Gamma[(m+1)/2]},$$

thus

$$g_{m+1}(y) = \begin{cases} \frac{1}{2^{(m+1)/2}\Gamma[(m+1)/2]} y^{(m-1)/2} \exp(-y/2) & , \quad y > 0, \\ 0 & , \quad y \leq 0, \end{cases}$$

which proves that the random variable η_{m+1} has a central chi-square probability distribution with $m+1$ degrees of freedom; thus the assertion is true for any finite integer n.

Definition A one-dimensional random variable ζ_n possesses a central σ^2 chi-square probability distribution with n degrees of freedom if its probability density is

$$f_n(z) = \begin{cases} \frac{1}{2^{n/2}\sigma^n\Gamma(n/2)} z^{(n/2)-1} \exp(-z/2\sigma^2) & , \quad z > 0, \\ 0 & , \quad z \leq 0, \end{cases} \tag{1.5.6}$$

where $\sigma > 0$.

The mean value and the dispersion of a random variable ζ_n with a central σ chi-square probability distribution with n degrees of freedom are

$$E[\zeta_n] = n\sigma^2 \quad \text{and} \quad D[\zeta_n] = 2n\sigma^4,$$

respectively.

---

[4] A function

$$B(p,q) = \int_0^1 x^{p-1}(1-x)^{q-1} dx, \; p > 0, \; q > 0,$$

is called a beta function.

The following relation is true between a gamma and a beta functions

$$B(p,q) = \Gamma(p)\Gamma(q)/\Gamma(p+q).$$

Let $\xi_1, \xi_2, \ldots, \xi_n$ be independent random variables such that $\xi_i \sim N(0, \sigma^2)$, $i = 1, 2, \ldots, n$. A random variable $\zeta_n = \sum_{i=1}^{n} \xi_i^2$ has the central σ^2 chi-square probability distribution with n degrees of freedom.

This assertion will be proved using the theorem of the probability density of a transformed random variable[5]. As the summands of the random variable η_n have the normal probability distribution $N(0,1)$ and the summands of the random variable ζ_n the normal probability distribution $N(0, \sigma^2)$, the transformation $\zeta_n = \sigma^2 \eta_n$ is taken into account, where the random variable η_n possesses the central chi-square probability distribution with n degrees of freedom whose probability density is of the form (1.5.5). The theorem on the probability density of a transformed random variable implies that the random variable ζ_n has the probability density $f_n(z) = g_n(z/\sigma^2)/\sigma^2$, thus $\zeta \sim \sigma^2 \chi^2(n)$.

A more general genesis of a central σ^2chi-square probability distribution is presented by the following Pearson lemma: Let $\xi \sim N_n(m, \sigma^2 H)$, where H is a known symmetric and positive definite matrix. The random variable $(\xi - m)' H^{-1} (\xi - m)$ has a central σ^2 chi-square probability distribution with n degrees of freedom.

The matrix H is assumed to be positive definite, thus it is possible to express it in the form $H = H^{1/2} H^{1/2}$. The random variable $\eta = H^{-1/2}(\xi - m)$ has the normal probability distribution $N_n(0, \sigma^2 I)$. The random variable $\sum_{i=1}^{n} \eta_i^2 = \eta' \eta$ has the central σ^2 chi-square probability distribution with n degrees of freedom. However, $\eta' \eta = (\xi - m)' H^{-1/2} H^{-1/2} (\xi - m) = (\xi - m)' H^{-1} (\xi - m)$, by which the Pearson theorem is proved.

Definition A random variable whose probability density is

$$g_{n,\delta}(y) = \begin{cases} \exp[-(y+\delta)/2] \sum_{r=0}^{\infty} \frac{1}{r!} \left(\frac{\delta^r}{2}\right) \frac{y^{r+(n/2)-1}}{2^{r+n/2}\Gamma(r+n/2)} & , \quad y > 0, \\ 0 & , \quad y \leq 0 \end{cases} \tag{1.5.7}$$

is said to have a non-central chi-square probability distribution with n degrees of freedom and the parameter of non-centrality δ.

---

[5]**Theorem of the probability density of a transformed random variable** Let η be a continuous random variable characterized by the probability density $g(\cdot)$, $y \in \mathcal{R}^1$. Consider a one-to-one transformation of the random variable η defined by a continuous and differentiable function $u(\cdot)$, $y \in \mathcal{R}^1$. Let $h(\cdot)$ denote the inverse function of the function $u(\cdot)$. The probability density $f(\cdot)$, $y \in \mathcal{R}^1$, of the random variable $\zeta = u(\eta)$ is of the form $f(z) = g[h(z)]|dh(z)/dz|$.

Let $\xi_1, \xi_2, \ldots, \xi_n$ be independent random variables such that $\xi_i \sim N(m_i, 1)$, $i = 1, 2, \ldots, n$. The random variable $\eta_n = \sum_{i=1}^{n} \xi_i^2$ has a non-central chi-square probability distribution with n degrees of freedom and the parameter of non-centrality is $\delta = \sum_{i=1}^{n} m_i^2$ ($\eta_n \sim \chi^2(n, \delta)$).

For the proof of this assertion cf., e.g., [10], p. 119.

Definition A random variable ζ_n is said to have a non-central σ^2 chi-square probability distribution with n degrees of freedom and the parameter of non-centrality if its probability distribution is

$$f_{n,\delta}(z) = \begin{cases} \exp\left[-\frac{1}{2}\left(\frac{z}{\sigma^2} + \delta\right)\right] \sum_{r=0}^{\infty} \frac{1}{r!} \left(\frac{\delta}{2}\right)^r \frac{z^{r+(n/2)-1}}{2^{r+n/2}\Gamma(r+n/2)\sigma^{2r+n}} & , \quad z > 0, \\ 0 & , \quad z \leq 0. \end{cases} \tag{1.5.8}$$

Let $\xi_1, \xi_2, \ldots, \xi_n$ be independent random variables such that $\xi_i \sim N(m_i, \sigma^2)$, $i = 1, 2, \ldots, n$. The random variable $\zeta_n = \sum_{i=1}^{n} \xi_i^2$ possesses a non-central σ^2 chi-square probability distribution with n degrees of freedom and with the non-centrality parameter $\delta = \sum_{i=1}^{n} m_i^2/\sigma^2$ ($\zeta \sim \sigma^2\chi^2(n, \delta)$).

Eq. (1.5.8) for the probability density of the random variable $\zeta_n = \sum_{i=1}^{n} \xi_i^2$, $\xi_i \sim N(m_i, \sigma^2)$, can be obtained from Eq. (1.5.7) that expresses the probability density of the random variable $\sum_{i=1}^{n} \xi_i^2/\sigma^2$ by applying the theorem of the probability density of a transformed random variable (see p. 84).

Definition A continuous random variable τ is said to possess the Student's t probability distribution with k degrees of freedom, if its probability density is

$$f_k(t) = \frac{\Gamma(\frac{k+1}{2})}{\sqrt{\pi k}\Gamma\left(\frac{k}{2}\right)} \left(1 + \frac{t^2}{k}\right)^{-(k+1)/2}, \quad t \in \mathcal{R}^1. \tag{1.5.9}$$

Let $\xi \sim N(0, \sigma^2)$ and $\eta \sim \sigma^2\chi^2(k)$ be independent random variables. The random variable

$$\tau = \frac{\xi}{(\eta/k)^{1/2}}$$

has the Student's t probability distribution with k degrees of freedom.

In accordance with this assumption, the random variables ξ and η are stochastically independent; therefore the probability density $f(\cdot)$, $(x,y)' \in \mathcal{R}^2$ of the random variable $(\xi,\eta)'$ equals the product of the densities of its components (see p. 41)

$$f(x,y) = f_\xi(x)f_\eta(y) = \frac{\exp[-(x^2+y)/2\sigma^2]y^{\frac{k}{2}-1}}{\sigma^{(k+1)/2}\sqrt{2\pi}2^{k/2}\Gamma(k/2)},$$

$$x \in (-\infty,\infty),\ y \in (0,\infty).$$

Based on the theorem of the probability density of a transformed random vector [6], the probability density $g(\cdot,\cdot\cdot)$ of the random vector $(\tau,\xi)'$, where $\tau = \xi/(\eta/k)^{1/2}$, and $\zeta = \eta$ is $g(t,z) = f[x(t,z),y(t,z)]|J(t,z)|$ and since the Jacobian $J(t,z)$ in our case is $J(t,z) = (z/k)^{1/2}$, we obtain

$$g(t,z) = \frac{z^{(k-1)/2}\exp\left[-\frac{z}{2\sigma^2}\left(\frac{t^2}{k}+1\right)\right]}{\sigma^{k+1}\sqrt{2k\pi}2^{k/2}\Gamma(k/2)},\ t \in (-\infty,\infty),\ z \in (0,\infty).$$

The probability density of the random variable τ that we seek is the marginal probability density of the first component of the random vector $(\tau,\xi)'$, thus

$$f_k(t) = \int_0^\infty g(t,z)dz,\ t \in (-\infty,\infty).$$

The problem is to determine the integral

$$\int_0^\infty \frac{z^{(k-1)/2}}{\sigma^{k+1}}\exp\left[-\frac{z}{2\sigma^2}\left(\frac{t^2}{k}+1\right)\right]dz.$$

The substitution $(z/\sigma^2)\left(\frac{t^2}{k}+1\right) = w$ yields

$$\int_0^\infty \frac{z^{\frac{k+1}{2}-1}}{\sigma^{k+1}}\exp\left[-\frac{z}{2\sigma^2}\left(\frac{t^2}{k}+1\right)\right]dz =$$

---

[6] **Theorem of the probability density of a transformed random vector** Let $(\xi,\eta)'$ be a two-dimensional random vector whose probability density is $g(\cdot,\cdot\cdot)$, $(x,y) \in \mathcal{R}^2$. Consider a one-to-one transformation of the vector $(\xi,\eta)'$ defined by a vector function $u(\cdot,\cdot\cdot) = (u_1(\cdot,\cdot\cdot),u_2(\cdot,\cdot\cdot))'$, $(x,y) \in \mathcal{R}^2$, whose components are continuous and differentiable. The probability density $f(\cdot,\cdot\cdot)$, $(u_1,u_2) \in \mathcal{R}^2$ of the random variable $(u_1(\xi,\eta),u_2(\xi,\eta))'$ is $f(u_1,u_2) = g[h_1(u_1,u_2),h_2(u_1,u_2)]|J|$, where

$$J = \det\begin{pmatrix} \partial h_1(u_1,u_2)/\partial u_1, & \partial h_1(u_1,u_2)/\partial u_2 \\ \partial h_2(u_1,u_2)/\partial u_1, & \partial h_2(u_1,u_2)/\partial u_2 \end{pmatrix}$$

and $h(\cdot,\cdot\cdot)$ is the inverse function of the function $u(\cdot,\cdot\cdot)$.

$$= 2^{\frac{k+1}{2}} \left(\frac{t^2}{k} + 1 \right)^{-(k+1)/2} \int_0^\infty w^{\frac{k+1}{2}-1} \exp(-w) dw =$$

$$= 2^{(k+1)/2} \left(\frac{t^2}{k} + 1 \right)^{-(k+1)/2} \Gamma\left(\frac{k+1}{2} \right),$$

whereby it has been proved that the probability density $f_k(\cdot)$, $t \in (-\infty, \infty)$, of the random variable $\tau = \xi/(\eta/k)^{1/2}$ is of the form (1.5.9).

Definition A continuous random variable τ is said to have a non-central Student's t probability distribution with k degrees of freedom and with the parameter of non-centrality m if its probability density $g_{k,m}(\cdot)$, $t \in (-\infty, \infty)$, is

$$g_{k,m}(t) = \frac{\left(1 + \frac{t^2}{k}\right)^{-(k+1)/2}}{\sqrt{k\pi} 2^{(k-1)/2} \Gamma(k/2)} \exp\left(-\frac{km}{2(k+t^2)} \right) \cdot$$

$$\cdot \int_0^\infty v^k \exp\left[-\frac{1}{2} \left(v - \frac{tm}{\sqrt{k+t^2}} \right)^2 \right] dv, \ t \in (-\infty, \infty). \tag{1.5.10}$$

By putting $m = 0$ in Eq. (1.5.10) we get Eq. (1.5.9) for the probability density of a central Student's t probability distribution with k degrees of freedom; it suffices to realize that

$$\int_0^\infty v^k \exp(-v^2/2) dv = 2^{(k-1)/2} \int_0^\infty \exp(-t) t^{\frac{k+1}{2}-1} dt = 2^{(k-1)/2} \Gamma\left(\frac{k+1}{2} \right).$$

Let $\xi \sim N(m, \sigma^2)$ and $\eta \sim \sigma^2 \chi^2(k)$ be independent random variables; the random variable

$$\tau = \frac{\xi}{(\eta/k)^{1/2}}$$

possesses a non-central probability distribution with k degrees of freedom and with the parameter of non-centrality m.

The idea of proving this assertion is the same as that of the previous one: the fact that $\xi \sim N(m, \sigma^2)$ has to be taken into account for determining the probability density of the random vector $(\xi, \eta)'$.

Definition The random variable ζ has the Fisher — Snedecor probability distribution with k, l degrees of freedom if its probability density is of the form

$$f_{k,l}(z) = \begin{cases} \frac{\Gamma(\frac{k+l}{2})(\frac{k}{l})^{k/2}}{\Gamma(k/2)\Gamma(l/2)} z^{\frac{k}{2}-1}\left(1+\frac{kz}{l}\right)^{-(k+l)/2} & , \quad z > 0, \\ 0 & , \quad z \le 0. \end{cases} \tag{1.5.11}$$

Let $\xi \sim \sigma^2\chi^2(k)$ and $\eta \sim \sigma^2\chi^2(l)$ be independent random variables; the random variable

$$\zeta = \frac{\xi/k}{\eta/l}$$

has the Fisher - Snedecor probability distribution with k, l degrees of freedom.

The random variables $\xi \sim \sigma^2\chi^2(k)$ and $\eta \sim \sigma^2\chi^2(l)$ are assumed to be independent, therefore the probability density of the random vector $(\xi, \eta)'$ is

$$f(x, y) = f_\xi(x) f_\eta(y) = \frac{\exp[-(kx + ly)/2\sigma^2] x^{(k/2)-1} y^{(l/2)-1}}{2^{(k+l)/2}\Gamma(k/2)\Gamma(l/2)\sigma^{k+l}},$$

$$x \in (0, \infty), \ y \in (0, \infty).$$

Let the components of the random vector $(\zeta, \nu)'$ be defined by the relations $\zeta = (\xi/k)/(\eta/l)$, $\nu = \eta$. The Jacobian of the transformation $z = (x/k)/(y/l)$, $u = y$ is

$$J(z, u) = \begin{vmatrix} ku/l, & kz/l \\ 0, & 1 \end{vmatrix} = \frac{k}{l}u.$$

Based on the theorem of the probability density of a transformed random vector, the probability density $g(\cdot, \cdot\cdot)$, $z > 0$, of the random vector $(\zeta, \nu)'$ is

$$g(z, u) = |J(z, u)| f[x(z, u), y(z, u)] =$$

$$= \frac{\left(\frac{k}{l}\right)^{k/2} z^{\frac{k}{2}-1} u^{\frac{k+l}{2}-1} \exp\left[-\frac{u}{2\sigma^2}\left(\frac{kz}{l}+1\right)\right]}{2^{(k+l)/2}\Gamma(k/2)\Gamma(l/2)\sigma^{k+l}}, \ z > 0, \ u > 0.$$

The sought probability density $f_{k,l}(\cdot)$ of the random variable $\zeta = (\xi/k)/(\eta/l)$ is, with respect to the probability density $g(\cdot, \cdot)$ of the random vector $(\zeta, \nu)'$, the marginal density of its first component, thus

$$f_{k,l}(z) = \int_0^\infty g(z,u)du =$$

$$= \frac{\left(\frac{k}{l}\right)^{(k/2)} z^{\frac{k}{2}-1}}{2^{(k+l)/2}\Gamma(k/2)\Gamma(l/2)} \int_0^\infty \frac{u^{\frac{k+l}{2}-1}}{\sigma^{k+l}} \exp\left[-\frac{u}{2\sigma^2}\left(\frac{kz}{l}+1\right)\right] du.$$

Using the substitution $\frac{u}{2\sigma^2}\left(\frac{kz}{l}+1\right) = w$ we get

$$\int_0^\infty \frac{u^{\frac{k+l}{2}-1}}{\sigma^{k+l}} \exp\left[-\frac{u}{2\sigma^2}\left(\frac{kz}{l}+1\right)\right] du =$$

$$= 2^{(k+l)/2}\left(1+\frac{k}{l}z\right)^{-(k+l)/2} \int_0^\infty w^{\frac{k+l}{2}-1}\exp(-w)dw =$$

$$= 2^{(k+l)/2}\left(1+\frac{k}{l}z\right)^{-(k+l)/2} \Gamma[(k+l)/2],$$

and the sought probability density

$$f_{k,l}(z) = \frac{\left(\frac{k}{l}\right)^{k/2}\Gamma\left(\frac{k+l}{2}\right) z^{\frac{k}{2}-1}}{\Gamma(k/2)\Gamma(l/2)} \left(1+\frac{k}{l}z\right)^{-(k+l)/2}, \quad z > 0,$$

of the random variable ζ is identical with Eq. (1.5.11).

Definition A random variable ζ possesses the non-central Fisher - Snedecor probability distribution with k, l degrees of freedom and with the parameter of non-centrality δ if its probability density is of the form

$$f_{k,l,\delta}(z) = \begin{cases} \exp(-\delta/2)\sum_{r=0}^\infty \frac{1}{r!}\left(\frac{\delta}{2}\right)^r \frac{\Gamma(\frac{k+l}{2}+r)(\frac{k}{l})^{\frac{k}{2}+r} z^{\frac{k}{2}+r-1}}{\Gamma(\frac{k}{2}+r)\Gamma(\frac{l}{2})(1+\frac{k}{l}z)^{[(k+l)/2]+r}}, & z > 0, \\ 0, & z \le 0. \end{cases} \tag{1.5.12}$$

Let $\xi \sim \sigma^2\chi^2(k,\delta)$ and $\eta \sim \sigma^2\chi^2(l)$ be independent random variables; the random variable $\zeta = (\xi/k)/(\eta/l)$ has the non-central Fisher — Snedecor probability distribution with k, l degrees of freedom and with the parameter of non-centrality δ. (This assertion is proved, e.g., in [10], p. 122.)

Let $\xi \sim \chi^2(k)$ and $\eta \sim \chi^2(l)$ be independent random variables; the random variable $\zeta = \xi + \eta$ has the central chi-square probability distribution with $k + l$ degrees of freedom.

In accordance with the assumption the random variables $\xi \sim \chi^2(k)$ and $\eta \sim \chi^2(l)$ are independent, the probability density $f(\cdot, \cdot\cdot)$ of the random vector $(\xi, \eta)'$ is defined as the product of the probability densities $g_k(\cdot)$ and $g_l(\cdot)$ of its components (see Eq. (1.5.5)), thus

$$f(x, y) = \frac{\exp[-(x+y)/2] x^{\frac{k}{2}-1} y^{\frac{l}{2}-1}}{2^{(k+l)/2}\Gamma(k/2)\Gamma(l/2)}, \; x > 0, \; y > 0,$$

which, after applying the theorem of the probability density of the sum of two random variables, yields

$$h(z) = \frac{\exp(-z/2)}{2^{(k+l)/2}\Gamma(k/2)\Gamma(l/2)} \int_0^z y^{\frac{k}{2}-1}(z-y)^{\frac{l}{2}-1} dy$$

(the integral domain is the interval $(0, z)$, since outside this interval at least one of the sub-integral functions is zero). Substituting $y = uz$ we obtain

$$\int_0^z y^{\frac{k}{2}-1}(z-y)^{\frac{l}{2}-1} dy =$$

$$= z^{\frac{k+l}{2}-1} \int_0^1 u^{\frac{k}{2}-1}(1-u)^{\frac{l}{2}-1} du =$$

$$= z^{\frac{k+l}{2}-1} B(k/2, l/2) = z^{\frac{k+l}{2}-1} \frac{\Gamma(k/2)\Gamma(l/2)}{\Gamma\left(\frac{k+l}{2}\right)},$$

thus

$$h(z) = \frac{\exp(-z/2) z^{\frac{k+l}{2}-1}}{2^{(k+l)/2}\Gamma\left(\frac{k+l}{2}\right)} \equiv g_{k+l}(z), \; z > 0,$$

where $g_{k+l}(\cdot)$ is of the form (1.5.5).

The assertion can be generalized in the following way: Let $\xi_1 \sim \chi^2(k_1)$, $\xi_2 \sim \chi^2(k_2), \ldots, \xi_n \sim \chi^2(k_n)$ be independent random variables, then $\sum_{i=1}^n \xi_i \sim \chi^2\left(\sum_{i=1}^n k_i\right)$.

The assertion: Let $\xi_1 \sim \sigma^2\chi^2(k_1), \xi_2 \sim \sigma^2\chi^2(k_2), \ldots, \xi_n \sim \sigma^2\chi^2(k_n)$ be independent random variables, then $\sum_{i=1}^n \xi_i \sim \sigma^2\chi^2\left(\sum_{i=1}^n k_i\right)$, can be proved in an analogous way.

Definition An $n \times n$ matrix Ξ is said to possess the Wishart probability distribution with N degrees of freedom ($N > n$) if its probability density is of the form

$$f(X) = \begin{cases} \dfrac{[\det(X)]^{(N-n-1)/2} \exp[(1/2)Tr(\Sigma^{-1}X)]}{2^{Nn/2}\pi^{N(N-1)/4}[\det(\Sigma)]^{N/2}\prod_{i=1}^{N}\Gamma(\frac{N+1-i}{2})} & , X \text{ is a positive definite } \\ & \quad n \times n \text{ matrix}, \\ 0 & , \text{otherwise}, \end{cases}$$

(1.5.13)

where Σ is an $n \times n$ positive definite matrix [7].

Let $\xi_1, \xi_2, \ldots, \xi_N$ be stochastically independent n-dimensional random vectors with the same probability distribution $N_n(0, \Sigma)$. The random matrix $\Xi = \sum_{i=1}^{N} \xi_i \xi_i'$ possesses the n-dimensional Wishart probability distribution with N degrees of freedom ($\Xi \sim W_n(N, \Sigma)$).

The assertion is proved, e.g., in [10], pp. 129-131.

If $n = 1$, the probability density given by (1.5.13) will be equivalent to (1.5.6) expressing the probability density of a random variable with a σ^2 chi-square probability density with N degrees of freedom (recall that in this case $\xi_i \sim N(0, \sigma^2)$, $i = 1, 2, \ldots, N$).

Let $\eta_1, \eta_2, \ldots, \eta_N$ be stochastically independent n-dimensional random vectors with the same normal probability distribution $N_n(m, \Sigma)$ and $N > n$. The $n \times n$-dimensional random matrix $\sum_{i=1}^{N}(\eta_i - \overline{\eta})(\eta_i - \overline{\eta})'$, where $\overline{\eta} = (1/N)\sum_{i=1}^{N} \eta_i \sim N_n(m, \Sigma/N)$, has the n-dimensional Wishart probability distribution with $N-1$ degrees of freedom, $\sum_{i=1}^{n}(\eta_i - \overline{\eta})(\eta_i - \overline{\eta})' \sim W_n(N-1, \Sigma)$.

The proof of this assertion can be found, e.g., in [2], pp.75, 213.

1.6 Random samples, the empirical distribution function, sample characteristics, and their probability distributions

1.6.1 Random samples, the empirical distribution function, fundamental theorems of mathematical statistics

We have already dealt with the basic notions from probability theory, needed to build reasonable stochastic models of various experiments.

---

[7] The trace of an $n \times n$ matrix A, denoted as $Tr(A)$, is defined as the sum of its diagonal elements $\{A\}_{i,i}$, thus $Tr(A) = \sum_{i=1}^{n}\{A\}_{i,i}$.

Mathematical statistics enables us to process the data resulting from various experiments and measurements.

A random variable ξ, which represents within an arbitrary stochastic model a measurement (observation) of a directly observable quantity, is defined on the probability space $(\Omega, \mathcal{S}, P)$, see p. 25. A stochastic model always operates on the set of all possible realizations $\{x = \xi(\omega) : \omega \in \Omega\}$ of the random variable ξ which models the measurement at the same time as a value of the distribution function $F(x) = P(\{\omega : \xi(\omega) < x\})$ is assigned, in an one-to-one way, to the value that is realized. In this way, all numerical characteristics of the random variable ξ are determined. The coincidence between the stochastic model and reality is verified by testing the results of a suitable experiment by means of methods of mathematical statistics. The scope of such an experiment can be very broad and heterogeneous — ranging from the experimental verification of the fact that a random variable really has an assumed probability distribution (e.g., that a die is homogeneous), through various sociological studies, where the characteristics are not studied on the whole population but only on certain representative samples and the results are processed in such a way that the conclusions apply to the entire population, to experiments for determining unknown parameters of the measured object based on some reasonable finite number of repeated measurements.

The random sample furnishes the input for mathematical statistics to begin its analyses.

Definition A sequence of n independent random variables or vectors with the same probability distribution is called a random sample. We are speaking of a random sample with size n from a given probability distribution (one- or multi-dimensional) characterized by a distribution function $F(\cdot)$, $x \in \mathcal{R}^i$, $i = 1, 2, \ldots, n$.

First, the elements of the random sample will be considered to be one-dimensional random variables. If $\{\xi(\omega) : \omega \in \Omega\} \subset \mathcal{R}^1$ is a set of all possible realizations of an arbitrary component of the random sample with the size n, the random sample, i.e., an n-dimensional random variable $(\xi_1, \xi_2, \ldots, \xi_n)'$, is realized on the set $\{\xi(\omega) : \omega \in \Omega\} \times \{\xi(\omega) : \omega \in \Omega\} \times \ldots \{\xi(\omega) : \omega \in \Omega\} \subset \mathcal{R}^n$ [8], called a sample space.

Example Let us investigate a sample space corresponding to the random sample with size n from a one-dimensional probability distribution

[8] A set of all ordered n-tuples $(a_1, a_2, \ldots, a_n)$, where $a_i \in A_i$, $i = 1, 2, \ldots, n$ is called the Cartesian product $A_1 \times A_2 \times \ldots A_n$ of the sets $A_1, A_2, \ldots, A_n$.

characterized by a distribution function $F(\cdot)$, $x \in \mathcal{R}^1$,

$$F(x) = \begin{cases} 0 & , x \in (-\infty, 1], \\ 0.16\dot{6} & , x \in (1, 2] \\ 0.33\dot{3} & , x \in (2, 3], \\ 0.500 & , x \in (3, 4], \\ 0.66\dot{6} & , x \in (4, 5], \\ 0.83\dot{3} & , x \in (5, 6], \\ 1 & , x \in (6, \infty) \end{cases}$$

(a random variable modelling a throw of a homogeneous die, p. 26). In this case the set of all possible realizations of the random variable ξ is of the form $\{\xi(\omega) : \omega \in \Omega\} = \{1, 2, 3, 4, 5, 6, \}$. If $n = 2$, the sample space consists of the set of points

$$\{1, 2, 3, 4, 5, 6\} \times \{1, 2, 3, 4, 5, 6\} =$$

$$= \{(1,1), (1,2), (1,3), (1,4), (1,5), (1,6), (2,1), (2,2), (2,3), (2,4), (2,5),$$

$$(2,6), (3,1), (3,2), (3,3), (3,4), (3,5), (3,6), (4,1), (4,2), (4,3), (4,4), (4,5),$$

$$(4,6), (5,1), (5,2), (5,3), (5,4), (5,5), (5,6), (6,1), (6,2), (6,3), (6,4), (6,5),$$

$$(6,6)\} \subset \mathcal{R}^2$$

(together 6^2 points); if $n = 3$, the sample space is a subset of the space $\mathcal{R}^3$ and consists of 6^3 points $\{1,2,3,4,5,6\} \times \{1,2,3,4,5,6\} \times \{1,2,3,4,5,6\} = \{(1,1,1), (1,2,1), (1,3,1), \ldots, (6,1,6), (6,2,6), (6,3,6), (6,4,6), (6,5,6,), (6,6,6)\}$.

If we return to the experiment performed by Meissner (see p. 20), we can state that in this experiment the sample space consists of 6^{1800} points belonging to the space $\mathcal{R}^{1800}$ (the coordinates of all points are the numbers 1,2,3,4,5,6).

If an n times replicated measurement of a distance between two points is taken into account and the measurement errors are assumed to possess a normal probability distribution, the sample space is $\mathcal{R}^n$ because $\{\xi(\omega) : \omega \in \Omega\} = \mathcal{R}^1$, where ξ is the random variable modelling the measurement of one directly observable distance.

Each realized random sample with size n is a point in an n-dimensional sample space, i.e., it is an n-dimensional real vector; indices of its components correspond to the order in which in the sampling of the random variable ξ was realized.

A simpler notion sufficient for many statistical considerations is the concept of a set of measured data. The set of measured data is a system of results of an experiment, where the order in which the outcomes were realized is not determinative. In other words, in contradistinction to a point in an n-dimensional sample space (an n-dimensional random sample is taken into account), which is an ordered n-tuple of measured results, the set of measured results is a disordered n-tuple of measured results. There exist many points of the sample space generating the same set of measured data (e.g., the points (1,1,2), (1,2,1) and (2,1,1) of a three-dimensional space generate the same set of measured data which contains two numbers one and one number two).

Within this detailed analysis of the problem it is necessary to realize the relation between the set of all possible realizations of the random variable ξ, the points of the sample space, and the set of measured data when an n-dimensional random sample remains at our disposal.

In essence three following situations occur:

1. ξ is a continuous random variable; the set $\{x = \xi(\omega) : \omega \in \Omega\}$ of all its realizations has a continuum power;

2. ξ is a discrete random variable that assumes a countable number of different values; the set $\{x = \xi(\omega) : \omega \in \Omega\}$ is countable;

3. ξ is a discrete random variable that assumes K different values; the set $\{x = \xi(\omega) : \omega \in \Omega\}$ contains K elements.

Let n be the size of the random sample; it has been realized by the values $x_{1s}, x_{2s}, \ldots, x_{ns}$ (the index s distiguishes the values realized in the sample from the values belonging to the set of possible realizations). The random sample considered has been realized at the point $(x_{1s}, x_{2s}, \ldots, x_{ns})'$ of the n-dimensional sample space. The definition of points of an n-dimensional sample space (they are defined as Cartesian products of n identical sets of all possible realizations of the random variable ξ) implies that the inequalities $x_{is} \neq x_{js}$, $i, j = 1, 2, \ldots, n$, $i \neq j$ need not hold, on the contrary the extremum case $x_{1s} = x_{2s} = \ldots = x_{ns}$ is admissible (the result of an n times repeated toss of a die is always the same number); if $K < n$, some realizations must be repeated. A multiple occurrence of the same realization within an n-dimensional random sample is possible even if a continuous random variable ξ or a discrete random variable with a countable or a finite number $K > n$ of realizations is considered.

It is not necessary to emphasize that a value which does not belong to the set of all possible realizations of the random variable considered cannot occur as an element of the random sample. If in an n-tuple $x_{1s}, x_{2s}, \ldots, x_{ns}$ of measured data only the values, without taking into account the order of their occurrence, are important, the system is called a set of measured data. If all possible permutations of components of the vector $(x_{1s}, x_{2s}, \ldots, x_{ns})'$ are considered, the same set of measured data is assigned to all the different points of the sample space generated in this way .

Let us order the set of measured data according to magnitude from the smallest to the largest; we then assign indices in parentheses to the elements ordered in this way, thus

$$x_{(1s)} \leq x_{(2s)} \leq \ldots \leq x_{(ns)}.$$

Furthermore, let this non-decreasing sequence of elements of the set of measured data satisfy the conditions

$$x_{(1s)} = x_{(2s)} = \ldots = x_{(k_1 s)} \leq x_{(k_1+1)s)} = \ldots =$$

$$= x_{(k_2 s)} \leq x_{((k_{N-1}+1)s)} = x_{(k_{N-1}+2)s} = \ldots = x_{(nv)}. \tag{1.6.1}$$

Let $x_{(1)}$ be a symbol used for denoting the first k_1, equalling values of the non-decreasing sequence,

$$x_{(1)} = x_{(1s)} = x_{(2s)} = \ldots = x_{(k_1 s)},$$

and analogously

$$x_{(2)} = x_{((k_1+1)s)} = x_{((k_1+2)s)} = \ldots = x_{(k_2 s)},$$

....

$$x_{(N)} = x_{((k_{N-1}+1)s)} = x_{((k_{N-1}+2)s)} = \ldots = x_{(k_N s)} = x_{(ns)}. \tag{1.6.2}$$

The non-decreasing sequence (1.6.1) forms the background for the definition of an empirical (sample) distribution function $S_n(\cdot)$ of a random variable ξ and other types of empirical moments. The symbols introduced by (1.6.2) are taken into account simultaneously.

Definition A function $S_n(\cdot)$, $x \in \mathcal{R}^1$, defined by the relation

$$S_n(x) = \begin{cases} 0, & x \in (-\infty, x_{(1)}], \\ k_{i-1}/n, & x \in (x_{(i-1)}, x_{(i)}],\ i = 2, \ldots, N, \\ 1, & x \in (x_{(N)}, \infty), \end{cases} \tag{1.6.3}$$

is said to be an empirical (sample) distribution function of the random variable ξ obtained on the basis of a random sample of size n.

Example The problem is to determine an empirical distribution function for the random variable modelling a toss of a die based on the results obtained by Meissner (see p. 20).

In this case $n = 1800$, $N = 6$, $x_{(1)} = 1$, $k_1 = 299$, $x_{(2)} = 2$, $k_2 = k_1 + 295 = 594$, $x_{(3)} = 3$, $k_3 = k_2 + 303 = 897$, $x_{(4)} = 4$, $k_4 = k_3 + 307 = 1204$, $x_{(5)} = 5$, $k_5 = k_4 + 289 = 1493$ and finally $x_{(6)} = 6$, $k_6 = 1800$, thus

$$S_{1800}(x) = \begin{cases} 0, & x \in (-\infty, 1], \\ 0.166, & x \in (1, 2], \\ 0.330, & x \in (2, 3], \\ 0.498, & x \in (3, 4], \\ 0.669, & x \in (4, 5], \\ 0.829, & x \in (5, 6], \\ 1, & x \in (6, \infty). \end{cases}$$

Comparing this empirical distribution function which results from a random 1800-dimensional sample with its theoretical form given on p. 93, we can observe a very good coincidence between them. We have verified the following important assertions called fundamental theorems of mathematical statistics. They are presented here without proofs, as the mathematical apparatus necessary for proving them goes beyond the framework of this publication.

The first fundamental theorem of mathematical statistics Let $S_n(\cdot)$, $x \in \mathcal{R}^1$, be an empirical distribution function of the random variable ξ, and let $F(\cdot)$, $x \in \mathcal{R}^1$, be its theoretical distribution function. Then, at each point $x \in \mathcal{R}^1$ with probability 1,

$$\lim_{n\to\infty} S_n(x) = F(x).$$

The second fundamental theorem of mathematical statistics Let $S_n(\cdot)$, $x \in \mathcal{R}^1$, be an empirical distribution function of a random variable ξ, and let $F(\cdot)$, $x \in \mathcal{R}^1$, be its theoretical distribution function. Then, with probability 1,

$$\lim_{n\to\infty} \sup_{-\infty<x<\infty} |S_n(x) - F(x)| = 0.$$

(For the proof of both fundamental theorems see, e.g., [6], p. 327.)

These two theorems form the foundation for the stochastic modelling and its ability to verify on the basis of random samples. They show that the empirical distribution function represents a good approximation of the theoretical distribution function, provided that the sample size is sufficiently large.

1.6.2 Sample random variables, empirical moments

The sample random variable represents the next important notion.

Definition A random variable $\eta = f(\xi)$, $\xi = (\xi_1, \xi_2, \ldots, \xi_n)'$ defined for each realization $(x_1, x_2, \ldots, x_n)'$ of a random sample $(\xi_1, \xi_2, \ldots, \xi_n)'$, is called a sample random variable.

The arithmetic mean $\bar{\xi} = (1/n)\sum_{i=1}^{n} \xi_i$, where ξ_i are elements of the vector ξ of the n-dimensional vector of the random sample, can serve as a typical example of a sample random variable. The number $\bar{x} = (1/n)\sum_{i=1}^{n} x_i$ is a realization of the sample random variable $\bar{\xi}$ corresponding to the realization $(x_1, x_2, \ldots, x_n)'$ of the random variable $(\xi_1, \xi_2, \ldots, \xi_n)'$.

The definition of a sample random variable enables us to define many different types of sample random variables. From the point of view of their applications in data processing, those types which give a good approximation of the fundamental moment characteristics of a random variable or a vector (the mean value, variance, excess, asymmetry, various correlation coefficients), based on random samples, are the most important.

The lth moment of a discrete random variable ξ at the point c is defined by Eq. (1.3.1). Without any loss of generality the set of all realizations $\{x_1, x_2, \ldots, x_K, \ldots\}$ of the random variable ξ is ordered in such a way that $x_1 < x_2 < \ldots < x_K < \ldots$. Using the relation $p_k = F(x_{k+1}) - F(x_k)$, Eq. (1.3.1) may be written in the following form

$$E[(\xi - c)^l] = \sum_k [F(x_{k+1}) - F(x_k)](x_k - c)^l. \tag{1.6.4}$$

This is the theoretical lth moment at the point c, defined on the basis of all elements from the set of all realizations of the random variable ξ and the corresponding probabilities p_k are $p_k = F(x_{k+1}) - F(x_k)$. After an experiment is made, we only have access to the realized system $\{x_{1s}, x_{2s}, \ldots, x_{ns}\}$ whose elements belong to the system of all realizations of the random variable ξ; thus, only an empirical distribution function $S_n(\cdot)$ can be determined; owing to the fundamental theorems of mathematical statistics, this yields a good approximation (for a sufficiently large n) of the corresponding theoretical distribution function (the assertion holds with probability 1). Thus, it is natural to approximate the right-hand side of Eq. (1.6.4) in the following way

$$\sum_k [F(x_{k+1}) - F(x_k)](x_k - c)^l \doteq \sum_{i=1}^{N} [S_n(x_{(i+1)}) - S_n(x_{(i)})](x_{(i)} - c)^l.$$

The definition of an empirical distribution function and the relation between $x_{(i)}$ and $x_{((k_{i-1}+1)s)}, x_{((k_{i-1}+2)s)}, \ldots, x_{(k_i,s)}$, $i = 1, 2, \ldots, N$, imply

$$\sum_{i=1}^{N} [S_n(x_{(i+1)}) - S_n(x_{(i)})](x_{(i)} - c)^l = \sum_{i=1}^{N} \frac{(k_{i+1} - k_i)}{n}(x_{(i)} - c)^l =$$

$$= (1/n) \sum_{j=1}^{n} (x_{js} - c)^l,$$

which means that the value of the expression $(1/n)\sum_{j=1}^{n}(x_{js} - c)^l$ can be considered to be a suitable approximation of the lth moment at the point c of the random variable ξ on the basis of a set $x_{1s}, x_{2s}, \ldots, x_{ns}$ of data, which is a realization of an n dimensional sample $\xi_1, \xi_2, \ldots, \xi_n$.

The same result may be obtained in the case of a continuous random variable ξ characterized by a probability density $f(\cdot)$, $x \in \mathcal{R}^1$, where we find a suitable approximation of the lth moment at the point c, based on a set of measured data $x_{1s}, x_{2s}, \ldots, x_{ns}$, which is a realization of an n dimensional random sample $\xi_{1s}, \xi_{2s}, \ldots, \xi_{ns}$ from a probability distribution characterized by a probability density $f(\cdot)$. Eq. (1.3.2) defining the lth theoretical moment considered can be writen in the form

$$E[(\xi - c)^l] = \int_{\mathcal{R}^1} (x - c)^l dF(x) \tag{1.6.5}$$ [9]

(the relation $f(x) = dF(x)/dx$ was taken into account). The right-hand side of this relation may be approximated by an integral sum based on the measured data and the theoretical distribution function by the empirical one; then

$$\int_{\mathcal{R}^1} (x - c)^l dF(x) \doteq \sum_{i=1}^{n} [S_n(x_{(i+1)}) - S_n(x_{(i)})](x_{(i)} - c)^l$$

and for the same reason as in the case of a discrete random variable, the value of the expression $(1/n)\sum_{j=1}^{n}(x_{js} - c)^l$ may be considered to be a good approximation of the lth moment of a continuous random variable ξ based on the given set of measured data.

Values of the expression $(1/n)\sum_{j=1}^{n}(x_{js} - c)^l$ differ for different samples with the same size from the same probability distribution, naturally (n-tuples of results obtained within repeated samples are different), the value of the expression $(1/n)\sum_{j=1}^{n}(x_{js} - c)^l$ is a realization of the random variable $(1/n)\sum_{j=1}^{n}(\xi_j - c)^l$, where ξ_j, $j = 1, 2, \ldots, n$ are components of the random vector $\xi = (\xi_1, \xi_2, \ldots, \xi_n)'$ realized in the random sample by the vector $x_s = (x_{1s}, x_{2s}, \ldots, x_{ns})'$.

The previous considerations enable us to give the following definitions of the empirical moments; simultaneously, we face the problem of investigating the probability distribution of each moment.

Definition Let $(\xi_1, \xi_2, \ldots, \xi_n)'$ be a random sample with size n from the population ξ. A realization of the random variable

$$\bar{\xi} = (1/n)\sum_{i=1}^{n} \xi_i \tag{1.6.6}$$

is called an empirical mean value of the random variable ξ; a realization of

---

[9] This form of the integral is called the Lebesgue — Riemann integral (cf., e.g., [15], p. 22).

the random variable

$$\hat{\sigma}^2 = (1/n)\sum_{i=1}^{n}(\xi_i - \bar{\xi})^2, \tag{1.6.7}$$

where $\bar{\xi}$ is given by (1.6.6) is called an empirical dispersion of the random variable ξ; a realization of the random variable

$$\hat{\gamma}_1 = \left[(1/n)\sum_{i=1}^{n}(\xi_i - \xi)^3\right] \Big/ \left[(1/n)\sum_{i=1}^{n}(\xi_i - \xi)^2\right]^{3/2} \tag{1.6.8}$$

is called an empirical coefficient of asymmetry of the random variable ξ, and a realization of the random variable

$$\hat{\gamma}_2 \left\{ \left[(1/n)\sum_{i=1}^{n}(\xi_i - \xi)^4\right] \Big/ \left[(1/n)\sum_{i=1}^{n}(\xi_i - \xi)^2\right]^2 \right\} - 3 \tag{1.6.9}$$

is called its empirical excess.

Definition Let $((\xi_1, \eta_1)', (\xi_2, \eta_2)', \ldots, (\xi_n, \eta_n)')'$ be a random sample with the size n from the population $(\xi, \eta)'$. A realization of the random variable

$$\hat{\rho}_{\xi,\eta} = \frac{\hat{\sigma}_{\xi,\eta}}{\hat{\sigma}_\xi \hat{\sigma}_\eta} = \frac{(1/n)\sum_{i=1}^{n}(\xi_i - \bar{\xi})(\eta_i - \bar{\eta})}{\left[(1/n)\sum_{i=1}^{n}(\xi_i - \bar{\xi})^2 (1/n)\sum_{i=1}^{n}(\eta_i - \bar{\eta})^2\right]^{1/2}} \tag{1.6.10}$$

is called an empirical correlation coefficient between the random variables ξ and η ($\bar{\xi}$ and $\bar{\eta}$ are defined by Eq. (1.6.6)).

Definition Let $((\xi_{11}, \xi_{21}, \ldots, \xi_{p1})', (\xi_{12}, \xi_{22}, \ldots, \xi_{p2})', \ldots, (\xi_{1n}, \xi_{2n}, \ldots, \xi_{pn})')'$ be a random sample with size n from a p-dimensional population $\xi = (\xi_1, \xi_2, \ldots, \xi_p)'$, $n > p$ [10]. A realization of a $p \times p$-dimensional random

[10] **Let $\xi_1, \xi_2, \ldots, \xi_n$ be a random sample from a continuous p-dimensional probability distribution whose covariance matrix Σ is regular ($R(\Sigma) = p$). If $n \geq p + 1$, the rank of the empirical covariance matrix $\hat{\Sigma}$ equals p with probability 1.**

matrix

$$\hat{\Sigma} = (1/n)\sum_{i=1}^{n}(\xi_i - \bar{\xi})(\xi_i - \bar{\xi})', \ \bar{\xi} = (1/n)\sum_{i=1}^{n}\xi_i, \tag{1.6.11}$$

the (r, s)th element of which is of the form

$$\{\hat{\Sigma}\}_{r,s} = (1/n)\sum_{i=1}^{n}(\xi_{ri} - \bar{\xi}_r)(\xi_{si} - \bar{\xi}_s),$$

is called an empirical covariance matrix of the random variable $(\xi_1, \xi_2, \ldots, \xi_p)'$.

Definition Let $((\eta_1, \xi_{11}, \xi_{21}, \ldots, \xi_{p1})', (\eta_2, \xi_{12}, \xi_{22}, \ldots, \xi_{p2})', \ldots, (\eta_n, \xi_{1n}, \xi_{2n}, \ldots, \xi_{pn})')'$, be a random sample with the size n from a $p+1$-dimensional population $(\eta, \xi')' = (\eta, \xi_1, \xi_2, \ldots, \xi_p)$, $n > p+1$. The empirical covariance matrix $\hat{\Sigma}_\xi$ of the random vector ξ is supposed to be regular. A realization of the random variable

$$\hat{\rho}_{\eta,\xi} = \left(\frac{\hat{\Sigma}_{\eta,\xi}\hat{\Sigma}_\xi^{-1}\hat{\Sigma}_{\xi,\eta}}{\hat{\sigma}_\eta^2}\right)^{1/2}, \tag{1.6.12}$$

where $\hat{\sigma}_\eta$ is given by Eq. (1.6.7), $\hat{\Sigma}_\xi$ by Eq. (1.6.11), and the p-dimensional random vector $\hat{\Sigma}_{\xi,\eta}$ is

$$\hat{\Sigma}_{\xi,\eta} = (1/n)\sum_{i=1}^{n}(\xi_i - \bar{\xi})(\eta_i - \bar{\eta}), \tag{1.6.13}$$

its realization is called an empirical coefficient of multiple correlation between the random variable η and the random vector ξ.

Definition Let $((\eta_1, \zeta_1, \xi_{11}, \xi_{21}, \ldots, \xi_{p1})', (\eta_2, \zeta_2, \xi_{12}, \xi_{22}, \ldots, \xi_{p2})', \ldots, (\eta_n, \zeta_n, \xi_{1n}, \xi_{2n}, \ldots, \xi_{pn})')'$, be a random sample with the size n from a $p+2$-dimensional population $(\eta, \zeta, \xi')' = (\eta, \zeta, \xi_1, \xi_2, \ldots, \xi_p)'$, then a realization of the random variable

$$\hat{\rho}_{\eta\cdot\zeta,\xi} = \frac{\hat{\sigma}_{\eta,\zeta} - \hat{\Sigma}_{\eta,\xi}\hat{\Sigma}_\xi^{-1}\hat{\Sigma}_{\xi,\eta}}{[(\hat{\sigma}_\eta^2 - \hat{\Sigma}_{\eta,\xi}\hat{\Sigma}_\xi^{-1}\hat{\Sigma}_{\xi,\eta})(\hat{\sigma}_\zeta^2 - \hat{\Sigma}_{\zeta,\xi}\hat{\Sigma}_\xi^{-1}\hat{\Sigma}_{\xi,\zeta})]^{1/2}},$$

where $\hat{\sigma}_{\eta,\zeta}$ is given by the relation (1.6.10), $\hat{\Sigma}_{\xi,\eta}$ and $\hat{\Sigma}_{\xi,\zeta}$ by the relation (1.6.13) and $\hat{\Sigma}_{\xi}$ by the relation (1.6.11), is called an empirical coefficient of partial correlation between the random variables η and ζ, when the influence of the random vector ξ is eliminated.

1.6.3 Probability distributions of some sample random variables

The probability distribution of a sample random variable depends on the probability distribution of the basic population, the definition of the sample random variable, and on the sample size n. Depending on the magnitude of the number n, the problem of determining the probability distribution of a sample random variable is solved in two ways:

1. by applying an exact procedure, if the random sample is small; in this case, n is a parameter of the probability distribution being determined;
2. if the size of the random sample is sufficiently large, one looks for an approximate probability distribution or its asymptotic (limit) probability distribution as $n \to \infty$.

It has to be realized that there does not exist any criterion which will enable us to judge unambiguously whether the size of a concrete sample is still small or has become already sufficiently large; the decision to apply an exact or a limit approach to derive the probability distribution of a sample random variable depends, except for the size of the random sample at hand, on the form of the function $f(\cdot)$, which transforms the random vector $(\xi_1, \xi_2, \ldots, \xi_n)'$ into the sample random variable.

Knowledge of the probability distribution of sample random variables is necessary when solving many concrete problems mainly concerning various analyses of the sets of measured data; however, their detailed derivation is not our main problem in all the following cases. For this reason, only the procedures for deriving the most frequently encountered probability distributions are given.

1. Exact probability distributions

Let $(\xi_1, \xi_2, \ldots, \xi_m)'$ be a random sample from a normal probability distribution $N(m, \sigma^2)$. The random variable $\overline{\xi} = (1/n)\sum_{i=1}^{n} \xi_i$ given by the relation (1.6.6) possesses the normal probability distribution $N(m, \sigma^2/n)$. (For the proof of this assertion see p. 77.)

Let $(\xi_1, \xi_2, \ldots, \xi_n)'$ be a random sample from the probability distribution $N(0, \sigma^2)$. The random variable $\hat{\sigma}^2 = (1/n)\sum_{i=1}^{n} \xi_i^2$ (see Eq. (1.6.7)) has a (σ^2/n) chi-square probability distribution with n degrees of freedom, and its probability density is of the form

$$f_n(s) = \begin{cases} \frac{1}{(2/n)^{n/2}\sigma^n\Gamma(n/2)} s^{(n/2)-1}\exp(-ns/2\sigma^2) & , \quad s > 0, \\ 0 & , \quad s \leq 0; \end{cases}$$

this assertion follows immediately from Eq. (1.5.6), which expresses the probability density of a σ^2chi-square probability distribution with n degrees of freedom; it suffices to rewrite it for the dispersion that is equal to σ^2/n.

Let $(\xi_1, \xi_2, \ldots, \xi_n)'$ be a random sample from a normal population $N(m, \sigma^2)$. The sample random variable $\hat{\sigma}^2 = (1/n)\sum_{i=1}^{n}(\xi_i - \overline{\xi})^2$, defined by Eq. (1.6.7) has a (σ^2/n)chi-square probability distribution with $n-1$ degrees of freedom. (The proof is given, e.g., in [1], p. 82.)

Let $((\xi_1, \eta_1)', (\xi_2, \eta_2)', \ldots, (\xi_n, \eta_n)')'$, $n > 3$, be a random sample from a two-dimensional probability distribution

$$N_2\left[(m_\xi, m_\eta)',\ \Sigma = \begin{pmatrix} \sigma_\xi^2, & \rho_{\xi,\eta}\sigma_\xi\sigma_\eta \\ \rho_{\xi,\eta}\sigma_\xi\sigma_\eta, & \sigma_\eta^2 \end{pmatrix}\right],$$

$$\sigma_\xi^2 > 0,\ \sigma_\eta^2 > 0,\ \rho_{\xi,\eta} \in [-1, 1].$$

The empirical correlation coefficient $\hat{\rho}_{\xi,\eta}$, defined by Eq. (1.6.10) has the probability density

$$f(x; \rho_{\xi,\eta}, n) = \begin{cases} \frac{2^{n-3}}{\pi\Gamma(n-2)}(1-\rho_{\xi,\eta})^{(n-1)/2}(1-x^2)^{(n-4)/2}\cdot \\ \quad \cdot\sum_{k=0}^{\infty}\left[\Gamma\left(\frac{n+k-1}{2}\right)\right]^2 \frac{(2\rho_{\xi,\eta}x)^k}{k!} & , \quad x \in (-1, 1), \\ 0 & , \quad x \notin (-1, 1). \end{cases} \tag{1.6.15}$$

If $\rho_{\xi,\eta} = 0$, then

$$f(x; , n) = \begin{cases} \frac{2^{n-3}\left[\Gamma\left(\frac{n-1}{2}\right)\right]^2}{\pi\Gamma(n-2)}(1-x^2)^{(n-4)/2} & , \quad x \in (-1, 1), \\ 0 & , \quad x \notin (-1, 1). \end{cases}$$

(For the proof, see, e.g., [23], p. 284.)

2. Approximate and asymptotic probability distributions

Let $(\xi_1, \xi_2, \ldots, \xi_n)'$ be a sample from a probability distribution such that $E[\xi_i] = m$ and $D[\xi_i] = \sigma^2 < \infty$, $i = 1, 2, \ldots, n$. The random variable

$$\nu_n = (1/n) \sum_{i=1}^{n} \xi_i$$

possesses an approximately normal probability distribution (the index n belonging to the sample random variable ν_n denotes the size of the random sample on the basis of which it is defined).

When we prove this assertion, we will explain in detail the difference between the notion of an approximate and asymptotic probability distribution.

Evidently, $E[\nu_n] = (1/n) \sum_{i=1}^{n} E[\xi_1] = m$ and $D[\nu_n] = E[(\nu_n - m)^2] = E\left[\left((1/n)\sum_{i=1}^{n}(\xi_i - m)\right)^2\right] = (1/n^2) \sum_{i=1}^{n} E[(\xi_i - m)^2] = \sigma^2/n$ (elements of the random sample are stochastically independent).

The problem is to investigate the probability distribution of the random variable

$$\zeta_n = \frac{\sqrt{n}(\nu_n - m)}{\sigma} = \frac{1}{\sqrt{n}\sigma} \sum_{i=1}^{n} (\xi_i - m) \tag{1.6.16}$$

(the random variable ζ_n is centred and normed — $E[\zeta_n] = 0$, $D[\zeta_n] = E[\zeta_n^2] = 1$).

Since the components of the random vector $(\xi_1, \xi_2, \ldots, \xi_n)'$ have the same probability distribution, their characteristic functions are identical. Let $\varphi_{\xi_c}(\cdot)$ be a characteristic function of the centred random variables $\xi_{ic} = \xi_i - m$, $i = 1, 2, \ldots, n$. The characteristic function of each of the summands in Eq. (1.6.16) is, in accordance with Eq. (1.4.2), of the form $\varphi_{\xi_c/(\sqrt{n}\sigma)}(t) = \varphi_{\xi_c}(t/(\sqrt{n}\sigma))$. As a consequence of the fact that the components of the random vector $(\xi_1, \xi_2, \ldots, \xi_n)'$ are stochastically independent, the characteristic function of the random variable defined by (1.6.16) is

$$\varphi_{\zeta_n}(t) = [\varphi_{\xi_c}(t/(\sqrt{n}\sigma))]^n, \ t \in \mathcal{R}^1.$$

An application of Eq. (1.4.6) yields

$$\varphi_{\xi_c}(t/(\sqrt{n}\sigma)) = 1 - \frac{t^2\sigma^2/(n\sigma^2)}{2!} + \mathcal{O}\left[\left(\frac{t}{\sqrt{n}\sigma}\right)^2\right], \ t \in \mathcal{R}^1,$$

where $\mathcal{O}[(t/\sqrt{n}\sigma))^2]$ is a remainder of the Taylor series. For a sufficiently large n

$$\left\{1-\frac{t^2}{2n}+\mathcal{O}\left[\left(\frac{t}{n}\right)^2\right]\right\}^n \doteq \left(1-\frac{t^2}{2n}\right)^n \doteq \exp(-t^2/2) = \lim_{n\to\infty}\left(1-\frac{t^2}{2n}\right)^n$$

and simultaneously $\lim_{n\to\infty} n\mathcal{O}[(t/(\sqrt{n}\sigma))^2] = 0$, thus

$$\varphi_{\xi_n}(t) \doteq \exp(-t^2/2).$$

After taking into account the remark on p. 77, concerning the one-to-one relation between the characteristic and distribution function of a random variable ξ, we can state the next conclusion:

If the size of the random sample $(\xi_1, \xi_2, \dots, \xi_n)'$ is sufficiently large, the random variable defined by Eq. (1.6.16) possesses an approximately normal probability distribution $N(0,1)$ (this probability distribution is approximate since the remainder of the Taylor series was omitted). As a consequence of this fact the random variable $\nu_n = \frac{\sigma}{\sqrt{n}}\zeta_n + m$ has an approximately normal probability distribution $N(m, \sigma^2/n)$.

Since

$$\lim_{n\to\infty}\varphi_{\zeta_n}(t) = \lim_{n\to\infty}\left\{1-\frac{t^2}{2n}+\mathcal{O}\left[\left(\frac{t}{n}\right)^2\right]\right\}^n =$$

$$= \lim_{n\to\infty}\left(1-\frac{t^2}{2n}\right)^n = \exp\left(-\frac{t^2}{2}\right),$$

which is the characteristic function of the normal probability ditribution $N(0,1)$, the random variable ζ_n, defined by Eq. (1.6.16), has approximately the normal probability distribution $N(0,1)$ and, as a consequence, the random variable ν_n possesses approximately the normal probability distribution $N(m, \sigma^2/n)$.

This is a proof of a very important central limit theorem (Lindeberg—Lévy): Let $\{\xi_n\}_{n=1}^{\infty}$ be a sequence of independent random variables having the same probability distribution, i.e., with the same mean value and the same finite dispersion. Then

$$\lim_{n\to\infty} F_n(z) = \frac{1}{\sqrt{2\pi}}\int_{-\infty}^{z}\exp(-t^2/2)dt,$$

where $\{F_n(\cdot)\}_{n=1}^{\infty}$ is a sequence of distribution functions corresponding to random variables of the form (1.6.16) (for more detail see [20], p. 97).

Let $(\xi_1, \xi_2, \ldots, \xi_n)'$ be a random sample from the normal probability distribution $N(m, \sigma^2)$. The empirical coefficient of asymmetry, defined Eq. (1.6.8), has approximately the normal probability distribution $N(0, 6/n)$ and the empirical excess, defined by Eq. (1.6.9), approximately the normal probability distribution $N(0, 24/n)$ ([15], p. 106).

The aim of this part was, in particular, to show methods of deriving probability distributions of sample random variables, whose knowledge is important for testing statistical hypotheses. Deriving the form of probability distributions of all sample random variables that appear in the third chapter of this book (which deals with testing statistical hypotheses) is a problem that extends beyond its framework; for this reason, only the fundamental assertions are given and further sources are stated for readers who want detailed derivations.

Chapter 2

ESTIMATION THEORY

2.1 The aim of data processing; a theoretical (deterministic), stochastic and statistical model of the object being investigated

The aim of data processing, from the point of view of estimation theory, is to determine both the optimum values and the fundamental statistical characteristics of those parameters which characterize the measured object.

Data processing is carried out in three stages:

Stage 1: Building a theoretical (deterministic) model stating relations between theoretical (errorless) values of unknown (determined) parameters and theoretical (errorless) values of directly observable values of the studied object, including conditions (if any exist) which the determined parameters have to satisfy.

Let $X_1, X_2, \ldots, X_p$ be theoretical values of directly observable values of the investigated object, $\Theta_1, \Theta_2, \ldots, \Theta_k$ theoretical values of the unknown parameters. The theoretical model consists of p relations

$$X_i = f_i(\Theta_1, \Theta_2, \ldots, \Theta_k) = f_i(\Theta),\ i = 1, 2, \ldots, p, \tag{2.1.1}$$

and, if any, q conditions

$$g_m(\Theta_1, \Theta_2, \ldots, \Theta_k) = 0,\ m = 1, 2, \ldots, q. \tag{2.1.2}$$

The relations (2.1.1) and (2.1.2) are written more frequently in their vector form

$$X = f(\Theta) \tag{2.1.1'}$$

and

$$g(\Theta) = 0, \tag{2.1.2'}$$

respectively, where $X = (X_1, X_2, \ldots, X_p)'$, $\Theta = (\Theta_1, \Theta_2, \ldots, \Theta_k)'$, $f(\Theta) = (f_1(\Theta), f_2(\Theta), \ldots, f_p(\Theta))'$, $g(\Theta) = (g_1(\Theta), g_2(\Theta), \ldots, g_q(\Theta))'$ are non-stochastic column vectors ($'$ denotes a transposition of a vector or a matrix) of the corresponding dimension, 0 is a zero vector, in our case q-dimensional.

Stage 2: Building a stochastic model corresponding to the theoretical model from the stage 1. A stochastic model respects the fact that the theoretical values of a directly observable value are not known immediately. The only possibility of obtaining information about magnitude consists in measuring the observable values. However, measuring errors are an organic component of each measurement (we know by our own experience that results of measurements repeated under the same conditions differ). A measurement of the same scalar (one-dimensional) quantity repeated n times under the same conditions results in an n-tuple of numerical values, each of which is a realization of the random variable which models the measurement of a directly observable quantity.

Let ξ_i be an n_i-dimensional random vector, modelling an n_i fold measurement of a directly observable quantity X_i of the theoretical model, $i = 1, 2, \ldots, p$; the number of measurements of the single components may differ. The experiment, which consists of measuring all p directly observable components of the vector X of the theoretical model of the measured object, is stochastically modelled by a random vector $\xi = (\xi_1', \xi_2', \ldots, \xi_p')'$, where each its n_i-dimensional block ξ_i models the measurement of the ith component of the theoretical model; self-evidently $n = \sum_{i=1}^{p} n_i$[1]. The random vector, which models stochastically the whole experiment, is fully characterized by its distribution function. We shall see below that full knowledge of this distribution function is not necessary for solving a great number of problems in estimation theory, mainly within the framework of linear estimation theory. Apparently, a priori knowledge of the distribution function enables us to use information contained in the measured data more effectively.

[1] Here we state only that the ith component X_i of the vector X is measured n_i times. For determining the optimum number of repetitions of a measurement of the concrete component, one should refer to a special statistical discipline, i.e., optimum experimental design, e.g., [25], [26] or [27].

The basic characteristics necessary for data processing are the mean value $E[\xi]$ and the covariance matrix $var(\xi) = \Sigma$ of the random vector ξ modelling the measurement.

Let ε denotes an n-dimensional vector modelling the random error of the measurement of a p-dimensional quantity. Its structure is the same as the structure of the random vector ξ, i.e., ε consists of p n_i-dimensional subvectors ε_i. According to the Gauss's second postulate on random errors, the probability distribution of random errors is symmetric around zero [7]. This implies that the mean value of an n-dimensional vector of random errors is zero, $E[\varepsilon] = 0$.

Now we are able to write the stochastic model which corresponds to the deterministic model (2.1.1′). The ith subvector ξ_i of the vector ξ, which models the n_i times repeated measurement of the ith component X_i of the vector X, is of the form

$$\xi_i = \begin{pmatrix} \xi_{i1} \\ \xi_{i2} \\ \vdots \\ \xi_{in_i} \end{pmatrix} = \begin{pmatrix} f_i(\Theta) \\ f_i(\Theta) \\ \vdots \\ f_i(\Theta) \end{pmatrix} + \begin{pmatrix} \varepsilon_{i1} \\ \varepsilon_{i2} \\ \vdots \\ \varepsilon_{in_i} \end{pmatrix} = \begin{pmatrix} 1 \\ 1 \\ \vdots \\ 1 \end{pmatrix} f_i(\Theta) + \begin{pmatrix} \varepsilon_{i1} \\ \varepsilon_{i2} \\ \vdots \\ \varepsilon_{in_i} \end{pmatrix} =$$

$$= i_{(n_i)} f_i(\Theta) + \varepsilon_i; \tag{2.1.3}$$

here $i_{(n_i)}$ is an n_i-dimensional vector, all the elements of which are units. The vector $\xi = (\xi_1', \xi_2', \ldots, \xi_p')'$ modelling the measurement of all components of the vector X has the following block structure

$$\xi = \begin{pmatrix} \xi_1 \\ \xi_2 \\ \vdots \\ \xi_p \end{pmatrix} + \begin{pmatrix} i_{(n_1)} f_1(\Theta) \\ i_{(n_2)} f_2(\Theta) \\ \vdots \\ i_{(n_p)} f_p(\Theta) \end{pmatrix} + \begin{pmatrix} \varepsilon_1 \\ \varepsilon_2 \\ \vdots \\ \varepsilon_p \end{pmatrix} =$$

$$= \begin{pmatrix} i_{(n_1)}, & 0_{(n_1)}, & \dots & 0_{(n_1)} \\ 0_{(n_2)}, & i_{(n_2)}, & \dots, & 0_{(n_2)} \\ \dots & \dots & \dots & \dots \\ 0_{(n_p)}, & 0_{(n_p)}, & \dots, & i_{(n_p)} \end{pmatrix} \begin{pmatrix} f_1(\Theta) \\ f_2(\Theta) \\ \vdots \\ f_p(\Theta) \end{pmatrix} + \begin{pmatrix} \varepsilon_1 \\ \varepsilon_2 \\ \vdots \\ \varepsilon_p \end{pmatrix} = Jf(\Theta) + \varepsilon, \tag{2.1.4}$$

where J denotes an $n \times p$-dimensional matrix of the above given form, $0_{(n_i)}$ are n_i-dimensional zero column vectors, $i = 1, 2, \dots, p$, and $f(\Theta)$ is a p-dimensional vector from the theoretical model (2.1.1′). Thus the stochastical model

$$\xi = Jf(\Theta) + \varepsilon, \; E[\xi] = Jf(\Theta), \; var(\xi) = var(\varepsilon) = \Sigma, \tag{2.1.5}$$

which reflects the method of measuring individual components X_i of the theoretical model, is assigned to the theoretical model (2.1.1′). The stochastic model is indispensable for determining the statistical properties of the optimum estimators of the unknown parameters obtained from a statistical model.

Stage 3: Consists in working out a statistical model. Applying methods of estimation theory, from the concrete realizations x of the random variable ξ modelling the experiment, we obtain the optimum (in the sense of exactly defined criteria) estimates $\hat{\Theta}(x)$ of the unknown parameters and determine their basic statistical characteristics.

The result of each experiment is an ordered n-tuple of real numbers $x = (x_1', x_2', \dots, x_p')'$. If the experiment is repeated N times, we get N different n-tuples $x_1 = (x_{11}', x_{12}', \dots, x_{1p}')'$, $x_2 = (x_{21}', x_{22}' \dots, x_{2p}')' \dots$, $x_N = (x_{N1}', x_{N2}', \dots, x_{Np}')'$, each of them being a realization of the n-dimensional random vector $\xi = (\xi_1', \xi_2' \dots, \xi_n')'$ from the stochastic model. Each realization x_j, $j = 1, 2, \dots, N$, of the random vector ξ is a point of the n-dimensional space $\mathcal{R}^n$; in estimation theory this space is called a sample space. Individual realizations of the experiment are assigned different points of the sample space; their dispersion depends on the distribution function $F_\xi(x, \Theta)$ of the random vector ξ. It implies that the theoretical value of the k-dimensional vector parameter $\Theta = (\Theta_1, \Theta_2, \dots, \Theta_k)'$ cannot be determined, since the only way that we can know it is through a realization (one or more) of the random variable ξ. It is evident that the variation of realizations x_j of the random vector ξ causes an analogous variation of values of the unknown parameters calculated from them. Thus, instead of the theoretical value of the vector Θ, we are able to determine a certain k-dimensional estimate $\hat{\Theta}(x_j)$ which is a realization of a k-dimensional random variable $\hat{\Theta}(\xi)$, called an estimator.

The probability distribution $F_\xi(x, \Theta)$ of an n-dimensional random vector ξ in the sample space $\mathcal{R}^n$ induces a probability distribution $F_{\hat{\Theta}}(t, \Theta)$ of the k-dimensional estimator in a k-dimensional parametric space $\mathcal{T} \subset \mathcal{R}^k$. The induced distribution function $F_{\hat{\Theta}}(t, \Theta) = P\{\omega : \hat{\Theta}_1(\xi(\omega)) < t_1, \hat{\Theta}_2(\xi(\omega)) < t_2, \ldots, \hat{\Theta}_k(\xi(\omega)) < t_k\}$ describes fully the statistical properties of the estimator $\hat{\Theta}(\xi)$ used.

The aim of estimation theory is to find a vector mapping $\hat{\Theta}(\cdot)$ of the n-dimensional sample space $\mathcal{R}^n$ to the k-dimensional parametric space $\mathcal{T}$ such that the resulting estimator will be

1. unbiased,
2. jointly efficient and
3. consistent.

An estimator $\hat{\Theta}(\cdot) : \mathcal{R}^n \to \mathcal{T}$ is said to be unbiased if the mean value of the estimator $\hat{\Theta}(\xi)$ equals the theoretical value of the estimated parameter Θ, i.e., if $E[\hat{\Theta}(\xi)] = \Theta, \ \Theta \in \mathcal{T}$.

An unbiased estimator $\hat{\Theta}(\cdot) : \mathcal{R}^n \to \mathcal{T}$ is said to be jointly efficient, if for each n-tuple of real numbers $a_1, a_2, \ldots, a_n$ and each other unbiased estimator $\tilde{\Theta}(\xi) \neq \hat{\Theta}(\xi)$, the dispersion of the expression $\sum_{i=1}^k a_i \hat{\Theta}_i(\xi)$ is not larger than the dispersion of the expression $\sum_{i=1}^k a_i \tilde{\Theta}(\xi)$, i.e., if

$$D\left[\sum_{i=1}^k a_i \hat{\Theta}_i(\xi)\right] \leq D\left[\sum_{i=1}^k a_i \tilde{\Theta}_i(\xi)\right]. \tag{2.1.6}$$

The relation (2.1.6) implies that the dispersion of any component of a jointly efficient estimator $\hat{\Theta}(\xi)$ cannot be larger than the dispersion of the same component of the unbiased estimator $\tilde{\Theta}(\xi)$. In order to prove this assertion, it suffices to choose $a_i = 1, \ a_j = 0, \ i \neq j, \ j = 1, 2, \ldots, k$.

The consistency of an estimator will be defined for a one-dimensional parametric space (single scalar parameter is being determined).

Let $\{\hat{\Theta}_r(\xi)\}_{r=1}^{\infty}$ be a sequence of estimators of the parameter Θ, such that the dimension of the sample space of the rth element of this sequence equals r. This sequence is said to be consistent if for an arbitrary $\varepsilon > 0$

$$\lim_{n\to\infty} P\{\omega : |\hat{\Theta}_n(\xi(\omega)) - \Theta| > \varepsilon\} = 0.$$

Thus, a sequence of estimators of the parameter Θ is consistent, if with an increasing number of performed measurements, the corresponding sequence of estimators with probability tending to 1 converges to its theoretical value.

Example The problem is to determine coefficients of a straight-line $X = a + bt$ on the basis of the measured values of the x-coordinate at the points t_1, t_2 and t_3 of the t axis.

The theoretical model of this situation is

$$X_1 = a + bt_1,\ X_2 = a + bt_2,\ X_3 = a + bt_3,$$

thus,

$$X = \begin{pmatrix} 1, & t_1 \\ 1, & t_2 \\ 1, & t_3 \end{pmatrix} \begin{pmatrix} a \\ b \end{pmatrix}.$$

The measurement is performed in such a way that at the side points of the t axis the measurement is carried out 3 times and at the center once. Under the formulated conditions the stochastical model is of the form

$$\xi = \begin{pmatrix} 1, & 0, & 0 \\ 1, & 0, & 0 \\ 1, & 0, & 0 \\ 0, & 1, & 0 \\ 0, & 0, & 1 \\ 0, & 0, & 1 \\ 0, & 0, & 1 \end{pmatrix} \begin{pmatrix} 1, & t_1 \\ 1, & t_2 \\ 1, & t_3 \end{pmatrix} \begin{pmatrix} a \\ b \end{pmatrix} + \varepsilon,\ E[\xi] = \begin{pmatrix} a + bt_1 \\ a + bt_1 \\ a + bt_1 \\ a + bt_2 \\ a + bt_3 \\ a + bt_3 \\ a + bt_3 \end{pmatrix},\ var(\xi) = \Sigma.$$

Now, it remains to determine the optimum (i.e., unbiased jointly efficient and consistent) estimators $\hat{a}(\xi)$ and $\hat{b}(\xi)$ of the unknown parameters a and b and to determine the covariance matrix $var[(\hat{a}(\xi), \hat{b}(\xi))']$ dependent on the covariance matrix $var(\xi)$ of the observation vector ξ.

Example The problem is to determine an unknown length of a rod interval; its theoretical length Θ will be estimated from an n times repeated direct measurement. The theoretical model of the formulated problem is $X = \Theta$ $(p = 1)$. The corresponding stochastic model is of the form $\xi = \Theta i + \varepsilon$ (i is an n-dimensional vector, all elements of which are units), the n-dimensional random vector ξ is characterized by its mean value $E[\xi] = \Theta i$ and the covariance matrix $var(\xi) = \Sigma$. The result of an n times repeated measurement of the rod interval is an n-tuple of real numbers generating an n-dimensional vector $x = (x_1, x_2, \ldots, x_n)'$. It is a realization of the random vector ξ which models the measurement. Let the estimator $\hat{\Theta}(\cdot) : \mathcal{R}^n \to \mathcal{T} \subset \mathcal{R}^1$ have the form of an arithmetic mean, $\hat{\Theta}(\xi) = (1/n)i'\Theta\xi$. The mean value of this estimator is $E[\hat{\Theta}(\xi)] = (1/n)i'E[\xi] = (1/n)i'\Theta i = \Theta$, which means that the chosen linear estimator of the parameter Θ is unbiased; the covariance matrix of the estimator $\hat{\Theta}(\xi)$ is $var[\hat{\Theta}(\xi)] = var[(1/n)i'\xi] = E[(1/n)^2(i\xi - \Theta)(i'\xi - \Theta)'] = (1/n)^2 E[i'(\xi - \Theta i)(\xi - \Theta i)'i] = (1/n)^2 i' var(\xi) i = (1/n)^2 i'\Sigma i$. If $\Sigma = \sigma^2 I$ (the measurements are not correlated), $var[\hat{\Theta}(\xi)] = (\sigma^2/n^2)i'Ii = \sigma^2/n$.

Further, we shall prove that under the conditions $E[\xi] = i\Theta$ and $var(\xi) = \sigma^2 I$, the unbiased estimator $\hat{\Theta}(\xi) = (1/n)i'\xi$ is an efficient estimator of the parameter Θ . According to the definition of efficiency it has to be proved that the dispersion of any other unbiased estimator is not smaller than the dispersion of an arithmetic mean.

Let $\tilde{\Theta}(\xi) = (a'i)^{-1}a'\xi$, where $a' = (a_1, a_2, \ldots, a_n)'$ is an n-dimensional vector of arbitrary real numbers such that at least one of them is non-zero. This estimator is unbiased, since $E[\tilde{\Theta}(\xi)] = (a'i)^{-1}a'E[\xi] = (a'i)^{-1}a'i\Theta = \Theta$. Its dispersion is $var[\tilde{\Theta}(\xi)] = \sigma^2(a'i)^{-2}a'Ia = \sigma^2(a'i)^{-2}a'a$. The value of the expression $(a'i)^{-2}a'a$ acquires its minimum value which equals $(1/n)$, if all the elements of the vector a are the same, i.e., for the estimator in the form of an arithmetic mean. If the covariance matrix of the vector ξ is not of the form $var(\xi) = \sigma^2 I$, the estimator of the parameter Θ in the form of an arithmetic mean being unbiased is not efficient.

Finally, we shall show that $\hat{\Theta}(\xi) = (1/n)i'\xi$, if $var(\xi) = \sigma^2 I$, is consistent. Let a sequence of the estimators

$$\hat{\Theta}_2(\xi) = (1/2)\sum_{i=1}^{2}\xi_i,\ \hat{\Theta}_3(\xi) = (1/3)\sum_{i=1}^{3}\xi_i, \ldots, \hat{\Theta}_n(\xi) = (1/n)\sum_{i=1}^{n}\xi_i, \ldots$$

be taken into account.

Due to the Chebyshev inequality[2],

$$P\{\omega : |\hat{\Theta}_n(\xi(\omega)) - E[\hat{\Theta}_n(\xi)]| \geq k\} \leq var[\hat{\Theta}_n(\xi)]/k^2, \ k > 0;$$

this, since $E[\hat{\Theta}_n(\xi)] = \Theta$ and $var[\hat{\Theta}(\xi)] = \sigma^2/n$, yields

$$P\{\omega : |\hat{\Theta}_n(\xi(\omega)) - \Theta| \geq k\} < \sigma^2/(nk^2),$$

therefore for any $k > 0$

$$\lim_{n\to\infty} P\{\omega : |\hat{\Theta}_n(\xi(\omega)) - \Theta| \geq k\} = 0,$$

which means that the sequence of the estimators which we considered is consistent.

If $\xi \sim N_n(\Theta i, \sigma^2 I)$, then $\hat{\Theta}(\xi) = (1/n)i'\xi \sim N(\Theta, \sigma^2/n)$ (see p. 103).

2.2 Basic linear measurement models, linear estimators

The previous section was devoted to methods and principles of determining optimum estimators in general. No type requirements were made for the functions $f(\cdot)$, $g(\cdot)$ and $\hat{\Theta}(\cdot)$. In this section all these functions will be assumed to be linear, which means that we will study linear estimators within the framework of the linear measurement model (where the relations between the directly observable quantities and the unknown parameters and the relations between the unknown parameters are linear).

This restriction to linear models of measurement does not mean a substantial reduction in the range of problems that are solvable within this framework. Rather, it follows from the fact that linear models of measurement involve all cases, where the functions $f(\cdot)$ and $g(\cdot)$ are not linear, and may however, still be linearized.

The usual method of linearizing nonlinear models consists in developing them into the Taylor series and neglecting the terms of the second and higher orders. In the case of an n-dimensional vector function $f(\cdot)$, $\Theta \in \mathcal{R}^k$,

$$f_{lin}(\Theta) = f(\Theta_0) + [\partial f(\Theta_0)/\partial\Theta']\Delta\Theta,$$

[2]**Chebyshev inequality** Let ξ be a random variable whose mean value equals m and dispersion is $\sigma^2 < \infty$. Then $P\{\omega : |\xi(\omega) - m| \geq \varepsilon\} \leq \sigma^2/\varepsilon^2$ for any $\varepsilon > 0$.

The assertion will be proved for a continuous random variable characterized by the probability density $f(\cdot)$, $x \in \mathcal{R}^1$. In this case $P\{\omega : |\xi(\omega) - m| \geq \varepsilon\} = \int_{|x-m|\geq\varepsilon} f(x)dx$ and since $|x-m|/\varepsilon \geq 1$ we get $\int_{|x-m|\geq\varepsilon} f(x)dx \leq (1/\varepsilon^2)\int_{|x-m|\geq\varepsilon} |x-m|^2 f(x)dx \leq (1/\varepsilon^2)\int_{\mathcal{R}^1} |x-m|^2 f(x)dx = \sigma^2/\varepsilon^2$, by which the Chebyshev inequality is proved.

where $\Theta_0 = (\Theta_{01}, \Theta_{02}, \ldots, \Theta_{0k})'$ is the vector of approximate values of the parameters being determined, $f(\Theta_0)$ is a p-dimensional vector of known function values of the vector function $f(\cdot)$ at the point $\Theta_0 \in \mathcal{R}^k$, $\partial f(\Theta_0)/\partial \Theta'$ is a $p \times k$-dimensional matrix of the first partial derivatives of the p-dimensional vector function $f(\cdot)$ at the point $\Theta_0 \in \mathcal{R}^k$ with respect to the row vector Θ', i.e.,

$$\partial f(\Theta_0)/\partial \Theta' = \begin{pmatrix} \partial f_1(\Theta_0)/\partial \Theta_1, & \partial f_1(\Theta_0)/\partial \Theta_2, & \ldots, & \partial f_1(\Theta_0)/\partial \Theta_k \\ \partial f_2(\Theta_0)/\partial \Theta_1, & \partial f_2(\Theta_0)/\partial \Theta_2, & \ldots, & \partial f_2(\Theta_0)/\partial \Theta_k \\ \cdots & \cdots & \cdots & \cdots \\ \partial f_p(\Theta_0)/\partial \Theta_1, & \partial f_p(\Theta_0)/\partial \Theta_2, & \ldots, & \partial f_p(\Theta_0)/\partial \Theta_k \end{pmatrix} =$$

$$= \begin{pmatrix} \partial f_1(\Theta_0)/\partial \Theta' \\ \partial f_2(\Theta_0)/\partial \Theta' \\ \cdots \\ \partial f_p(\Theta_0)/\partial \Theta' \end{pmatrix};$$

the index lin denotes the linearized function $f(\cdot)$, $\Theta \in \mathcal{T} \subset \mathcal{R}^k$. If the approximate value of the unknown vector parameter Θ in its zero approximation is not known with sufficient accuracy, the method of successive approximations has to be applied, which after a finite number of steps ensures that the remainder of the Taylor series becomes sufficiently small and may be neglected.

The $p \times k$-dimensional matrix $A = \partial f(\Theta_0)\partial \Theta'$, the elements of which are known numerical coefficients, is called a design matrix.

The relation between the theoretical (errorless) value X of the directly observable quantity and the theoretical value $\Theta = \Theta_0 + \Delta\Theta$ of the unknown vector parameter is

$$X = f(\Theta_0) + A\Delta\Theta. \tag{2.2.1}$$

Linearizing in the same way the q-dimensional vector function $g(\cdot)$, $\Theta \in \mathcal{T} \subset \mathcal{R}^k$, expressing q conditions that the values of the parameters have to fulfil, we obtain

$$g_{lin}(\Theta) = g(\Theta_0) + [\partial g(\Theta_0)/\partial \Theta']\Delta\Theta = 0,$$

where the index lin denotes the linearized form of the function $g(\cdot)$, $\Theta \in \mathcal{T} \subset \mathcal{R}^k$, again. Denoting $g(\Theta_0) = b$ (a q-dimensional vector of known values of the vector function $g(\cdot)$ at the point Θ_0) and $\partial g(\Theta_0)/\partial \Theta' = B$ (a $q \times k$-dimensional matrix of values of the first derivatives of the q-dimensional vector function $g(\cdot)$ at the same point), we can write the linearized condition binding the values of the unknown parameters in the following way

$$b + B\Delta\Theta = 0. \tag{2.2.2}$$

It is evident that $f_{lin}(\cdot)$ and $g_{lin}(\cdot)$ at any point in the parametric space will be the better approximation of the functions $f(\cdot)$ and $g(\cdot)$, respectively, the smaller the difference is between the approximate value Θ_0 and the actual theoretical value Θ of the unknown vector parameter.

The relation (2.2.1) expresses the linear relation between the theoretical value X of the directly observable vector quantity of the measured object and the theoretical value of the difference $\Delta\Theta = \Theta - \Theta_0$ of the unknown parameter. Analogously, the relation (2.2.2) expresses q linear conditions, which the theoretical values of the increments $\Delta\Theta_1, \Delta\Theta_2, \ldots, \Delta\Theta_k$ of the parameters have to satisfy. We speak of the linear model for determining $\Delta\Theta$.

The unknown vector parameter Θ is estimated in the form $\hat{\Theta}(\eta) = \Theta_0 + \widehat{\Delta\Theta}(\eta)$, using the realizations of a random vector $\eta = \xi - Jf(\Theta_0)$, $E[\eta] = JA\Delta\Theta$, $var(\eta) = var(\xi) = \Sigma$. The increment $\Delta\Theta$ is the immediate subject of estimating. The estimate $\widehat{\Delta\Theta}(y)$ must fulfil the system of conditions $b + B\widehat{\Delta\Theta}(y) = 0$ (if it exists).

The requirement of linear estimators means that we are restricted to the estimators of the form $\hat{\Theta}(\eta) = \Theta_0 + t + T\eta$, where t is a k-dimensional real vector and T a $k \times n$-dimensional matrix. The problem is to determine a k-dimensional vector t and a $k \times n$-dimensional matrix T such that the resulting estimators of the unknown parameters will be unbiased and jointly efficient.

The majority of situations occurring in data processing can be described by six basic linear models: 1) A linear model for processing data from direct measurements, 2) from indirect measurements, 3) from direct measurements with conditions, 4) from incomplete direct measurements with conditions, 5) from indirect measurements with conditions, and 6) from incomplete indirect measurements with conditions. Individual cases are presented in the following review in which the first row describes the theoretical model, the second and third row the corresponding stochastic model, and the fourth row contains the regularity conditions of the model; they ensure the existence of all matrix inverses necessary for computing the numerical value of the optimum estimate of the unknown parameters (e.g., if the design matrix in the model of indirect measurements did not have the full rank in columns, i.e., if $R(A) < k$, where k is the number of unknown parameters, not each parameter would be estimable separately, only a certain linear combination of the unknown parameters would be estimable; in order to convince ourselves of this assertion, it suffices to take into account an arbitrary simple theoretical model in model 2 with a design matrix A such that $R(A) < k$; see also p. 238).

In principle, we take into account in all the models a p-dimensional vector X of theoretical values for directly observable quantities, an n-dimensional random vector ξ of measured data, a k-dimensional vector of unknown parameters, and the number of conditions (if they exist) that equal q. Such data determines uniquely the dimensions of vectors and matrices occurring in individual linear models of measurement.

Model 1 (a direct measurement of a scalar parameter)

$$X = \Theta_0 + \Delta\Theta, \ \Theta \in \mathcal{R}^1,$$
$$\xi = i(\Theta_0 + \Delta\Theta) + \varepsilon,$$
$$E[\xi] = i(\Theta_0 + \Delta\Theta), \ E[\varepsilon] = 0, \ var(\xi) = var(\varepsilon) = \Sigma,$$
$$R(\Sigma) = n.$$

A measurement of the height of an object repeated n times under the same conditions can serve as an example of Model 1.

Model 2 (an indirect measurement of a vector parameter)

$$X = f(\Theta_0) + A\Delta\Theta, \ X \in \mathcal{R}^p, \ \Theta \in \mathcal{R}^k,$$
$$\xi = J[f(\Theta_0) + A\Delta\Theta] + \varepsilon,$$
$$E[\xi] = J[f(\Theta_0) + A\Delta\Theta], \ E[\varepsilon] = 0, \ var(\xi) = var(\varepsilon) = \Sigma,$$
$$R(A) = k \le p, \ R(J) = p \le n, \ R(\Sigma) = n.$$

The problem of determining coefficients a_1, a_2 and a_3 in the relation $y(t) = a_1 + a_2t + a_3t^2$, which models a certain physical dependence on the basis of the measured values of $y(t)$ at the given points $t_1, t_2, \ldots, t_n$, may serve as an example of Model 2.

Model 3 (a direct measurement of a vector parameter with a system of conditions)

$$
\begin{array}{|l|}
\hline
X = \Theta_0 + \Delta\Theta,\ b + B\Delta\Theta = 0,\ X \in \mathcal{R}^k,\ \Theta \in \mathcal{T} \subset \mathcal{R}^k,\ (p = k) \\
\xi = J(\Theta_0 + \Delta\Theta) + \varepsilon,\ b + B\Delta\Theta = 0, \\
E[\xi] = J(\Theta_0 + \Delta\Theta),\ E[\varepsilon] = 0,\ var(\xi) = var(\varepsilon) = \Sigma, \\
R(B) = q \leq k,\ R(J) = k \leq n,\ R(\Sigma) = n. \\
\hline
\end{array}
$$

A repeated measurement of the angles α_1, α_2 and α_3 in a plane triangle bound by the condition $\alpha_1 + \alpha_2 + \alpha_3 = \pi$ may serve as an example of Model 3.

Model 4 (an incomplete direct measurement of a vector parameter with a system of conditions)

The problem is to determine the optimum estimators of the elements of a k-dimensional vector parameter Θ if some of its components are directly observable and the rest of them are not directly observable. The k-dimensional parameter vector Θ is split into two subvectors: Θ_1 is a p-dimensional subvector consisting of directly measured parameters and Θ_2 is a $k-p$-dimensional subvector of those parameters which are interesting from the point of view of the problem solved but not directly observable. They can be determined on the basis of a system of q $(k \geq q \geq k-p)$ conditions $g(\Theta_1) + h(\Theta_2) = 0$, which express the relation between them and the directly observable parameters Θ_1 of the unknown parameter Θ. The linearized form of this system of conditions can be obtained by applying the above mentioned procedure for developing them into the Taylor series and neglecting terms of the second and higher orders. Then the model of an incomplete direct measurement of a vector parameter with a system of conditions can be written in the following way

$$X = \Theta_{10} + \Delta\Theta_1,\ \Theta = \begin{pmatrix} \Theta_1 \\ \Theta_2 \end{pmatrix},\ b + (B, C)\begin{pmatrix} \Delta\Theta_1 \\ \Delta\Theta_2 \end{pmatrix} = 0,\ X \in \mathcal{R}^p,$$
$$\Theta_1 \in \mathcal{T}_1 \subset \mathcal{R}^p,\ \Theta_2 \in \mathcal{T}_2 \subset \mathcal{R}^{k-p},$$
$$\xi = J(\Theta_{10} + \Delta\Theta_1) + \varepsilon,\ b + (B, C)\begin{pmatrix} \Delta\Theta_1 \\ \Delta\Theta_2 \end{pmatrix} = 0,$$
$$E[\xi] = J(\Theta_{10} + \Delta\Theta_1),\ E[\varepsilon] = 0,\ var(\xi) = var(\varepsilon) = \Sigma,$$
$$R[(B, C)] = q \le k,\ R(C) = k - p \le q,\ R(J) = p \le n,\ R(\Sigma) = n.$$

One of the simplest examples of this model is a direct measurement of the angles α_1 and α_2 in a plane triangle where the angle α_3, which is not directly observable, has to be determined from the condition $\alpha_1 + \alpha_2 + \alpha_3 = 2\pi$.

Model 5 (an indirect measurement of a vector parameter with a system of conditions)

$$X = f(\Theta_0) + A\Delta\Theta,\ b + B\Delta\Theta = 0,\ X \in \mathcal{R}^p,\ \Theta \in \mathcal{T} \subset \mathcal{R}^k,$$
$$\xi = J[f(\Theta_0) + A\Delta\Theta] + \varepsilon,\ b + B\Delta\Theta = 0,$$
$$E[\xi] = J[f(\Theta_0) + A\Delta\Theta],\ E[\varepsilon] = 0,\ var(\xi) = var(\varepsilon) = \Sigma,$$
$$R(A) = k \le p,\ R(B) = q \le k,\ R(J) = p \le n,\ R(\Sigma) = n.$$

The problem of determining the coefficients a_1, a_2, b_1, b_2 and b_3 in the relations $y_a(t) = a_1 + a_2 t$, $t \in [T_1, T_2]$ and $y_b(t) = b_1 + b_2 t + b_3 t^2$, $t \in [T_2, T_3]$, based of the measured values of $y_a(t)$ at certain points of the interval $[T_1, T_2]$ and measured values of $y_b(t)$ at some points of the interval $[T_2, T_3]$ if we know that the functions $y_a(\cdot)$ and $y_b(\cdot)$, models a physical relation which is continuous at the point T_2, i.e., $y_a(T_2) = y_b(T_2)$.

Model 6 (an incomplete indirect measurement of a vector parameter with a system of conditions)

A k-dimensional vector parameter, analogous to Model 4, is of the form $\Theta = (\Theta_1', \Theta_2')'$; there exists a functional relation between l elements of the vector Θ_1 and the theoretical values of directly observable quantities of the measured object. Moreover, there exist q conditions expressing relations between the elements of the vector Θ. Thus

$$X = f(\Theta_{10}) + A\Delta\Theta_1,\ \Theta = \begin{pmatrix} \Theta_1 \\ \Theta_2 \end{pmatrix},\ b + (B, C)\begin{pmatrix} \Delta\Theta_1 \\ \Delta\Theta_2 \end{pmatrix} = 0,$$
$$X \in \mathcal{R}^p,\ \Theta_1 \in \mathcal{T}_1 \subset \mathcal{R}^l,\ \Theta_2 \in \mathcal{T}_2 \subset \mathcal{R}^{k-l},$$

$$\xi = J[f(\Theta_{10}) + A\Delta\Theta_1] + \varepsilon,\ b + (B, C)\begin{pmatrix} \Delta\Theta_1 \\ \Delta\Theta_2 \end{pmatrix} = 0,$$
$$E[\xi] = J[f(\Theta_{10}) + A\Delta\Theta_1],\ E[\varepsilon] = 0,\ var(\xi) = var(\varepsilon) = \Sigma,$$
$$R(A) = l \le p,\ R[(B, C)] = q \le k,\ R(C) = k - l,\ R(J) = p \le n,$$
$$R(\Sigma) = n.$$

Example In a plane triangle whose side a is known, the magnitudes of all angles are to be determined on the basis of the measured angles α and β and the side b. The problem is to build a theoretical and stochastic model of this situation.

The parameters being determined are $\Theta_1 = \alpha$, $\Theta_2 = \beta$ and $\Theta_3 = \gamma$. The theoretical values of the directly observable quantities are $X_1 = \alpha$, $X_2 = \beta$ and $X_3 = b$, i.e., the first two parameters are the directly observable parameters to be determined, the relation between the directly observable side b and the parameter α and β, which is also being determined, is expressed by the law of sines $b(\alpha, \beta) = a \sin\beta / \sin\alpha$. The linearized form of this relation can be obtained by developing it in the Taylor series at the point (α_0, β_0), which is sufficiently close to the point (α, β), and then neglecting its terms of orders higher than 1

$$b(\alpha, \beta) = a(\sin\beta_0 / \sin\alpha_0) - a(\sin\beta_0 \cos\alpha_0 / \sin_2\alpha_0)\Delta\alpha +$$
$$+ a(\cos\beta_0 / \sin\alpha_o)\Delta\beta + \ldots.$$

The linearized theoretical model which describes this problem is

$$\begin{pmatrix} X_1 \\ X_2 \\ X_3 \end{pmatrix} = \begin{pmatrix} \alpha_0 \\ \beta_0 \\ a\sin\beta_0/\sin\alpha_0 \end{pmatrix} +$$
$$\begin{pmatrix} 1, & 0 \\ 0, & 1 \\ -a\sin\beta_0\cos\alpha_0/sin^2\alpha_0, & a\cos\beta_0/\sin\alpha_0 \end{pmatrix}\begin{pmatrix} \Delta\alpha \\ \Delta\beta \end{pmatrix}$$

under the condition

$$(\alpha_0 + \beta_0 - \pi) + (1, 1)\begin{pmatrix} \Delta\alpha \\ \Delta\beta \end{pmatrix} + \gamma = 0.$$

Let the matrix J reflecting the measurement method (its form has to conform to our knowledge of the optimum experimental design) have the form

$$J' = \begin{pmatrix} 1, & 1, & 1, & 0, & 0, & 0, & 0, & 0 \\ 0, & 0, & 0, & 1, & 1, & 1, & 0, & 0 \\ 0, & 0, & 0, & 0, & 0, & 0, & 1, & 1 \end{pmatrix},$$

which means that both angles are measured three times and the side b twice. In this way we obtain the stochastic model of the form

$$\begin{pmatrix} \xi_1 \\ \xi_2 \\ \xi_3 \\ \xi_4 \\ \xi_5 \\ \xi_6 \\ \xi_7 \\ \xi_8 \end{pmatrix} = \begin{pmatrix} 0 \\ 0 \\ 0 \\ 0 \\ 0 \\ 0 \\ a \sin \beta_0 / \sin \alpha_0 \\ a \sin \beta_0 / \sin \alpha_0 \end{pmatrix} +$$

$$+ \begin{pmatrix} 1, & 0 \\ 1, & 0 \\ 1, & 0 \\ 0, & 1 \\ 0, & 1 \\ 0, & 1 \\ -a \sin \beta_0 \cos \alpha_0 / \sin^2 \alpha_0, & a \cos \beta_0 / \sin \alpha_0 \\ -a \sin \beta_0 \cos \alpha_0 / \sin^2 \alpha_0, & a \cos \beta_0 / \sin \alpha_0 \end{pmatrix} \begin{pmatrix} \Delta\alpha \\ \Delta\beta \end{pmatrix} + \begin{pmatrix} \varepsilon_1 \\ \varepsilon_2 \\ \varepsilon_3 \\ \varepsilon_4 \\ \varepsilon_5 \\ \varepsilon_6 \\ \varepsilon_7 \\ \varepsilon_8 \end{pmatrix} =$$

$$= Jf(\alpha_0, \beta_0) + JA(\Delta\alpha, \Delta\beta)' + \varepsilon$$

under the condition

$$(\alpha_0 + \beta_0 - \pi) + \Delta\alpha + \Delta\beta + \gamma = 0,$$

where

$$E[\varepsilon] = 0, \ E[\xi] = Jf(\alpha_0, \beta_0) + JA(\Delta\alpha, \Delta\beta)', \ var(\xi) = var(\varepsilon) = \Sigma.$$

If the measurements are stochastically independent and σ_1^2 and σ_2^2 the dispersion of the length and angle measurements, respectively, the covariance matrix Σ of the vector ξ, which models the described measurement, is

$$\Sigma = \begin{pmatrix} \sigma_1^2, & 0, & 0, & 0, & 0, & 0, & 0, & 0 \\ 0, & \sigma_1^2, & 0, & 0, & 0, & 0, & 0, & 0 \\ 0, & 0, & \sigma_1^2, & 0, & 0, & 0, & 0, & 0 \\ 0, & 0, & 0, & \sigma_1^2, & 0, & 0, & 0, & 0 \\ 0, & 0, & 0, & 0, & \sigma_1^2, & 0, & 0, & 0 \\ 0, & 0, & 0, & 0, & 0, & \sigma_1^2, & 0, & 0 \\ 0, & 0, & 0, & 0, & 0, & 0, & \sigma_2^2, & 0 \\ 0, & 0, & 0, & 0, & 0, & 0, & 0, & \sigma_2^2 \end{pmatrix},$$

thus $D[\xi_1] = D[\xi_2] = D[\xi_3] = D[\xi_4] = D[\xi_5] = D[\xi_6] = \sigma_1^2$ and $D[\xi_7] = D[\xi_8] = \sigma_2^2$.

Model 2 has a privileged position among all linear models of measurements. Each of the other models may be transformed into the Model 2 form. A very important consequence of this statement is that any theorem proved for Model 2 is automatically true for all six basic linear models.

The problem of transforming any model to the Model 2 form lies in determining the form of the design matrix after the transformation using the matrices and vectors of the model before the transformation and in verifying if the matrix A_{T_i} (T shows that this matrix was generated by the transformation, the index i denotes the model undergoing the transformation) is regular, i.e., if it possesses the full rank in columns.

The simplest situation occurs in Model 1 of a direct measurement of a scalar parameter, where $A_{T_1} = 1,\ R(1) = 1,\ k = p = 1$.

In Model 2, evidently, $A_{T_2} = A,\ R(A_{T_2}) = R(A) = k \leq p$.

Consider Model 3 (direct measurements of a vector parameter with a system of conditions). Didactical reason, if necassary, leads to introducing indices in brackets denoting dimensions of vectors (a simple index) and matrices (a double index). In our case the subject of a transformation is the theoretical model $X_{(k)} = \Theta_{0(k)} + \Delta\Theta_{(k)}$, $b_{(q)} + B_{(q,k)}\Delta\Theta_{(k)} = 0_{(k)}$, $R(B_{(q,k)}) = q \leq k$. The number q of conditions binding the function values of the components of the q-dimensional parameter vector Θ enables us to express q of them ($q \leq k$) by the aid of the remaining $k-q$. The term $B_{(q,k)}\Delta_{(k)}$ occurring in the condition equation can be expressed by using a suitable way of renumbering the order of parameters in the form

$$B_{(q,k)}\Delta\Theta_{(k)} = B^{(1)}_{(q,q)}\Delta\Theta^{(1)}_{(q)} + B^{(2)}_{(q,k-q)}\Delta\Theta^{(2)}_{(k-q)},$$

where

$$\Delta\Theta_{(k)} = \begin{pmatrix} \Delta\Theta^{(1)}_{(q)} \\ \Delta\Theta^{(2)}_{(k-q)} \end{pmatrix}, \quad B_{(q,k)} = (B^{(1)}_{(q,q)}, B^{(2)}_{(q,k-q)}),$$

such that $R(B^{(1)}_{(q,q)}) = q$. The equation of conditions,

$$b_{(q)} + B^{(1)}_{(q,q)}\Delta\Theta^{(1)}_{(q)} + B^{(2)}_{(q,k-q)}\Delta\Theta^{(2)}_{(k-q)} = 0_{(q)},$$

rewritten using the previous procedure enables us to calculate $\Delta\Theta^{(1)}_{(q)}$:

$$\Delta\Theta^{(1)}_{(q)} = -(B^{(1)}_{(q,q)})^{-1}(b_{(q)} + B^{(2)}_{(q,k-q)}\Delta\Theta^{(2)}_{(k-q)}); \tag{2.2.3}$$

the inverse $(B^{(1)}_{(q,q)})^{-1}$ exists since the splitting of the matrix $B_{(q,k)}$ into its blocks was performed in such a way that $R(B^{(1)}_{(q,q)}) = q$. Finally the relation

$$X_{(k)} = \Theta_{0(k)} + \begin{pmatrix} \Delta\Theta^{(1)}_{(q)} \\ \Delta\Theta^{(2)}_{(k-q)} \end{pmatrix}$$

enables us to express the vector $\Delta\Theta^{(1)}_{(q)}$, by using the relation (2.2.3) in the form

$$X_{(k)} = \Theta_{0(k)} \begin{pmatrix} -(B^{(1)}_{(q,q)})^{-1}b_{(q)} - (B^{(1)}_{(q,q)})^{-1}B^{(2)}_{(q,k-q)}\Delta\Theta^{(2)}_{(k-q)} \\ \Delta\Theta^{(2)}_{(k-q)} \end{pmatrix} =$$

$$= \Theta_{0(k)} + \begin{pmatrix} -(B_{(q,q)})^{-1}b_{(q)} \\ 0_{(k-q)} \end{pmatrix} + \begin{pmatrix} -(B^{(1)}_{(q,q)})^{-1}B^{(2)}_{(q,k-q)} \\ I_{(k-q,k-q)} \end{pmatrix}\Delta\Theta^{(2)}_{(k-q)}.$$

Thus Model 3 $X_{(k)} = \Theta_{0(k)} + \Delta\Theta_{(k)}$, $b_{(q)} + B_{(q,k)}\Delta\Theta_{(k)} = 0_{(q)}$, $R(B) = q \leq k$, is equivalent with Model 2 of the form

$$X_{T_3(k)} = \Theta_{0(k)} + A_{T_3(k,k-q)}\Delta\Theta^{(2)}_{(k-q)}, \ X_{T_3} \in \mathcal{R}^k, \ \Delta\Theta^{(2)} \in \mathcal{T} \subset \mathcal{R}^{k-q},$$

where

$$X_{T_3(k)} = X_{(k)} + \begin{pmatrix} -(B^{(1)}_{(q,q)})^{-1}b_{(q)} \\ 0_{(k-q)} \end{pmatrix},$$

$$A_{T3(k,k-q)} = \begin{pmatrix} -(B^{(1)}_{(q,q)})^{-1} B^{(2)}_{(q,k-q)} \\ I_{(k-q,k-q)} \end{pmatrix},$$

and simultaneously $R(A_{T_3}) = k - q$, which means that the design matrix after the transformation has the full rank in columns. It is a consequence of the fact that the columns of the matrix $I_{(k-q,k-q)}$ are linearly independent, which implies $R(A_{T_3}) = k - q$.

The equivalence of Model 4 (an incomplete direct measurement of a vector parameter with a system of conditions) with Model 2 will be proved by means of Model 3. By way of illustration we shall prove in detail a concrete case for $p = 6$, $q = 4$ and $k = 7$. The parameters being determined are $\Delta\Theta_1, \Delta\Theta_2, \Delta\Theta_3, \Delta\Theta_4, \Delta\Theta_5, \Delta\Theta_6$ and $\Delta\Theta_7$. The first six are directly observable, they generate the six-dimensional vector $\Delta\Theta_1 = (\Delta\Theta_1, \Delta\Theta_2, \Delta\Theta_3, \Delta\Theta_4, \Delta\Theta_5, \Delta\Theta_6)'$, the parameter $\Delta\Theta_7$ is not directly observable, it possesses the meaning of the vector $\Delta\Theta_2$.

The four condition equations written in their vector form

$$b + B\Delta\Theta_1 + C\Delta\Theta_2 = 0,$$

where b is a four-dimensional real vector, B is a 4×6-dimensional matrix of real coefficients, 0 denotes a four-dimensional zero vector, can be written in the form of a system of the following linear equations

$$\begin{aligned}
b_1 + b_{11}\Delta\Theta_1 + b_{12}\Delta\Theta_2 + b_{13}\Delta\Theta_3 + b_{14}\Delta\Theta_4 + b_{15}\Delta\Theta_5 + & \\
+ b_{16}\Delta\Theta_6 + c_{11}\Delta\Theta_7 &= 0 \\
b_2 + b_{21}\Delta\Theta_1 + b_{22}\Delta\Theta_2 + b_{23}\Delta\Theta_3 + b_{24}\Delta\Theta_4 + b_{25}\Delta\Theta_5 + & \\
+ b_{26}\Delta\Theta_6 + c_{21}\Delta\Theta_7 &= 0 \\
b_3 + b_{31}\Delta\Theta_1 + b_{32}\Delta\Theta_2 + b_{33}\Delta\Theta_3 + b_{34}\Delta\Theta_4 + b_{35}\Delta\Theta_5 + & \\
+ b_{36}\Delta\Theta_6 + c_{31}\Delta\Theta_7 &= 0 \\
b_4 + b_{41}\Delta\Theta_1 + b_{42}\Delta\Theta_2 + b_{43}\Delta\Theta_3 + b_{44}\Delta\Theta_4 + b_{45}\Delta\Theta_5 + & \\
+ b_{46}\Delta\Theta_6 + c_{41}\Delta\Theta_7 &= 0.
\end{aligned} \tag{2.2.4}$$

Let the conditions in the system (2.2.4) be ordered in such a way that the coefficient $c_{11} \neq 0$ (such an ordering is possible, since the condition of regularity of Model 4 - $R(C) = k - p$ in our concrete case when $k - p = 1$ ensures that at least one of the coefficients c_{i1}, $i = 1, 2, 3, 4$ must be nonzero).

Expressing the parameter $\Delta\Theta_7$ from the first equation of the system (2.2.4)

$$\Delta\Theta_7 = -c_{11}^{-1}(b_1+b_{11}\Delta\Theta_1+b_{12}\Delta\Theta_2+b_{13}\Delta\Theta_3+b_{14}\Delta\Theta_4+b_{15}\Delta\Theta_5+b_{16}\Delta\Theta_6) \tag{2.2.5}$$

and inserting it into all the other equations of the system, we get a system which consists of the following three conditions

$$\begin{gathered} b_i - c_{i1}c_{11}^{-1}b_1 + (b_{i1} - c_{i1}c_{11}^{-1}b_{11})\Delta\Theta_1 + (b_{i2} - c_{i1}c_{11}^{-1}b_{12})\Delta\Theta_2 + \\ +(b_{i3} - c_{i1}c_{11}^{-1}b_{13})\Delta\Theta_3 + (b_{i4} - c_{i1}c_{11}^{-1}b_{14})\Delta\Theta_4 + \\ +(b_{i5} - c_{i1}c_{11}^{-1}b_{15})\Delta\Theta_5 + (b_{i6} - c_{i1}c_{11}^{-1}b_{16})\Delta\Theta_6 = 0, \ i = 2,3,4, \end{gathered}$$

expressing the relations between the values of the directly observable parameters. Denoting $b_{T_4} = (b_2 - c_{21}c_{11}^{-1}b_1, b_3 - c_{31}c_{11}^{-1}b_1, b_4 - c_{41}c_{11}^{-1}b_1)'$ and

$$B_{T_4} = \begin{pmatrix} b_{21} - c_{21}c_{11}^{-1}b_{11}, & b_{22} - c_{21}c_{11}^{-1}b_{12}, & b_{23} - c_{21}c_{11}^{-1}b_{13}, & b_{24} - c_{21}c_{11}^{-1}b_{14}, & b_{25} - c_{21}c_{11}^{-1}b_{15}, & b_{26} - c_{21}c_{11}^{-1}b_{16}, \\ b_{31} - c_{31}c_{11}^{-1}b_{11}, & b_{32} - c_{31}c_{11}^{-1}b_{12}, & b_{33} - c_{31}c_{11}^{-1}b_{13}, & b_{34} - c_{31}c_{11}^{-1}b_{14}, & b_{35} - c_{31}c_{11}^{-1}b_{15}, & b_{36} - c_{31}c_{11}^{-1}b_{16}, \\ b_{41} - c_{41}c_{11}^{-1}b_{11}, & b_{42} - c_{41}c_{11}^{-1}b_{12}, & b_{43} - c_{41}c_{11}^{-1}b_{13}, & b_{44} - c_{41}c_{11}^{-1}b_{14}, & b_{45} - c_{41}c_{11}^{-1}b_{15}, & b_{46} - c_{41}c_{11}^{-1}b_{16} \end{pmatrix},$$

we get the model $X = \Theta_{10} + \Delta\Theta_1$, $b_{T_4} + B_{T_4}\Delta\Theta_1$, $R(B_{T_4}) = 3$ (due to the regularity assumptions of Model 4, $R[(B,C)] = 4$ and $R(C) = 1$, as a consequence of which the considered transformation of the matrix (B,C) into the matrix B_{T_4} could not cause a linear dependence between originally independent rows); the parameter $\Delta\Theta_7$, not being directly observable, is determined by Eq. (2.2.5).

Generally, the subject of the transformation is Model 4 of the form $X_{(p)} = \Theta_{10(p)} + \Delta\Theta_{1(p)}$, $\Theta_{(k)} = (\Theta'_{1(p)}, \Theta'_{2(k-p)})'$, $b_{(q)} + B_{(q,p)}\Delta\Theta_{1(p)} + C_{(q,k-q)}\Delta\Theta_{2(k-p)} = 0_{(q)}$, $R[(B_{(q,p)}, C_{(q,k-p)})] = q \leq k$, $R(C_{(q,k-p)}) = k-p \leq q$. As $R(C) = k-p$ (the number of parameters which are not directly observable), the matrix $C_{(q,k-p)}$ can be divided horizontally into two submatrices $C^{(1)}_{(k-p,k-p)}$ and $C^{(2)}_{(q-k+q,k-p)}$ in such a way that $R(C^{(1)}_{(k-p,k-p)}) = k-p$. After dividing in an analogous way the matrix $B_{(q,p)}$ and the vector $b_{(q)}$, we may write the system of conditions

$$b_{(q)} + (B_{(q,p)}, C_{(q,k-p)})\begin{pmatrix} \Delta\Theta_{1(p)} \\ \Delta\Theta_{2(k-p)} \end{pmatrix} = 0_{(q)}$$

in the following way

$$\begin{pmatrix} b^{(1)}_{(k-p)} \\ b^{(2)}_{(q-k+p)} \end{pmatrix} + \begin{pmatrix} B^{(1)}_{(k-p,p)}, & C^{(1)}_{(k-p,k-p)} \\ B^{(2)}_{(q-k+p,p)}, & C^{(2)}_{(q-k+p,k-p)} \end{pmatrix} \begin{pmatrix} \Delta\Theta_{1(p)} \\ \Delta\Theta_{2(k-p)} \end{pmatrix} =$$

$$\begin{pmatrix} 0_{(k-p)} \\ 0_{(q-k+p)} \end{pmatrix}.$$

The first system of conditions enables us to determine $\Delta\Theta_2$ (the inverse of the matrix $C^{(1)}$ exists, since $R(C^{(1)}) = k - p$). Inserting it into the second system of conditions, we obtain

$$\Delta\Theta_{2(k-p)} = -(C^{(1)}_{(k-p,k-p)})^{-1}(b^{(1)}_{(k-p)} + B^{(1)}_{(k-p,p)}\Delta\Theta_{1(p)}) \tag{2.2.6}$$

with a system of conditions

$$b^{(2)}_{(q-k+p)} - C^{(2)}_{(q-k+p,k-p)}(C^{(1)}_{(k-p,k-p)})^{-1}b^{(1)}_{(k-p)} +$$

$$+[B^{(2)}_{(q-k+p,p)} - C^{(2)}_{(q-k+p,k-p)}(C^{(1)}_{(k-p,k-p)})^{-1}B^{(1)}_{(k-p,p)}]\Delta\Theta_{1(p)} = 0_{(q-k+p)},$$

expressing the conditions between the directly observable parameters only.

The rows of the matrix $B_{T_4} = B^{(2)}_{(q-k+p,p)} - C^{(2)}_{(q-k+p,k-p)}(C^{(1)}_{(k-p,k-p)})^{-1}$ $B^{(1)}_{(k-p,p)}$ are linearly independent (the rows of the matrix (B, C) are independent as a consequence of the assumption $R[(B, C)] = q$ and the transformation of this matrix we used could not cause any mutual dependence between them, cf. [17], p. 131), therefore $R(B_{T_4}) = q - k + p < p$. Denoting, moreover, $b_{T_4} = b^{(2)}_{(q-k+p)} - C^{(2)}_{(q-k+p,k-p)}(C^{(1)}_{(k-p,k-p)})^{-1}b^{(1)}_{(k-p)}$, we can see that we have obtained Model 3 in the form $X_{(p)} = \Theta_{10(p)} + \Delta\Theta_{1(p)}$, $b_{T_4(q-k+p)} + B_{T_4(q-k+p,p)}\Delta\Theta_{1(p)} = 0_{(q-k+p)}$, $R(B_{T_4}) = q - k + p < p$ when simultaneously $\Delta\Theta_2$ is given by the equation (2.2.6). The way of transforming Model 3 into Model 2 has already been shown.

Consider now Model 5 (indirect measurements with a condition) $X_{(p)} = f_{(p)}(\Theta_{0(k)}) + A_{(p,q)}\Delta\Theta_{(k)}$, $b_{(q)} + B_{(q,k)}\Delta\Theta_{(k)} = 0_{(q)}$, $R(A_{(p,k)}) = k \leq p$, $R(B_{(q,k)}) = q \leq k$. As a consequence of the assumption $R(B_{(q,k)}) = q \leq k$, the system of conditions $b_{(q)} + B_{(q,k)}\Delta\Theta_{(k)} = 0_{(q)}$ can be written in the form

$$b_{(q)} + B^{(1)}_{(q,q)}\Delta\Theta^{(1)}_{(q)} + B^{(2)}_{(q,k-q)}\Delta\Theta^{(2)}_{(k-q)} = 0_{(q)}$$

(the matrix $B_{(q,k)}$ is divided in the vertical direction into two matrices $B^{(1)}_{(q,q)}$ and $B^{(2)}_{(q,k-q)}$ in such a way that $R(B^{(1)}_{(q,q)}) = q$), which makes it possible to express the q-dimensional vector parameter $\Delta\Theta^{(1)}_{(q)}$ in the form

$$\Delta\Theta^{(1)}_{(q)} = -(B^{(1)}_{(q,q)})^{-1} b_{(q)} - (B^{(1)}_{(q,q)})^{-1} B^{(2)}_{(q,k-q)} \Delta\Theta^{(2)}_{(k-q)}; \tag{2.2.7}$$

this means that

$$A_{(p,q)}\Delta\Theta_{(q)} = A_{(p,q)} \begin{pmatrix} \Delta\Theta^{(1)}_{(q)} \\ \Delta\Theta^{(2)}_{(k-q)} \end{pmatrix} =$$

$$= A_{(p,q)} \left[\begin{pmatrix} -(B^{(1)}_{(q,q)})^{-1} b_{(q)} \\ 0_{(k-q)} \end{pmatrix} + \begin{pmatrix} -(B^{(1)}_{(q,q)})^{-1} B^{(2)}_{(q,k-q)} \\ I_{(k-q,k-q)} \end{pmatrix} \Delta\Theta^{(2)}_{(k-q)} \right]$$

and the columns of the matrix

$$A_{T_5(q,k-q)} = A_{(p,k)} \begin{pmatrix} -(B^{(1)}_{(q,q)})^{-1} B^{(2)}_{(q,k-q)} \\ I_{(k-q,k-q)} \end{pmatrix}$$

are linearly independent, as a consequence of which its rank is $R(A_{T_5}) = k - q$. Denoting, furthermore,

$$X_{T_5(p)} = X_{(p)} + A_{(p,k)} \begin{pmatrix} -(B^{(1)}_{(q,q)})^{-1} b_{(q)} \\ 0_{(k-q)} \end{pmatrix},$$

we see that Model 2 of the form $X_{T_5} = f(\Theta_0) + A_{T_5}\Delta\Theta^{(2)}$, $R(A_{T_5}) = k - q \leq p$, in which $\Delta\Theta^{(1)}$ is given simultaneously by the relation (2.2.7) and is equivalent to the original Model 5.

The equivalence of Model 6 (an incomplete indirect measurement of a vector parameter with a system of conditions) with Model 2 will be proved by means of Model 5.

Let us consider the model $X_{(p)} = f_{(p)}(\Theta_{10(l)}) + A_{(p,l)}\Delta\Theta_{1(l)}$, $\Theta_{(k)} = (\Theta'_{1(l)} + \Delta\Theta'_{1(l)}, \Theta'_{2(k-l)} + \Delta\Theta'_{2(k-l)})'$, $b_{(q)} + B_{(q,l)}\Delta\Theta_{1(l)} + C_{(q,k-l)}\Delta\Theta_{2(k-l)} = 0_{(q)}$, $R(A_{(p,l)}) = l \leq p$, $R[(B_{(q,l)}, C_{(q,k-l)})] = q \leq k$, $R(C_{(q,k-l)}) = k - l \leq q$. The same method of modification as in Model 4 can be applied now: The system of conditions $b_{(q)} + B_{(q,l)}\Delta\Theta_{1(l)} + C_{(q,k-l)}\Delta\Theta_{2(k-l)} = 0_{(q)}$ can be split into the following two systems:

$$b^{(1)}_{(k-l)} + B^{(1)}_{(k-l,l)}\Delta\Theta_{1(l)} + C^{(1)}_{(k-l,k-l)}\Delta\Theta_{1(k-l)} = 0_{(k-l)}$$

and

$$b^{(2)}_{(q-k+l)} + B^{(2)}_{(q-k+l,l)}\Delta\Theta_{1(l)} + C^{(2)}_{(q-k+l,k-l)}\Delta\Theta_{2(k-l)} = 0_{(q-k+l)}$$

such that the square $k-l$ matrix $C^{(1)}$ will be regular (this is possible because of the assumption $R(C) = k-l$). The first system of conditions yields

$$\Delta\Theta_{2(k-l)} = -(C^{(1)}_{(k-l,k-l)})^{-1}(b^{(1)}_{(k-l)} + B^{(1)}_{(k-l,l)})\Delta\Theta_{1(l)},$$

which, being inserted into the second, gives the system of $q-k+l$ conditions

$$b^{(2)}_{(q-k+l)} - C^{(2)}_{(q-k+l,k-l)}(C^{(1)}_{(k-l,k-l)})^{-1}b^{(1)}_{(k-l)} + [B^{(2)}_{(q-k+l,l)} -$$

$$-C^{(2)}_{(q-k+l,k-l)}(C^{(1)}_{(k-l,k-l)})^{-1}B^{(1)}_{(k-l,l)}]\Delta\Theta_{1(l)} = 0_{(q-k+l)},$$

which expresses the relations between the elements of the vector parameter $\Delta\Theta_1$, only. Analogous to Model 4, we obtain finally the conclusion that the rank of the matrix $B_{T_6} = B^{(2)}_{(q-k+l,l)} - C^{(2)}_{(q-k+l,k-l)}(C^{(1)}_{(k-l,k-l)})^{-1}B^{(1)}_{(k-l,l)}$ is $R(B_{T_6}) = q-k+l$, which proves the equivalence of Model 6 to Model 5, and thus, simultaneously, to Model 2.

2.3 Method of least squares, theorem on equivalence of estimators

The method of least squares enables us to solve the problem of determining an optimum estimator of a vector parameter $\Delta\Theta \in \mathcal{T} \subset \mathcal{R}^k$ based on a realization y of an n-dimensional random vector η, $E[\eta] = JA\Delta\Theta$, $var(\eta) = \Sigma$, $R(JA) = k \leq n$ (where $\mathcal{T}$ is the set of all possible values of the parameter $\Delta\Theta$, A is a design matrix of the experiment, and J is a matrix reflecting the method of performing the experiment).

The principle of least squares in linear estimation theory consists of the following: The element of the k-dimensional manifold[3] $\mathcal{T} \subset \mathcal{R}^k$, which minimizes, for each realization y of the random variable η, the quadratic form

$$Q(\Delta\Theta, y) = (JA\Delta\Theta - y)'\Sigma^{-1}(JA\Delta\Theta - y),\ \Delta\Theta \in \mathcal{T},$$

[3] A linear manifold $\mathcal{T} \subset \mathcal{R}^k$ in the case of a vector parameter $\Delta\Theta$ which must satisfy a system of conditions $B\Delta\Theta + b = 0$ has the form $\mathcal{T} = \{u : Bu + b = 0\}$. Realize that for $b \neq 0$ the zero element of the k-dimensional space $\mathcal{R}^k$ is not an element of the linear manifold, i.e., $\mathcal{T}$ is a subset of the space $\mathcal{R}^k$, not its subspace.

represents the best (in the sense of this method) estimate of the parameter $\Delta\Theta$. We denote the element belonging to each realization y of the random vector η by $\widehat{\Delta\Theta}(y)$. The set of realizations of the random vector η is assigned the set of realizations of the random variable $\widehat{\Delta\Theta}(\eta)$. A random variable $\widehat{\Delta\Theta}(\eta)$ is called an LS estimator of a k-dimensional random vector parameter $\Delta\Theta$.

F.C. Gauss is the originator of the least squares method. The least squares method, whose large-scale applicability extends well beyond estimation theory, has a very simple geometric interpretation. The problem of finding out for any $y \in \mathcal{R}^n$ a point of the set $\{JA\Delta\Theta : \Delta\Theta \in \mathcal{T}\}$ which minimizes the quadratic form $Q(\Delta\Theta, y)$ is an analog of the simple problem in analytic geometry of determining the point on the straight line $y = a + bx$, which is closest to an arbitrary point B on the plane $\mathcal{R}^2$. For any concrete point B, the point B_p on the straight line which is a projection of the point B along the perpendicular to it, represents the solution of the problem.

The quadratic form $Q(\Delta\Theta, y)$ represents the square of the distance between the point $y \in \mathcal{R}^n$ and the set $\{JA\Delta\Theta : \Delta\Theta \in \mathcal{T}\} \subset \mathcal{R}^n$ in an n-dimensional vector space in which the distance between two points x and y is defined by the relation $\rho^2(x, y) = (x - y)'\Sigma^{-1}(x - y)$. If $\Sigma = I$, the Euclidean metric $\rho^2_{\mathcal{E}}(x, y) = (x - y)'(x - y) = \sum_{i=1}^{n}(x_i - y_i)^2$ has been treated [14].

The great importance of the least squares method seen from the point of view of estimation theory is emphasized by the following theorem on the equivalence of estimators.

Theorem of the equivalence of estimators Linear estimators of unknown parameters in basic linear models, optimum in the sense of the least squares principle (the LS estimators), are unbiased and jointly efficient.

The proof of the theorem on the equivalence of estimators will be presented in the following section.

2.4 Linear unbiased and jointly efficient estimators of unknown parameters in basic linear models of measurement

The problem is to derive in an explicit form the linear unbiased and jointly efficient estimators of unknown parameters in all the situations described by the six fundamental linear models of measurement.

In all cases, the input for data processing is a realization x of an n-dimensional random variable ξ transformed by means of the value $f(\Theta_0)$ of the vector function $f(\cdot)$ at the approximate point Θ_0 to the realization y of the random variable $\eta = \xi - Jf(\Theta_0)$. From a realization y of the random variable we get the estimate $\widehat{\Delta\Theta}(y)$ of the difference $\Theta - \Theta_0$; therefore the resulting estimator of the unknown parameter Θ is of the form $\hat{\Theta}(\eta) = \Theta_0 + \widehat{\Delta\Theta}(\eta)$.

Model 1 (a direct measurement of a scalar parameter)

The problem is to find a linear unbiased and efficient estimator of a scalar parameter $\Theta = \Theta_0 + \Delta\Theta$ within a stochastic model

$$\xi = i\Theta_0 + i\Delta\Theta + \varepsilon,\ \eta = \xi - i\Theta_0,$$
$$E[\eta] = i\Delta\Theta,\ var(\eta) = var(\xi) = var(\varepsilon) = \Sigma,\ R(\Sigma) = n.$$

Because we require linearity in the estimator, we look for it in the form

$$\widehat{\Delta\Theta}(\eta) = L'\eta + l,\ L \in \mathcal{R}^n,\ l \in \mathcal{R}^1.$$

Due to the definition, this estimator is unbiased if $E[\widehat{\Delta\Theta}(\eta)] = \Delta\Theta$ for an arbitrary $\Delta\Theta \in \mathcal{R}^1$. Since $E[L'\eta + l] = L'E[\eta] + l = L'i\Delta\Theta + l$, the relation for determining the coefficients L and l is

$$L'i\Delta\Theta + l = \Delta\Theta,$$

which (since it has to be fulfilled for any $\Delta\Theta \in \mathcal{R}^1$) implies

$$l = 0,\ L'i = 1.$$

In accordance with the definition, an unbiased linear estimator $\widehat{\Delta\Theta} = L'\eta$, $L'i = 1$, of a scalar parameter is efficient if its dispersion is minimum. The dispersion of a random variable $\widehat{\Delta\Theta}(\eta) = L'\eta$ is $D[\widehat{\Delta\Theta}(\eta)] = L'\Sigma L$. This means that in the class of the n-dimensional vectors which satisfy the condition $L'i = 1$ we must seek an L_0 which minimizes the value of the expression $L'\Sigma L$. This becomes a problem of determining the extremum of a function with a subsidiary condition. It can be determined using the method of the Lagrange undetermined multipliers. The auxiliary Lagrange function is

$$\Phi(L) = L'\Sigma L - 2\lambda(L'i - 1),$$

λ is the Lagrange multiplier. Its partial derivative with respect to L is $\partial\Phi(L)/\partial L = 2\Sigma L - 2\lambda i$. A solution of the equation $\Sigma L - \lambda i = 0$ is $L_0 = \Sigma^{-1}i\lambda$, where λ can be determined from the condition of unbiasedness which must be satisfied by each L, i.e., by $L_0 = \Sigma^{-1}i\lambda$ as well. Thus $L_0'i = 1 \Rightarrow i'\Sigma^{-1}i\lambda = 1 \Rightarrow \lambda = (i'\Sigma^{-1}i)^{-1}$. At the point $L_0 = \Sigma^{-1}i(i'\Sigma^{-1}i)^{-1}$ the function $\Phi(\cdot)$ and, consequently, the dispersion of the random variable $L_0'\eta$ acquire the extremum value equal to $L_0'\Sigma L_0$. In order to prove that this extremum equals the minimum of the dispersion, it is necessary to prove that every other linear unbiased estimator $L\eta$, $L \neq L_0$, satisfies the inequality $D[L'\eta] = L'\Sigma L > D[L_0'\eta] = L_0'\Sigma L_0$. Due to the definition of the positive definiteness of the matrix Σ, each non-zero n-dimensional vector $L - L_0$ must have the property

$$(L - L_0)'\Sigma(L - L_0) > 0.$$

Since $(L - L_0)'\Sigma(L - L_0) = L'\Sigma L - 2L_0'\Sigma L + L_0\Sigma L_0$ and $L_0\Sigma L = = (i'\Sigma i)^{-1}i'\Sigma^{-1}\Sigma L = (i'\Sigma^{-1}i)^{-1}i'L = (i'\Sigma^{-1}i)^{-1} = L_0'\Sigma L_0$,

$$(L - L_0)'\Sigma(L - L_0) = L'\Sigma L - L_0'\Sigma L_0$$

and this difference is positive as the matrix Σ has to be positive definite. The function $\Phi(\cdot)$ and, simultaneously, the dispersion of an unbiased linear estimator $\widehat{\Delta\Theta}(\eta) = L_0'\eta$ are minimal.

The linear unbiased and efficient estimator of a scalar parameter $\Theta = \Theta_0 + \Delta\Theta$ within the Model 1 is

$$\hat{\Theta}(\eta) = \Theta_0 + (i'\Sigma^{-1}i)^{-1}\Sigma^{-1}i'\Sigma^{-1}\eta \tag{2.4.1}$$

and its dispersion is

$$D[\hat{\Theta}(\eta)] = (i'\Sigma^{-1}i)^{-1}. \tag{2.4.2}$$

The next problem is to prove that the estimator (2.4.1) minimizes for each realization y of the random variable η the quadratic form

$$Q(\Delta\Theta, y) = (i\Delta\Theta - y)'\Sigma^{-1}(i\Delta\Theta - y), \ \Delta\Theta \in \mathcal{R}^1.$$

The extremum of this form is a point $\widetilde{\Delta\Theta} \in \mathcal{R}^1$ (at the point $\widetilde{\Delta\Theta}$ the first derivative is zero). Because of $\partial Q(\Delta\Theta, y)/\partial\Delta\Theta = 2i'\Sigma^{-1}(i\Delta\Theta - y)$, $\widetilde{\Delta\Theta} =$

$(i'\Sigma^{-1}i)^{-1}i'\Sigma^{-1}y$ is a solution of the equation $i'\Sigma^{-1}i\Delta\Theta - i'\Sigma^{-1}y = 0$, which is a realization of the estimator of the form (2.4.1). This point is a point of the minimum of the quadratic form $Q(\Delta\Theta, y)$ as $\partial^2 Q(\Delta\Theta, y)/\partial\Delta\Theta^2 = 2i'\Sigma^{-1}i$, and this is a positive number (which equals the dispersion of a random variable $\mu = i'\zeta$, where $var(\zeta) = \Sigma^{-1}$ and Σ^{-1} is a positive definite matrix). Thus, the theorem on equivalence for Model 1 is proved.

Let us analyse, in more detail, the estimator (2.4.1) for special and frequently occurring forms of the covariance matrix $var(\eta)$.

a) Let $var(\eta) = \sigma^2 H$, where $H = P^{-1}$ is a diagonal matrix of inverse weights. The linear unbiased and invariant estimator of the parameter Θ in Model 1 has the form of a weighted mean

$$\widehat{\Delta\Theta}(\eta) = (i'Pi)^{-1}i'P\eta = \left(\sum_{i=1}^{n} p_i\right)^{-1} \sum_{i=1}^{n} p_i\eta_i,$$

and the dispersion of the estimator $\hat{\Theta}(\eta) = \Theta_0 + \left(\sum_{i=1}^{n} p_i\right)^{-1} \sum_{i=1}^{n} p_i\eta_i$, according to (2.4.2), is

$$D[\hat{\Theta}(\eta)] = \sigma^2 / \left(\sum_{i=1}^{n} p_i\right).$$

Let $var(\eta) = \sigma^2 I$. The linear unbiased and efficient estimator of the scalar parameter $\Delta\Theta$ within Model 1 is

$$\widehat{\Delta\Theta}(\eta) = (i'i)^{-1}i'\eta = (1/n)\sum_{i=1}^{n} \eta_i$$

and

$$D\left[\Theta_0 + (1/n)\sum_{i=1}^{n} \eta_i\right] = \sigma^2/n.$$

Here it is necessary to caution against applying es .nators that have the form of an arithmetic or weighted mean if the measured data are not realizations of a random vector whose covariance matrix is of the form $var(\eta) = \sigma^2 I$ or $var(\eta) = \sigma^2 P^{-1}$, respectively. The estimates obtained by using them deviate substantially from reality, [18].

Model 2 (an indirect measurement of a vector parameter)

A linear unbiased and jointly efficient estimator of a vector parameter $\Theta = \Theta_0 + \Delta\Theta$ within the model

$$\xi = Jf(\Theta_0) + JA\Delta\Theta + \varepsilon,\ \eta = \xi - Jf(\Theta_0),$$
$$E[\eta] = JA\Delta\Theta,\ var(\xi) = var(\eta) = var(\varepsilon) = \Sigma,$$
$$R(JA) = k \le n,\ R(\Sigma) = n$$

is to be derived.

A linear estimator $\widehat{\Delta\Theta}(\eta) = L'\eta + l$, where L is an $n \times k$-dimensional matrix and l a k-dimensional vector will be unbiased if $L'JA = I$ (a unit matrix of the type $k \times k$) and $l = 0$ (a k-dimensional zero vector), since only under these conditions, for an arbitrary $\Delta\Theta \in \mathcal{R}^k$, $E[L'\eta + l] = JA\Delta\Theta$.

An unbiased linear estimator

$$\widehat{\Delta\Theta}(\eta) = L'\eta,\ L'JA = I,$$

is jointly efficient if for any k-tuple of real numbers $a_1, a_2, \ldots, a_k$ and any other linear unbiased estimator $\widetilde{\Delta\Theta}(\eta) \neq \widehat{\Delta\Theta}(\eta)$ the dispersion of the random variable $\sum_{i=1}^{n} a_i\widehat{\Delta\Theta}_i(\eta)$ is not greater than the dispersion of the random variable $\sum_{i=1}^{n} a_i\widetilde{\Delta\Theta}_i(\eta)$.

In accordance with this definition of the joint efficiency, we are looking for an n-dimensional vector $m = La$ which ensures that the dispersion of the unbiased estimator $a'L'\eta = m'\eta$ will be minimum for each $a \in \mathcal{R}^k$. Since $D[m'\eta] = m'\Sigma m$ and the condition of unbiasedness is $m'JA = a'$ (the original condition of unbiasedness $L'JA = I$ was multiplied from the left by a'), the Lagrange auxiliary function for determining the extremum point of the quadratic form $m'\Sigma m$ with a subsidiary condition $m'JA = a'$ is $\Phi(m) = m'\Sigma m - 2\lambda'(A'J'm - a)$ (λ is a k-dimensional Lagrange undetermined multiplier). Its first derivative with respect to m is $\partial\Phi(m)/\partial m = 2\Sigma m - 2JA\lambda$. The extremum point of the function $\Phi(\cdot)$ is the point m_0 being a solution of the equation $\Sigma m - JA\lambda = 0$, thus $m_0 = \Sigma^{-1}JA\lambda$. The subsidiary condition $a' = m'JA$ ensuring the unbiasedness of the estimator implies $\lambda'A'J'\Sigma^{-1}JA = a$, thus $\lambda = (A'J'\Sigma^{-1}JA)^{-1}a$ and $m_0' = a'(A'J'\Sigma^{-1}JA)^{-1}A'J'\Sigma^{-1}$. The dispersion $D[m_0'\eta] = m_0'\Sigma m_0 = a'(A'J'\Sigma^{-1}JA)^{-1}a$ of the linear unbiased estimator $m_0'\eta = a'(A'J'\Sigma^{-1}JA)^{-1}A'J'\Sigma^{-1}\eta$ is extremum in the class of all estimators fulfilling the condition of unbiasedness $m'JA = a'$.

It remains to prove that this extremum is the minimum. The covariance matrix Σ is positive definite, in accordance with the definition of the positive definiteness for an arbitrary $\alpha \in \mathcal{R}^n$, $\alpha \neq 0$; it satisfies the inequality $\alpha'\Sigma\alpha > 0$. If α is chosen in the form $\alpha = m - m_0$, where m ($m \neq m_0$) fulfils the condition $m'JA = a'$, then $(m - m_0)'\Sigma(m - m_0) > 0$. Because of

$$(m - m_0)'\Sigma(m - m_0) = m'\Sigma m - 2m_0\Sigma m + m_0'\Sigma m_0$$

and

$$m_0'\Sigma m = a'(A'J'\Sigma^{-1}JA)^{-1}A'J'm = a'(A'J'\Sigma^{-1}JA)^{-1}a = m_0'\Sigma m_0,$$

we get

$$(m - m_0)'\Sigma(m - m_0) = m'\Sigma m - m_0'\Sigma m_0$$

and this relation expressing the difference of dispersions of the linear unbiased estimators $m'\eta$ and $m_0'\eta$ is always greater than zero, which implies that the dispersion of the random variable $m_0'\eta = a'L'\eta = a'\widehat{\Delta\Theta}(\eta)$ is minimum in the class of all linear unbiased estimators, which means that the estimator $\hat{\Theta}(\eta) = \Theta_0 + \widehat{\Delta\Theta}(\eta)$ of the parameter $\Theta = \Theta_0 + \Delta\Theta$, where

$$\widehat{\Delta\Theta}(\eta) = (A'J'\Sigma^{-1}JA)^{-1}A'J'\Sigma^{-1}\eta, \tag{2.4.3}$$

is linear unbiased and jointly efficient. Its variance is

$$var[\Theta_0 + \widehat{\Delta\Theta}(\eta)] = (A'J'\Sigma^{-1}JA)^{-1}. \tag{2.4.4}$$

Each linear unbiased and jointly efficient estimate

$$\widehat{\Delta\Theta}(y) = (A'J'\Sigma^{-1}JA)^{-1}A'J'\Sigma^{-1}y$$

of the vector parameter $\Delta\Theta$ minimizes the quadratic form $Q(\Delta\Theta, y) = (JA\Delta\Theta - y)'\Sigma^{-1}(JA\Delta\Theta - y)$, $\Delta\Theta \in \mathcal{R}^k$, for each realization y of the random variable η, i.e., the estimator (2.4.3) is the LS-estimator.

The quadratic form $Q(\cdot, y)$ acquires its extremum for a fixed $y \in \mathcal{R}^n$ at the point $\widetilde{\Delta\Theta}(y)$ which is a solution of the equation

$$(\partial Q(\Delta\Theta, y)/\partial\Delta\Theta =)\ 2A'J'\Sigma^{-1}(JA\Delta\Theta - y) = 0,$$

i.e., at the point $\widetilde{\Delta\Theta}(y) = (A'J'\Sigma^{-1}JA)^{-1}A'J'\Sigma^{-1}y$ being a realization of the estimator (2.4.3), as a consequence of which $\widetilde{\Delta\Theta}(y) = \widehat{\Delta\Theta}(y)$ for each realization y of the random variable η. Futhermore, since $\partial^2 Q(\Delta\Theta, y)/ /\partial\Delta\Theta\partial\Delta\Theta' = 2A'J'\Sigma^{-1}JA$ and the matrix $A'J'\Sigma^{-1}JA$ is positive definite, the quadratic form $Q(\cdot, y)$ acquires its minimum at the point $\widetilde{\Delta\Theta}(y) = \widehat{\Delta\Theta}(y)$.

As a consequence of the universal position of Model 2 among other linear models of measurement, this proof of the theorem of the equivalence of the LS estimator and the linear unbiased and jointly efficient estimator may be considered to be general, valid for each of the models transformable to Model 2.

This theorem enables us to obtain explicit formulas for the linear unbiased and jointly efficient estimator by using the least squares method (the reader can convince himself of the difficulties which arise when we derive them because we require unbiasedness and joint efficiency in the case where the parameter space is a certain linear manifold $\mathcal{T} \subset \mathcal{R}^k$ as a consequence of the system of conditions binding the values of the unknown parameters).

Model 3 (a direct measurement of a vector parameter with a system of conditions)

The relations for the linear unbiased and jointly efficient estimator of the vector parameter $\Theta = \Theta_0 + \Delta\Theta$ in the stochastic model

$$\xi = J\Theta_0 + J\Delta\Theta + \varepsilon, \ \eta = \xi - J\Theta_0, \ b + B\Delta\Theta = 0$$
$$E[\eta] = J\Delta\Theta, \ var(\xi) = var(\eta) = var(\varepsilon) = \Sigma,$$
$$R(B) = q \leq k, \ R(J) = k \leq n, \ R(\Sigma) = n$$

will be derived using the least squares method. A random vector is sought, each realization of which minimizes the quadratic form

$$Q(\Delta\Theta, y) = (J\Delta\Theta - y)'\Sigma^{-1}(J\Delta\Theta - y), \ \Delta\Theta \in \mathcal{T} \subset \mathcal{R}^k,$$

satisfying simultaneously the condition $b + B\Delta\Theta = 0$. This condition restricts the definition domain of the k-dimensional vector parameter $\Delta\Theta$ to the set $\mathcal{T} = \{\Delta\Theta : b + B\Delta\Theta = 0\}$.

The problem is solved using the method of the Lagrange undetermined multipliers. The first partial derivative of the Lagrange auxiliary function $\Phi(\Delta\Theta, y) = (J\Delta\Theta - y)'\Sigma^{-1}(J\Delta\Theta - y) - 2\lambda'(b + B\Delta\Theta)$, where λ is a q-dimensional Lagrange multiplier, yields $\partial\Phi(\Delta\Theta, y)/\partial\Delta\Theta = 2J'\Sigma^{-1}(J\Delta\Theta - y) - 2B'\lambda$, which means that the sought point $\widehat{\Delta\Theta}(y)$, at which the quadratic form acquires an extremum bound by a condition, is a solution of the equation $J'\Sigma^{-1}(J\Delta\Theta - y) - 2B'\lambda = 0$ of the form

$$\widehat{\Delta\Theta}(y) = (J'\Sigma^{-1}J)^{-1}[J'\Sigma^{-1}y + B'\lambda]. \tag{2.4.5}$$

Inserting (2.4.5) into the condition $b + B\Delta\Theta = 0$ we get the equation $b + B(J'\Sigma^{-1}J')^{-1}J'\Sigma^{-1}y + B(J'\Sigma^{-1}J)^{-1}B'\lambda = 0$ for determining λ. It implies

$$\lambda = -[B(J'\Sigma^{-1}J)^{-1}B']^{-1}[b + B(J'\Sigma^{-1}J)^{-1}J'\Sigma^{-1}y],$$

as a consequence of which

$$\widehat{\Delta\Theta}(y) = (J'\Sigma^{-1}J)^{-1}J'\Sigma^{-1}y - (J'\Sigma^{-1}J)^{-1}B'[B(J'\Sigma^{-1}J)^{-1}B']^{-1}\cdot$$
$$\cdot[b - B(J'\Sigma^{-1}J)^{-1}J'\Sigma^{-1}y]. \tag{2.4.6}$$

The next problem is to show that the estimator of the form (2.4.6) is a minimum of the quadratic form $Q(\cdot, y)$ under the subsidiary condition, i.e., to prove that for any other linear unbiased estimator $\widetilde{\Delta\Theta}(y)$ such that $b + B\widetilde{\Delta\Theta}(y) = 0$

$$(J\widetilde{\Delta\Theta}(y) - y)'\Sigma^{-1}(J\widetilde{\Delta\Theta}(y) - y) - (J\widehat{\Delta\Theta}(y) - y)'\Sigma^{-1}(J\widehat{\Delta\Theta}(y) - y) > 0.$$

We will use the same procedure to prove this inequality as in the previous model. It consists in utilizing the fact that the matrix Σ^{-1} is positive definite. In accordance with the definition, the positive definite matrix Σ^{-1} must satisfy the inequality $a'\Sigma^{-1}a > 0$ for any non-zero $a \in \mathcal{R}^n$. For an a chosen in the form $a = (J\widetilde{\Delta\Theta} - y) - (J\widehat{\Delta\Theta} - y)$ (for the sake of simplicity, the argument of the estimates considered is omitted in the next part)

$$[(J\widetilde{\Delta\Theta} - y) - (J\widehat{\Delta\Theta} - y)]'\Sigma^{-1}[(J\widetilde{\Delta\Theta} - y) - (J\widehat{\Delta\Theta} - y)] > 0. \tag{2.4.7}$$

Since

$$[(J\widetilde{\Delta\Theta} - y) - (J\widehat{\Delta\Theta} - y)]'\Sigma^{-1}[(J\widetilde{\Delta\Theta} - y) - (J\widehat{\Delta\Theta} - y)] =$$

$$= (J\widetilde{\Delta\Theta} - y)'\Sigma^{-1}(J\widetilde{\Delta\Theta} - y) - 2(J\widetilde{\Delta\Theta} - y)'\Sigma^{-1}(J\widehat{\Delta\Theta} - y) +$$
$$+ (J\widehat{\Delta\Theta} - y)'\Sigma^{-1}(J\widehat{\Delta\Theta} - y)$$

the problem now consists in investigating the expression $(J\widetilde{\Delta\Theta} - y)'\Sigma^{-1}$ $[(J\widehat{\Delta\Theta} - y)$. Expressing $\widehat{\Delta\Theta}$ here by means of (2.4.5) (for simplicity in writing we shall not use the explicit expression (2.4.6)), we obtain

$$(J\widetilde{\Delta\Theta} - y)'\Sigma^{-1}(J\widehat{\Delta\Theta} - y) = (J\widetilde{\Delta\Theta} - y)'\Sigma^{-1}[J(J'\Sigma^{-1}J)^{-1}(J\Sigma^{-1}y + B'\lambda) - y].$$

Since

$$\widetilde{\Delta\Theta}' J'\Sigma^{-1}(J\widehat{\Delta\Theta} - y) = \widetilde{\Delta\Theta}' J'\Sigma^{-1}J(J'\Sigma^{-1}y + B'\lambda) - y] =$$

$$= \widetilde{\Delta\Theta}(J'\Sigma^{-1}y + B'\lambda - J'\Sigma^{-1}y) = \widetilde{\Delta\Theta}' B'\lambda,$$

when simultaneously $b' + \widetilde{\Delta\Theta}' B' = 0$ and $b' + \widehat{\Delta\Theta}' B' = 0$, we see that $\widetilde{\Delta\Theta}' B' = \widehat{\Delta\Theta}' B'$. Hence, the equalities

$$\widetilde{\Delta\Theta}' J'\Sigma^{-1}(J\widehat{\Delta\Theta} - y) = \widetilde{\Delta\Theta}' B' = \widehat{\Delta\Theta}' B' = \widehat{\Delta\Theta}' J'\Sigma^{-1}(J\widehat{\Delta\Theta} - y)$$

implying immediately that

$$(J\widetilde{\Delta\Theta} - y)'\Sigma^{-1}(J\widehat{\Delta\Theta} - y) = (J\widehat{\Delta\Theta} - y)'\Sigma^{-1}(J\widehat{\Delta\Theta} - y)$$

are proved, thus

$$(J\widetilde{\Delta\Theta} - y)'\Sigma^{-1}(J\widetilde{\Delta\Theta} - y) - 2(J\widetilde{\Delta\Theta} - y)'\Sigma^{-1}(J\widehat{\Delta\Theta} - y) +$$

$$+ (J\widehat{\Delta\Theta} - y)'\Sigma^{-1}(J\widehat{\Delta\Theta} - y) =$$
$$= (J\widetilde{\Delta\Theta} - y)'\Sigma^{-1}(J\widetilde{\Delta\Theta} - y) - (J\widehat{\Delta\Theta} - y)'\Sigma^{-1}(J\widehat{\Delta\Theta} - y)$$

and this difference is positive definite as a consequence of the relation (2.4.7).

The estimate of the form (2.4.6) which minimizes for each realization $y \in \mathcal{R}^n$ the quadratic form $Q(\cdot, y)$ and fulfils simultaneously the condition $b + B\widehat{\Delta\Theta}(y) = 0$ is an LS-estimate and due to the theorem of the equivalence, it is a linear unbiased and jointly efficient estimator of the parameter $\Delta\Theta$. Thus the linear unbiased and jointly efficient estimator of the vector parameter $\Theta = \Theta_0 + \Delta\Theta$ within Model 3 is

$$\hat{\Theta}(\eta) = \Theta_0 + (J'\Sigma^{-1}J)^{-1}\Sigma^{-1}\eta -$$

$$-(J'\Sigma^{-1}J)^{-1}B'[B(J'\Sigma^{-1}J)^{-1}B']^{-1}[b+B(J'\Sigma^{-1}J)^{-1}J'\Sigma^{-1}\eta] \tag{2.4.8}$$

and its variance is

$$var[\hat{\Theta}(\eta)] = var[\widehat{\Delta\Theta}(\eta)] =$$
$$= (J'\Sigma^{-1}J)^{-1} - (J'\Sigma^{-1}J)^{-1}B'[B(J'\Sigma^{-1}J)^{-1}B']^{-1}B(J'\Sigma^{-1}J)^{-1}. \tag{2.4.9}$$

If we observe in detail the structure of the relationship (2.4.8), we see that it is split into two parts, namely $(J'\Sigma^{-1}J)\Sigma^{-1}\eta$, which is an estimator of the parameter $\Delta\Theta$ without taking into account the conditions to be fulfilled by its components, and the term $-(J'\Sigma^{-1}J)^{-1}B'[B(J'\Sigma^{-1}J)^{-1}B']^{-1}\cdot [b+B(J'\Sigma^{-1}J)^{-1}J'\Sigma^{-1}\eta$ expressing a correction for the condition $b + B\Delta\Theta = 0$ to be satisfied.

If each directly observable parameter of the vector Θ is measured just once, $J = I$ (a unit matrix of the type $k \times k$) and the relations (2.4.8) and (2.4.9) for the LS-estimator of the parameter Θ and its variance assume the following simpler form

$$\hat{\Theta}(\eta) = \Theta_0 + \eta - B'(B\Sigma B')^{-1}(b + B\eta)$$

and

$$var[\hat{\Theta}(\eta)] = var[\widehat{\Delta\Theta}(\eta)] = \Sigma - \Sigma B'(B\Sigma B')^{-1}B\Sigma.$$

Model 4 (an incomplete direct measurement of a vector parameter with a system of conditions)

The relations for the linear unbiased and jointly efficient estimator of the vector parameter $(\Theta_1', \Theta_2')' = (\Theta_{10}', \Theta_{20}')' + (\Delta\Theta_1', \Delta\Theta_2')'$ in the stochastic model

$$\xi = J\Theta_{10} + J\Delta\Theta_1 + \varepsilon,\ \eta = \xi - J\Theta_{10},\ b + B\Delta\Theta_1 + C\Delta\Theta_2 = 0$$
$$E[\eta] = J\Delta\Theta_1,\ var(\xi) = var(\eta) = var(\varepsilon) = \Sigma,$$
$$R[(B, C)] = q \le k,\ R(C) = k - p \le q,\ R(J) = k \le n,\ R(\Sigma) = n$$

will be determined using the least squares method. In accordance with it, a point $\Delta\Theta_1$ of the manifold $\mathcal{T} = \{\Delta\Theta_1 : b + B\Delta\Theta_1 + C\Delta\Theta_2 = 0\}$ of the space $\mathcal{R}^p$ is sought, which for each realization y of the random vector η minimizes the quadratic form

$$Q(\Delta\Theta_1, y) = (J\Delta\Theta_1 - y)'\Sigma^{-1}(J\Delta\Theta_1 - y), \; \Delta\Theta_1 \in \mathcal{T} \subset \mathcal{R}^p.$$

It is a problem of determining the extremum of a function with a subsidiary condition, therefore the method of the Lagrange undetermined multipliers is applied. The Lagrange auxiliary function is of the form $\Phi(\Delta\Theta_1, \Delta\Theta_2, y) = (J\Delta\Theta_1 - y)'\Sigma^{-1}(J\Delta\Theta_1 - y) - 2\lambda'(b + B\Delta\Theta_1 + C\Delta\Theta_2)$, where λ is a q-dimensional Lagrange multiplier, its first partial derivatives with respect to $\Delta\Theta_1$ and $\Delta\Theta_2$ are $\partial\Phi(\Delta\Theta_1, \Delta\Theta_2, y)/\partial\Delta\Theta_1 = 2J'\Sigma^{-1}(J\Delta\Theta_1 - y) - 2B'\lambda$, $\partial\Phi(\Delta\Theta_1, \Delta\Theta_2, y)/\partial\Delta\Theta_2 = -2C'\lambda$; the bound extremum of the quadratic form $Q(\cdot, \cdot\cdot, y)$ occurs at the point

$$\widehat{\Delta\Theta}_1(y) = (J'\Sigma^{-1}J)^{-1}(J'\Sigma^{-1}y + B'\lambda), \tag{2.4.10}$$

which is a solution of the equation $J'\Sigma^{-1}(J\Delta\Theta_1 - y) - B'\lambda = 0$. The q-dimensional Lagrange coefficient λ and the parameter $\Delta\Theta_2$ can be determined from the system of equations

$$\begin{gathered} B(J'\Sigma^{-1}J)^{-1}B'\lambda + C\Delta\Theta_2 = -[b + B(J'\Sigma^{-1}J)^{-1}J'\Sigma^{-1}y], \\ C'\lambda = 0 \end{gathered} \tag{2.4.11}$$

the former being obtained by inserting $\Delta\Theta_1$ of the form (2.4.10) into the condition $b + B\Delta\Theta_1 + C\Delta\Theta_2 = 0$ and the latter is the annulled first derivative of the Lagrange auxiliary function with respect to the parameter $\Delta\Theta_2$. In order to get a solution for this system of equations we have to invert the square matrix

$$\begin{pmatrix} B(J'\Sigma^{-1}J)^{-1}B', & C \\ C', & 0 \end{pmatrix} \tag{2.4.12}$$

of the type $q + k - p$ (0 in (2.4.12) is a zero square matrix of the type $k - p$). First, we shall convince ourselves that this matrix is regular. The assumptions $R[(B,C)] = q$ and $R(C) = k - p$ imply that the $q \times q$ matrix B contains $q-(k-p)$ independent columns; since $R(J) = p$ and $R(\Sigma) = n$, the square matrix $(J'\Sigma^{-1}J)^{-1}$ of the type p is regular, the square matrix $B(J'\Sigma^{-1}J)^{-1}B'$ of the type $q \times q$ contains $q-(k-p)$ independent columns or rows (this matrix is not regular); since $R[(B,C)] = q$ and simultaneously $R(C) = k - p$, the columns of the matrix $B(J'\Sigma^{-1}J)^{-1}B'$ are independent of the columns of the matrix C, thus all the columns and the rows of the matrix (2.4.12) are independent, thus it is regular. Its inverse

$$\begin{pmatrix} B(J'\Sigma^{-1}J)^{-1}B', & C \\ C', & 0 \end{pmatrix}^{-1} = \begin{pmatrix} Q_{11}, & Q_{12} \\ Q_{21}, & Q_{22} \end{pmatrix} \tag{2.4.13}$$

exists, however there does not exist a possibility that we could enumerate it in an explicit form using the rules for an inverse of a block matrix[4], since neither the matrix $B(J'\Sigma^{-1}J)^{-1}B'$ nor the zero matrix 0 are regular. A solution of the equation system (2.4.11) yields

$$\begin{pmatrix} \lambda \\ \widehat{\Delta\Theta}_2(y) \end{pmatrix} = \begin{pmatrix} -Q_{11}[b + B(J'\Sigma^{-1}J)^{-1}J'\Sigma^{-1}y] \\ -Q_{21}[b + B(J'\Sigma^{-1}J)^{-1}J'\Sigma^{-1}y] \end{pmatrix},$$

which finally yields the point of extremum of the quadratic form $Q(\Delta\Theta_1, y) = (J\Delta\Theta_1 - y)'\Sigma^{-1}(J\Delta\Theta_1 - y)$ satisfying simultaneously the condition $b + B\Delta\Theta_1 + C\Delta\Theta_2 = 0$ in the form

$$\begin{pmatrix} \widehat{\Delta\Theta}_1(y) \\ \widehat{\Delta\Theta}_2(y) \end{pmatrix} = \begin{pmatrix} -(J'\Sigma^{-1}J)^{-1}B'Q_{11} \\ -Q_{21} \end{pmatrix} b +$$

$$+ \begin{pmatrix} (J'\Sigma^{-1}J)^{-1}[I - B'Q_{11}B(J'\Sigma^{-1}J)^{-1}] \\ -Q_{21}B(J'\Sigma^{-1}J)^{-1} \end{pmatrix} J'\Sigma^{-1}y.$$

In order to prove that this is a minimum point of the quadratic form with the subsidiary conditions, it has to be shown that every other estimator $(\widetilde{\Delta\Theta}_1{}', \widetilde{\Delta\Theta}_2{}')'$ such that $b + B\widetilde{\Delta\Theta}_1{}' + C\widetilde{\Delta\Theta}_2 = 0$ satisfies the inequality

$$(J\widehat{\Delta\Theta}_1 - y)'\Sigma^{-1}(J\widehat{\Delta\Theta}_1 - y) < (J\widetilde{\Delta\Theta}_1 - y)'\Sigma^{-1}(J\widetilde{\Delta\Theta}_1 - y).$$

[4]For the inverse of a regular block matrix see p. 50, where the conditions for its existence are stated.

This assertion will be proved using an analogous procedure, as in the previous case of Model 3: The covariance matrix Σ is positive definite, therefore for each $\widetilde{\Delta\Theta}_1 \neq \widehat{\Delta\Theta}_1$ at an arbitrary but fixed y

$$[(J\widetilde{\Delta\Theta}_1 - y) - (J\widehat{\Delta\Theta}_1 - y)]'\Sigma^{-1}[(J\widetilde{\Delta\Theta}_1 - y) - (J\widehat{\Delta\Theta}_1 - y)] > 0.$$

Specifying the left-hand side of this inequality for single quadratic forms, we see that the problem is to investigate the term $(J\widetilde{\Delta\Theta}_1 - y)'\Sigma^{-1}(J\widehat{\Delta\Theta}_1 - y)$. Expressing $\widehat{\Delta\Theta}_1$ here by means of (2.4.10) (for the simplicity of the record again the relatively complicated explicit previous expression for $\widehat{\Delta\Theta}_1(y)$ is not used), we obtain $(J\widetilde{\Delta\Theta}_1 - y)'\Sigma^{-1}(J\widehat{\Delta\Theta}_1 - y) = (J\widetilde{\Delta\Theta}_1 - y)\Sigma^{-1}[J(J'\Sigma^{-1}J)^{-1}\ (J'\Sigma^{-1}y + B'\lambda) - y]$. Since $\widetilde{\Delta\Theta}_1{}'J'\Sigma^{-1}(J\widehat{\Delta\Theta}_1 - y) = \widetilde{\Delta\Theta}_1{}'J'\Sigma^{-1}[J(J'\Sigma^{-1}J)^{-1}(J'\Sigma^{-1}y + B'\lambda) - y] = \widetilde{\Delta\Theta}_1{}'J'\Sigma^{-1}J(J'\Sigma^{-1}J)^{-1}$ $(J'\Sigma^{-1}y + B'\lambda) - \widetilde{\Delta\Theta}_1 J'\Sigma^{-1}y = \widetilde{\Delta\Theta}_1{}' \cdot B'\lambda$ when simultaneously $b + B\widehat{\Delta\Theta}_1 + C\widehat{\Delta\Theta}_2 = 0$ and $b + B\widetilde{\Delta\Theta}_1 + C\widetilde{\Delta\Theta}_2 = 0$, we obtain $\widetilde{\Delta\Theta}_1{}'B'\lambda = \{\widehat{\Delta\Theta}_1{}'B' + [\widehat{\Delta\Theta}_2{}' - \widetilde{\Delta\Theta}_2{}']C'\}\lambda$; moreover, since $C'\lambda = 0$, we see that $\widetilde{\Delta\Theta}_1{}'B'\lambda = \widehat{\Delta\Theta}_1{}'B'\lambda$ and simultaneously

$$(J\widetilde{\Delta\Theta}_1 - y)'\Sigma^{-1}(J\widehat{\Delta\Theta}_1 - y) = (J\widehat{\Delta\Theta}_1 - y)'\Sigma^{-1}(J\widehat{\Delta\Theta}_1 - y).$$

By this procedure the proof of the assertion that in Model 4 the estimators of the unknown parameters $\Theta_1 = \Theta_{10} + \Delta\Theta_1$ and $\Theta_2 = \Theta_{20} + \Delta\Theta_2$ of the form

$$\begin{pmatrix} \hat{\Theta}_1(\eta) \\ \hat{\Theta}_2(\eta) \end{pmatrix} = \begin{pmatrix} \Theta_{10} \\ \Theta_{20} \end{pmatrix} + \begin{pmatrix} (J'\Sigma^{-1}J)^{-1} \\ 0 \end{pmatrix} J'\Sigma^{-1}\eta - \\ - \begin{pmatrix} (J'\Sigma^{-1}J)^{-1}B'Q_{11} \\ Q_{21} \end{pmatrix} [b + B(J'\Sigma^{-1}J)^{-1}J'\Sigma^{-1}\eta] \tag{2.4.14}$$

are the LS-estimators is concluded. The variance is

$$var\begin{pmatrix} \hat{\Theta}_1(\eta) \\ \hat{\Theta}_2(\eta) \end{pmatrix} =$$

$$= \begin{pmatrix} (J'\Sigma^{-1}J)^{-1} - (J'\Sigma^{-1}J)^{-1}B'Q_{11}B(J'\Sigma^{-1}J)^{-1} \\ -Q_{21}B(J'\Sigma^{-1}J)^{-1} \end{pmatrix} J'\Sigma^{-1}J\cdot$$

$$\cdot[(J'\Sigma^{-1}J)^{-1} - (J'\Sigma^{-1}J)^{-1}B'Q_{11}B(J'\Sigma^{-1}J)^{-1}, -(J'\Sigma^{-1}J)^{-1}B'Q_{12}] =$$

$$= \begin{pmatrix} var[\hat{\Theta}_1(\eta)], & cov[\hat{\Theta}_1(\eta), \hat{\Theta}_2(\eta)] \\ cov[\hat{\Theta}_2(\eta), \hat{\Theta}_1(\eta)], & var[\hat{\Theta}_2(\eta)] \end{pmatrix},$$

where

$$var[\hat{\Theta}_1(\eta)] = (J'\Sigma^{-1}J)^{-1}[I - B'Q_{11}B(J'\Sigma^{-1}J)^{-1}]J'\Sigma^{-1}J\cdot$$

$$\cdot[I - (J'\Sigma^{-1}J)^{-1}B'Q'_{11}B](J'\Sigma^{-1}J)^{-1} =$$

$$= (J'\Sigma^{-1}J)^{-1} - 2(J'\Sigma^{-1}J)^{-1}B'Q_{11}B(J'\Sigma^{-1}J)^{-1} +$$

$$+(J'\Sigma^{-1}J)^{-1}B'Q_{11}B(J'\Sigma^{-1}J)^{-1}B'Q'_{11}B(J'\Sigma^{-1}J)^{-1},$$

$$cov[\hat{\Theta}_1(\eta), \hat{\Theta}_2(\eta)] =$$

$$-(J'\Sigma^{-1}J)^{-1}B'Q'_{21} + (J'\Sigma^{-1}J)^{-1}B'Q_{11}B(J'\Sigma^{-1}J)^{-1}B'Q'_{21},$$

$$cov[\hat{\Theta}_2(\eta), \hat{\Theta}_1(\eta)] =$$

$$-Q_{21}B(J'\Sigma^{-1}J)^{-1} + Q_{21}B(J'\Sigma^{-1}J)^{-1}B'Q'_{11}B(J'\Sigma^{-1}J)^{-1},$$

and

$$var[\hat{\Theta}_2(\eta)] = Q_{21}B(J'\Sigma^{-1}J)^{-1}B'Q'_{21}.$$

The expression for the variance can be simplified by the use of relations following from the fact that the matrix $\begin{pmatrix} Q_{11}, & Q_{12} \\ Q_{21}, & Q_{22} \end{pmatrix}$ is an inverse of the matrix $\begin{pmatrix} B(J'\Sigma^{-1}J)^{-1}B', & C \\ C', & 0 \end{pmatrix}$, i.e., that

$$\begin{pmatrix} Q_{11}, & Q_{12} \\ Q_{21}, & Q_{22} \end{pmatrix} \begin{pmatrix} B(J'\Sigma^{-1}J)^{-1}B', & C \\ C', & 0 \end{pmatrix} = \begin{pmatrix} I, & 0 \\ 0, & I \end{pmatrix}.$$

The system of equations $Q_{11}B(J'\Sigma^{-1}J)^{-1}B' + Q_{12}C' = I$ and $Q_{11}C = 0$, the former multiplied from the left by Q_{11}, yields

$$Q_{11}B(J'\Sigma^{-1}J)^{-1}B'Q_{11} + Q_{21}C'Q_{11} = Q_{11}$$

and (utilizing the latter)

$$Q_{11}B(J'\Sigma^{-1}J)^{-1}B'Q_{11} = Q_{11}$$

(simultaneously the fact that the input matrix and its inverse are symmetric has been taken into account). In an analogous way the system of equations

$Q_{21}B(J'\Sigma^{-1}J)^{-1}B'+Q_{22}C' = 0$ and $Q_{21}C = I$, the former being multiplied from the right by the matrix Q'_{21}, yields

$$Q_{21}B(J'\Sigma^{-1}J)^{-1}B'Q'_{21} = -Q_{22}.$$

Similarly the system of equations $Q_{21}B(J'\Sigma^{-1}J)^{-1}B' + Q_{22}C' = 0$ and $Q_{11}C = 0$ gives

$$Q_{21}B(J'\Sigma^{-1}J)^{-1}B'Q_{11} = 0.$$

Inserting these relations into the expression for $var[\hat{\Theta}_1(\eta)]$, $cov[\hat{\Theta}_1(\eta), \hat{\Theta}_2(\eta)]$, $cov[\hat{\Theta}_2(\eta), \hat{\Theta}_1(\eta)]$ and $var[\hat{\Theta}_2(\eta)]$, we obtain the following final expressions for the variance of the linear unbiased and jointly efficient estimator (2.4.14) of the unknown parameters within Model 4 in the form

$$var\begin{pmatrix} \hat{\Theta}_1(\eta) \\ \hat{\Theta}_2(\eta) \end{pmatrix} =$$
$$\begin{pmatrix} (J'\Sigma^{-1}J)^{-1} - (J'\Sigma^{-1}J)^{-1}B'Q_{11}B(J'\Sigma^{-1}J)^{-1}, & -(J'\Sigma^{-1}J)^{-1}B'Q_{12} \\ -Q_{21}B(J'\Sigma^{-1}J)^{-1}, & -Q_{22} \end{pmatrix} \quad (2.4.15)$$

Model 5 (an indirect measurement of a vector parameter with a system of conditions)

For determining the linear unbiased and jointly efficient estimator of the vector parameter $\Theta = \Theta_0 + \Delta\Theta$ in the stochastic model

$$\xi = Jf(\Theta_0) + JA\Delta\Theta + \varepsilon,\ \eta = \xi - Jf(\Theta_0),\ b + B\Delta\Theta_0 = 0,$$
$$E[\eta] = JA\Delta\Theta,\ var(\xi) = var(\eta) = var(\varepsilon) = \Sigma,$$
$$R(A) = k \leq p,\ R(B) = q \leq k,\ R(J) = k \leq n,\ R(\Sigma) = n,$$

we apply the least squares method. The resulting expressions need not to be derived in detail, they are a simple modification of the corresponding relations from Model 3. It suffices to substitute the matrix JA whose rank equals $R(JA) = k$ for the matrix J in all relations true for Model 3. In such a way the expressions (2.4.8) and (2.4.9) yield the sought linear unbiased and jointly efficient estimator of the parameter Θ in the form

$$\hat{\Theta}(\eta) = \Theta_0 + (A'J'\Sigma^{-1}JA)^{-1}A'J'\Sigma^{-1}\eta - (A'J'\Sigma^{-1}JA)^{-1}B' \cdot [B(A'J'\Sigma^{-1}JA)^{-1}B']^{-1} \cdot \cdot [b + B(A'J'\Sigma^{-1}JA)^{-1}A'J'\Sigma^{-1}\eta], \tag{2.4.16}$$

and its variance is

$$var[\hat{\Theta}(\eta)] = (A'J'\Sigma^{-1}JA)^{-1} - (A'J'\Sigma^{-1}JA)^{-1} \cdot \cdot B'[B(A'J'\Sigma^{-1}JA)^{-1}B']^{-1}B(A'J'\Sigma^{-1}JA)^{-1}. \tag{2.4.17}$$

Model 6 (an incomplete indirect measurement of a vector parameter with a system of conditions)

In the same way as in the previous case of Model 5, the relations for the linear unbiased and jointly efficient estimator of the vector parameter $\Theta = (\Theta_1', \Theta_2')' = (\Theta_{10}', \Theta_{20}')' + (\Delta\Theta_1', \Delta\Theta_2')'$ in the stochastic model

$$\xi = J[f(\Theta_{10}) + A\Delta\Theta_1] + \varepsilon,\ \eta = \xi - Jf(\Theta_{10}),\ b + B\Delta\Theta_1 + C\Delta\Theta_2 = = 0,$$
$$E[\eta] = JA\Delta\Theta_1,\ var(\xi) = var(\eta) = var(\varepsilon) = \Sigma,$$
$$R(A) = l \le p,\ R[(B, C)] = q \le k,\ R(C) = k - l \le q,$$
$$R(J) = p \le n,\ R(\Sigma) = n,$$

are a simple modification of the corresponding expressions from Model 4; it suffices to substitute the matrix JA for the matrix J, $R(JA) = l$. The mentioned modification of relations (2.4.14) and (2.4.15) yields the expressions for the LS-estimators of the parameters Θ_1 and Θ_2 in the form

$$\begin{pmatrix} \hat{\Theta}_1(\eta) \\ \hat{\Theta}_2(\eta) \end{pmatrix} = \begin{pmatrix} \Theta_{10} \\ \Theta_{20} \end{pmatrix} + \begin{pmatrix} (A'J'\Sigma^{-1}JA)^{-1} \\ 0 \end{pmatrix} A'J'\Sigma^{-1}\eta -$$

$$-\begin{pmatrix} (A'J'\Sigma^{-1}JA)^{-1}B'Q_{11} \\ Q_{21} \end{pmatrix} [b + B(A'J'\Sigma^{-1}JA)^{-1}A'J'\Sigma^{-1}\eta] \tag{2.4.18}$$

and its variance matrix

$$var\begin{pmatrix} \hat{\Theta}_1(\eta) \\ \hat{\Theta}_2(\eta) \end{pmatrix} = \begin{pmatrix} var[\hat{\Theta}_1(\eta)], & cov[\hat{\Theta}_1(\eta), \hat{\Theta}_2(\eta)] \\ cov[\hat{\Theta}_2(\eta), \hat{\Theta}_1(\eta)], & var[\hat{\Theta}_2(\eta)] \end{pmatrix} =$$
$$= \begin{pmatrix} (A'J'\Sigma^{-1}JA)^{-1} - (A'J'\Sigma^{-1}JA)^{-1}B'Q_{11}B(A'J'\Sigma^{-1}JA)^{-1}, \\ -Q_{21}B(A'J'\Sigma^{-1}JA)^{-1}, \end{pmatrix.}$$
$$\begin{matrix} -(A'J'\Sigma^{-1}JA)^{-1}B'Q_{12} \\ -Q_{22} \end{matrix}\Bigg), \tag{2.4.19}$$

where

$$\begin{pmatrix} Q_{11}, & Q_{12} \\ Q_{21}, & Q_{22} \end{pmatrix} = \begin{pmatrix} B(A'J'\Sigma^{-1}JA)^{-1}B', & C \\ C', & 0 \end{pmatrix}^{-1}.$$

This inverse is unique, even if it cannot be determined by an explicit formula.

2.5 Mixed linear models

Many problems in the natural sciences and technology must be modelled by using a mixed linear model. In comparison with stochastic models described in Sections 2.2 and 2.4, the mixed model is general in the sense that results of measurement are assumed to be perturbed, not only by the error component ε, but also by a further random component φ, whose mean value can but need not be zero. The random vector φ is bound with the measured object and is its typical and non-separable part. This is a reason for denoting the sum of the trend component (i.e., the mean value of the observed random vector ζ) and the random vector φ as the useful component of the measurement (the useful signal) in contradistinction to ε, denoted as the disturbing component (the noise). A mixed model, as contrasted with the simple linear models from Sections 2.2 and 2.4, enables us to determine not only the optimum estimators of the mean value parameters, including their characteristics for accuracy, but enables us to get an optimum prediction for the useful signal both at the set of observation points and outside of it by using some a priori information about the observed random variable ζ and its useful component.

Before dealing with mathematical problems associated with deriving the necessary relations, we shall discuss several situations typically encountered when generating the mixed model (in geodetic practice the term collocation model and in geophysics the term kriging is often used).

Data processing can start at the point that the theoretical model represents the relations between the actual values of the unknown parameters and the theoretical values of the directly measured quantity. This model is, of course, only an approximation of reality, the degree to which it reflects the level of our knowledge of the measured object. The result of a measurement, in addition to the measurement error, contains information about all sorts of influences connected with circumstances not involved in the theoretical model (the main reasons for difficulty with model design can be traced to inexpedient complexity in the resulting model or to the fact that not all data necessary for it are available).

In other words, simple linear models presume a good knowledge of the theory about the subject being investigated, i.e., a good knowledge of the functions $f(\cdot)$ and $g(\cdot)$, see p. 108, which is not the case in many situations; in the extreme case, the purpose of measuring is to obtain preliminary information concerning them.

Thus, the results of the measurement of a directly observable quantity may be affected not only by the measurement errors but by a whole series of further factors whose influence on the measured value is not known sufficiently; for this reason the effect becomes random and as such must be studied.

An investigation of the procedure and the rules for settling large buildings by means of geodetic methods can serve as an example of a mixed linear model. Let a theoretically reasonable curve representing the settling corresponding to a concrete type of the subsoil and a given way of loading it be known. The model situation cannot be identical with the concrete structure of the subsoil in the given terrain, moreover it cannot be mapped in three-dimensional space. The theoretically reasonable trend component of the settling of the foundations, whose parameters are optimumly estimable from the measured data set(the experiment consists in replicated levelling measurements) yields a global view only of the settlement. Moreover, the theoretical model is often not known a priori with a sufficient amount of detail, which leads to seeking the form of the settlement curve which fits as well as possible the measured data. The trend component cannot be used for the optimum prediction of the foundation settlement of the building at the given moment. For this purpose, a useful signal has to be applied, since it reflects the concrete properties of the concrete subsoil under the concrete way of loading it.

The mixed model simulates the problems which arise within an optimum interconnection of geodetic networks, where the stochastic term φ indicates that in the joined or supplementary networks the coordinates of the previous stage are the optimum estimates of the theoretical values (i.e., they are realizations of random variables, not deterministic values).

As another example, the problem of interpolating or extrapolating time or space spread continuous phenomena on the basis of measurements performed at discrete points in their definition domains can be taken into account. Here, moreover, the maximum suppression of the noise component of the measurement can be required.

Summarizing these considerations, we see that the theoretical model $X = f(\Theta)$, $g(\Theta) = 0$ (see the relations (2.1.1′) and (2.1.2′)) is taken into account again. If the measurement is affected only by the measurement errors, the corresponding stochastic model is of the form

$$\xi = Jf(\Theta) + \varepsilon,\ g(\Theta) = 0,\ E[\xi] = Jf(\Theta),\ var(\xi) = var(\varepsilon) = \Sigma.$$

If the measurement is assumed to be perturbed also by a stochastic component φ, $E[\varphi] = s(\vartheta)$, where $s(\cdot)$ is an n-dimensional vector function of known form, ϑ is an s-dimensional vector of the unknown parameters of the mean value of the random vector $\varphi = s(\vartheta) + \varphi_0$, we get a stochastic model

$$\zeta = Jf(\Theta) + s(\vartheta) + \varphi_0 + \varepsilon,\ g(\Theta) = 0,\ E[\zeta] = Jf(\Theta) + s(\vartheta), \tag{2.5.1}$$

$$var(\zeta) = var(\varphi_0) + var(\varepsilon) = var(\varphi) + var(\varepsilon),$$

$$cov(\varphi_0, \varepsilon) = cov(\varphi, \varepsilon) = 0,$$

called a mixed linear model. The sum $Jf(\Theta) + \varphi$ represents the useful component and ε the noise of the random vector modelling the measurement. If the random component φ vanishes, the model (2.5.1) reverts to its above mentioned simpler version. We assume no function at dependence either between the components of the vector parameter Θ or between the components of the vector parameters Θ and ϑ mutually.

If the functions $f(\cdot)$, $s(\cdot)$ and $g(\cdot)$ are not linear, they can be linearized using the procedures described in Section 2.2:

$$f_{lin}(\Theta) = f(\Theta_0) + (\partial f(\Theta_0)/\partial\Theta')\Delta\Theta = f(\Theta_0) + A\Delta\Theta,$$
$$g_{lin}(\Theta) = g(\Theta_0) + (\partial g(\Theta_0)/\partial\Theta')\Delta\Theta = b + B\Delta\Theta,$$
$$s_{lin}(\vartheta) = s(\vartheta_0) + (\partial s(\vartheta_0)/\partial\vartheta')\Delta\vartheta = s(\vartheta_0) + S\Delta\vartheta;$$

the matrices A and B and the vector b are explained in detail in Sections 2.2 and 2.4; the matrix S is an $n \times s$-dimensional matrix such that $R(S) = s \leq n$ and simultaneously $R(JA, S) = k + s \leq n$.

These modifications result in the model

$$\begin{aligned}
&\zeta = Jf(\Theta_0) + s(\vartheta_0) + JA\Delta\Theta + S\Delta\vartheta + \varphi_0 + \varepsilon,\ b + B\Delta\Theta = 0,\\
&E[\zeta] = Jf(\Theta_0) + s(\vartheta_0) + JA\Delta\Theta + S\Delta\vartheta,\ E[\varphi_0] = 0,\ E[\varepsilon] = 0,\\
&var(\zeta) = var(\varphi_0) + var(\varepsilon),\ cov(\varphi_0, \varepsilon) = 0,\\
&R(A) = k \leq p,\ R(S) = s \leq n,\ R(J) = p \leq n,\\
&R[(JA, S)] = k + s \leq n,\ R(B) = q \leq k,
\end{aligned} \tag{2.5.1'}$$

in which the regularity conditions ensure estimability of the parameters Θ and ϑ and, simultaneously, the existence of all matrix inverses necessary for computing the optimum estimates.

Within a statistical model the problem is to determine the linear unbiased and jointly efficient estimator of the $(k+s)$-dimensional vector parameter $(\Theta', \vartheta')'$ and the optimum (i.e., a linear unbiased with the minimum residual dispersion) approximation[5], if need be, of the useful signal at any point in its definition domain, thus also at the points lying outside of the set of observation points.

2.5.1 A linear unbiased and jointly efficient estimator of the vector $(\Theta', \vartheta')'$

The next problem is to represent explicit formulas of linear unbiased and jointly efficient estimators of the vector parameter $(\Theta', \vartheta')'$ within the

[5]The notions "an approximation of a random variable" and "an estimator of an unknown parameter" will be distinguished on principle.

six basic linear models of measurement and to determine its fundamental statistical characteristics of accuracy, i.e., $var(\hat{\Theta})$, $var(\hat{\vartheta})$ and $cov(\hat{\Theta}, \hat{\vartheta})$. The relations for the optimum estimators of the hypervector $(\Theta', \vartheta')'$ in all linear models of measurement are a generalization of the corresponding relations derived in Section 2.4.

A comparison of structures of the corresponding stochastic models in their simple version from Section 2.4 and a mixed version of Section 2.5 imply (the comparison is made for measurement Model 2 as a consequence of its universal position among the other models of measurement) that instead of a simple model

$$\eta = JA\Delta\Theta + \varepsilon, \ \eta = \xi - Jf(\Theta_0), \ E[\eta] = JA\Delta\Theta,$$

$$var(\xi) = var(\eta) = var(\varepsilon) = \Sigma,$$

the model

$$\nu = JA\Delta\Theta + S\Delta\vartheta + \varphi_0 + \varepsilon, \ \nu = \zeta - Jf(\Theta_0) - s(\vartheta_0),$$

$$E[\nu] = JA\Delta\Theta + S\Delta\vartheta = (A, S)\begin{pmatrix} \Delta\Theta \\ \Delta\vartheta \end{pmatrix},$$

$$var(\zeta) = var(\nu) = var(\varphi_0) + var(\varepsilon),$$

in which, instead of the vector parameter Θ, the hypervector $(\Theta', \vartheta')'$ of unknown parameters occurs, and instead of the matrix JA (reflecting the way of measuring the directly observable quantities of the investigated object), the hypermatrix (JA, S) is considered. The set of measured data consists now of realizations of the random variable ζ. The relation (2.4.4) expressing the linear unbiased and jointly efficient estimator of the vector of unknown parameters in the simple form of Model 2 can be rewritten for the corresponding mixed model in the following way

$$\begin{pmatrix} \widehat{\Delta\Theta}(\nu) \\ \widehat{\Delta\vartheta}(\nu) \end{pmatrix} = [(JA, S)'var^{-1}(\nu)(JA, S)]^{-1}(JA, S)'var^{-1}(\nu)\nu. \tag{2.5.2}$$

In order to obtain explicit formulas for $\widehat{\Delta\Theta}(\nu)$ and $\widehat{\Delta\vartheta}(\nu)$, we must invert the matrix $(JA, S)'var^{-1}(\nu)(JA, S)$ with the following block structure

$$(JA, S)'var^{-1}(\nu)(JA, S) = \begin{pmatrix} A'J'var^{-1}(\nu)JA, & A'J'var^{-1}(\nu)S \\ S'var^{-1}(\nu)JA, & S'var^{-1}(\nu)S \end{pmatrix}.$$

Here the following theorem on the inverse of a block matrix is of great importance.

Let

$$C = \begin{pmatrix} C_{11}, & C_{12} \\ C_{21}, & C_{22} \end{pmatrix}$$

be a square regular matrix such that C_{11} and C_{22} are square matrices and at least one of them is regular. Then

$$\begin{pmatrix} C_{11}, & C_{12} \\ C_{21}, & C_{22} \end{pmatrix}^{-1} = \begin{pmatrix} C^{11}, & C^{12} \\ C^{21}, & C^{22} \end{pmatrix},$$

where

$$C^{11} = (C_{11} - C_{12}C_{22}^{-1}C_{21})^{-1} = C_{11}^{-1} + C_{11}^{-1}C_{12}(C_{22} - C_{21}C_{11}^{-1}C_{12})^{-1}C_{21}C_{11}^{-1},$$

$$C^{12} = -(C_{11} - C_{12}C_{22}^{-1}C_{21})^{-1}C_{12}C_{22}^{-1} = -C_{11}^{-1}C_{12}(C_{22} - C_{21}C_{11}^{-1}C_{12})^{-1},$$

$$C^{21} = -(C_{22} - C_{21}C_{11}^{-1}C_{12})^{-1}C_{21}C_{11}^{-1} = -C_{22}^{-1}C_{21}(C_{11} - C_{12}C_{22}^{-1}C_{21})^{-1},$$

$$C^{22} = (C_{22} - C_{21}C_{11}^{-1}C_{12})^{-1} = C_{22}^{-1} + C_{22}^{-1}C_{21}(C_{11} - C_{12}C_{22}^{-1}C_{21})^{-1}C_{12}C_{22}^{-1}.$$

These alternative expressions of submatrices of the inverse matrix enable us to choose that instance in which one of the matrices C_{11} and C_{22} is regular. If they are both regular, the alternative expressions represent a good check of the computational correctness of the inverse of matrix C.

The proof of this assertion is based on the definition of an inverse of an arbitrary regular matrix. Due to it,

$$\begin{pmatrix} C^{11}, & C^{12} \\ C^{21}, & C^{22} \end{pmatrix} \begin{pmatrix} C_{11}, & C_{12} \\ C_{21}, & C_{22} \end{pmatrix} = \begin{pmatrix} I, & 0 \\ 0, & I \end{pmatrix},$$

therefore it suffices to convince ourselves that

$$C^{11}C_{11} + C^{12}C_{21} = I, \; C^{21}C_{11} + C^{22}C_{21} = 0,$$

$$C^{11}C_{12} + C^{12}C_{22} = 0, \; C^{21}C_{12} + C^{22}C_{22} = I.$$

This system of matrix equations determines the block matrices C^{11}, C^{12}, C^{21} and C^{22} uniquely.

Now we shall review the expressions for linear unbiased and jointly efficient estimators of the hypervector $(\Theta', \vartheta')'$ of the unknown parameters and their fundamental statistical characteristics within individual basic linear measurement models.

Model 1 (a direct measurement of a scalar parameter)

Model 2 (an indirect measurement of a vector parameter)

The expressions for an unbiased and jointly efficient linear estimator of the parameters Θ and ϑ in the stochastic model

$$\begin{aligned}
&\zeta = i\Theta_0 + s(\vartheta_0) + i\Delta\Theta + S\Delta\vartheta + \varphi_0 + \varepsilon,\ \nu = \zeta - i\Theta_0 - s(\vartheta_0),\\
&E[\nu] = i\Delta\Theta + S\Delta\vartheta,\ E[\varepsilon] = 0,\ E[\varphi_0] = 0,\\
&var(\zeta) = var(\nu) = var(\varphi_0) + var(\varepsilon),\ cov(\varphi_0, \varepsilon) = 0,\\
&R[(i, S)] = 1 + s \leq n,\ R[var(\nu)] = n
\end{aligned}$$

represent a special case of the corresponding expressions in the stochastic model

$$\begin{aligned}
&\zeta = Jf(\Theta_0) + s(\vartheta_0) + JA\Delta\Theta + S\Delta\vartheta + \varphi_0 + \varepsilon,\\
&\nu = \zeta - Jf(\Theta_0) - s(\vartheta_0),\\
&E[\nu] = JA\Delta\Theta + S\Delta\vartheta,\ E[\varphi_0] = 0, E[\varepsilon] = 0\\
&var(\zeta) = var(\nu) = var(\varphi_0) + var(\varepsilon),\ cov(\varepsilon, \varphi_0) = 0,\\
&R[(JA, S)] = k + s \leq n,\ R(A) = k \leq p,\ R(J) = p \leq n,\\
&\quad R(S) = s \leq n,\ R[var(\nu)] = n
\end{aligned}$$

for $JA = i$. This is the reason for presenting their general form only:

$$\begin{gathered}
\hat{\Theta}(\nu) = \Theta_0 +\\
+\{A'J'var^{-1}(\nu)JA - A'J'var^{-1}(\nu)S[S'var^{-1}(\nu)S]^{-1}S'var^{-1}(\nu)JA\}^{-1}\cdot\\
\cdot\{A'J' - A'J'var^{-1}(\nu)S[S'var^{-1}(\nu)S]^{-1}S'\}var^{-1}(\nu)\nu
\end{gathered} \tag{2.5.3}$$

$$\begin{gathered}
\hat{\vartheta}(\nu) = \vartheta_0 +\\
+\{S'var^{-1}(\nu)S - S'var^{-1}(\nu)JA[A'J'var^{-1}(\nu)JA]^{-1}A'J'var^{-1}(\nu)S\}^{-1}\cdot\\
\cdot\{S' - S'var^{-1}(\nu)JA[A'J'var^{-1}(\nu)JA]^{-1}A'J'\}var^{-1}(\nu)\nu;
\end{gathered} \tag{2.5.4}$$

their variance and covariance matrices are

$$var[\hat{\Theta}(\nu)] =$$
$$\{A'J'var^{-1}(\nu)JA - A'J'var^{-1}(\nu)S[S'var^{-1}(\nu)S]^{-1}S'var^{-1}(\nu)JA\}^{-1} =$$
$$[A'J'var^{-1}(\nu)JA]^{-1} + [A'J'var^{-1}(\nu)JA]^{-1}A'J'var^{-1}(\nu)S\{S'var^{-1}(\nu)S-$$
$$-S'var^{-1}(\nu)JA[A'J'var^{-1}(\nu)JA]^{-1}A'J'var^{-1}(\nu)S\}^{-1}S'var^{-1}(\nu)JA\cdot$$
$$\cdot[A'J'var^{-1}(\nu)JA]^{-1}, \tag{2.5.5}$$

$$var[\hat{\vartheta}(\nu)] =$$
$$= \{S'var^{-1}(\nu)S - S'var^{-1}(\nu)JA[A'J'var^{-1}(\nu)JA]^{-1}A'J'var^{-1}(\nu)S\}^{-1}$$
$$= [S'var^{-1}(\nu)S]^{-1} + [S'var^{-1}(\nu)S]^{-1}S'var^{-1}(\nu)JA\{A'J'var^{-1}(\nu)JA-$$
$$-A'J'var^{-1}(\nu)S[S'var^{-1}(\nu)S]^{-1}S'var^{-1}(\nu)JA\}^{-1}A'J'var^{-1}(\nu)S\cdot$$
$$\cdot[S'var^{-1}(\nu)S]^{-1}, \tag{2.5.6}$$

$$cov[\hat{\Theta}(\nu), \hat{\vartheta}(\nu)] = -\{A'J'var^{-1}(\nu)JA - A'J'var^{-1}(\nu)S[S'var^{-1}(\nu)S]^{-1}\cdot$$
$$\cdot S'var^{-1}(\nu)JA\}^{-1}A'J'var^{-1}(\nu)S[S'var^{-1}(\nu)S]^{-1} =$$
$$= -[A'J'var^{-1}(\nu)JA]^{-1}A'J'var^{-1}(\nu)S\{S'var^{-1}(\nu)S - S'var^{-1}(\nu)JA\cdot$$
$$\cdot[A'J'var^{-1}(\nu)JA]^{-1}A'J'var^{-1}(\nu)S\}^{-1}. \tag{2.5.7}$$

All alternative expressions given provide a good check of the computation correctness.

The complexity of the resulting expressions which we meet in the following sections, particularly in connection with problems of interpolation and extrapolation of the useful part of the observed random vector and in connection with the problem of the optimum elimination of a systematic deterministic effect, requires that we introduce simplified symbols. We shall explain this, using the example of Model 2.

Let us denote

$$\begin{pmatrix} A'J'var^{-1}(\nu)JA, & A'J'var^{-1}(\nu)S \\ S'var^{-1}(\nu)JA, & S'var^{-1}(\nu)S \end{pmatrix}^{-1} = \begin{pmatrix} N^{11}, & N^{12} \\ N^{21}, & N^{22} \end{pmatrix},$$

$$N^{11} =$$

$$= \{A'J'var^{-1}(\nu)JA - A'J'var^{-1}(\nu)S[S'var^{-1}(\nu)S]^{-1}S'var^{-1}(\nu)JA\}^{-1},$$

$$N^{11} =$$

$$= \{S'var^{-1}(\nu)S - S'var^{-1}(\nu)JA[A'J'var^{-1}(\nu)JA]^{-1}A'J'var^{-1}(\nu)S\}^{-1},$$

$$N^{12} = -N^{11}A'J'var^{-1}(\nu)S[S'var^{-1}(\nu)S]^{-1},$$

$$N^{21} = -N^{22}S'var^{-1}(\nu)JA[A'J'var^{-1}(\nu)JA]^{-1}; \tag{2.5.8}$$

all inverse matrices that occur here exist as a consequence on the assumptions on the regularity of Model 2; the alternative expressions are not given here, however their existence required for checking the correctness of the computation has been noted; evidently $(N^{12})' = N^{21}$.

Using the auxiliary relations (2.5.8) which have a relatively simple and mutually correlated structure, we can write the expressions for the linear unbiased and jointly efficient estimator of the supervector $(\Theta', \vartheta')'$ and its covariance matrix in a relatively (from its structure point of view) simple form

$$\begin{pmatrix} \hat{\Theta}(\nu) \\ \hat{\vartheta}(\nu) \end{pmatrix} = \begin{pmatrix} \Theta_0 \\ \vartheta_0 \end{pmatrix} + \begin{pmatrix} N^{11}, & N^{12} \\ N^{21}, & N^{22} \end{pmatrix} \begin{pmatrix} A'J' \\ S' \end{pmatrix} var^{-1}(\nu)\nu =$$

$$= \begin{pmatrix} \Theta_0 \\ \vartheta_0 \end{pmatrix} + \begin{pmatrix} (N^{11}A'J' + N^{12}S')var^{-1}(\nu)\nu \\ (N^{21}A'J' + N^{22}S')var^{-1}(\nu)\nu \end{pmatrix}, \tag{2.5.9}$$

and

$$var \begin{pmatrix} \hat{\Theta}(\nu) \\ \hat{\vartheta}(\nu) \end{pmatrix} = \begin{pmatrix} var[\hat{\Theta}(\nu)], & cov[\hat{\Theta}(\nu), \hat{\vartheta}(\nu)] \\ cov[\hat{\vartheta}(\nu), \hat{\Theta}(\nu)], & var[\hat{\vartheta}(\nu)] \end{pmatrix} =$$

$$= \begin{pmatrix} N^{11}, & N^{12} \\ N^{21}, & N^{22} \end{pmatrix}. \tag{2.5.10}$$

Model 3 (a direct measurement of a vector parameter with a system of conditions)

Model 5 (an indirect measurement of a vector parameter with a system of conditions)

Similar to the previous case, Model 3

$$\begin{array}{l} \zeta = J\Theta_0 + s(\vartheta_0) + J\Delta\Theta + S\Delta\vartheta + \varphi_0 + \varepsilon, \nu = \zeta - J\Theta_0 - s(\vartheta_0), \\ B\Delta\Theta + b = 0, \\ E[\nu] = J\Delta\Theta + S\Delta\vartheta,\ E[\varphi_0] = 0,\ E[\varepsilon] = 0, \\ var(\zeta) = var(\nu) = var(\varphi_0) + var(\varepsilon),\ cov(\varphi_0, \varepsilon) = 0, \\ R[(J, S)] = k + s \leq n,\ R(S) = s \leq n,\ R(J) = k \leq n, \\ R(B) = q \leq k,\ R[var(\nu)] = n \end{array}$$

is a special case of Model 5

$$\begin{array}{l} \zeta = Jf(\Theta_0) + s(\vartheta_0) + JA\Delta\Theta + S\Delta\vartheta + \varphi_0 + \varepsilon, \\ \nu = \zeta - Jf(\Theta_0) - s(\vartheta_0), \\ B\Delta\Theta + b = 0, \\ E[\nu] = JA\Delta\Theta + S\Delta\vartheta,\ E[\varphi_0] = 0,\ E[\varepsilon] = 0, \\ var(\zeta) = var(\nu) = var(\varphi_0) + var(\varepsilon),\ cov(\varphi_0, \varepsilon) = 0, \\ R[(JA, S)] = k + s \leq n,\ R(S) = s \leq n,\ R(JA) = k \leq n, \\ R(B) = q \leq k,\ R[var(\nu)] = n \end{array}$$

for $A = I$. For this reason the relations for the linear unbiased and jointly efficient estimator of the parameter hypervector $(\Theta', \vartheta')'$ and its covariance matrix are given for the general case of Model 5 only.

The condition $B\Delta\Theta + b = 0$ binds only the values of the parameter vector Θ, therefore before applying the relation (2.4.16) as a starting point for deriving the form of the linear unbiased and jointly efficient estimator of the hypervector $(\Theta_0' + \Delta\Theta', \vartheta_0' + \Delta\vartheta')'$ it must be rewritten to a form formally binding the elements of the hypervector $(\Delta\Theta', \vartheta)'$. We can directly convince ourselves that the condition of the form

$$(B_{(q,k)}, 0_{(q,s)}) \begin{pmatrix} \Delta\Theta_{(k)} \\ \vartheta_{(s)} \end{pmatrix} + b_{(q)} = 0_{(q)}$$

(similar to Section 2.2, the simple indices in parentheses denote the dimension of a vector and the double indices denote the dimension of a matrix), which binds in a formal way the elements of the hypervector $(\Delta\Theta', \vartheta')'$, is identical with the condition $B_{(q,k)}\Delta\Theta_{(k)} + b_{(q)} = 0_{(q)}$, concerning the elements of the parameter vector $\Delta\Theta$. Substituting now a $(k+s)$-dimensional vector $(\widehat{\Delta\Theta}', \widehat{\Delta\vartheta}')'$ in the relation (2.4.16) for the k-dimensional vector $\widehat{\Delta\Theta}$, the $n \times (k+s)$-dimensional hypermatrix (JA, S) for the $n \times k$-dimensional matrix JA, as a consequence of this, the block matrix

$$\begin{pmatrix} A'J'var^{-1}(\nu)JA, & A'J'var^{-1}(\nu)S \\ S'var^{-1}(\nu)JA, & S'var^{-1}(\nu)S \end{pmatrix}^{-1} = \begin{pmatrix} N^{11}, & N^{12} \\ N^{21}, & N^{22} \end{pmatrix}$$

(its individual blocks are defined by (2.5.8)) for the matrix $[A'J'var^{-1}(\nu)JA]^{-1}$, we obtain

$$\begin{pmatrix} \widehat{\Delta\Theta}_{(k)}(\nu) \\ \widehat{\Delta\vartheta}_{(s)}(\nu) \end{pmatrix} = \begin{pmatrix} N^{11}_{(k,k)}, & N^{12}_{(k,s)} \\ N^{21}_{(s,k)}, & N^{22}_{(s,s)} \end{pmatrix} \begin{pmatrix} (JA)'_{(k,n)} \\ S'_{(s,n)} \end{pmatrix} var^{-1}_{(n,n)}(\nu)\nu_{(n)} -$$

$$- \begin{pmatrix} N^{11}_{(k,k)}, & N^{12}_{(k,s)} \\ N^{21}_{(s,k)}, & N^{22}_{(s,s)} \end{pmatrix} \begin{pmatrix} B'_{(k,q)} \\ 0_{(s,q)} \end{pmatrix} \left[(B_{(q,k)}, 0_{(q,s)}) \begin{pmatrix} N^{11}_{(k,k)}, & N^{12}_{(k,s)} \\ N^{21}_{(s,k)}, & N^{22}_{(s,s)} \end{pmatrix} \cdot \right.$$

$$\left. \cdot \begin{pmatrix} B'_{(k,q)} \\ 0_{(s,q)} \end{pmatrix} \right]^{-1} b_{(q)} + (B_{(q,k)}, 0_{(q,s)}) \begin{pmatrix} N^{11}_{(k,k)}, & N^{12}_{(k,s)} \\ N^{21}_{(s,k)}, & N^{22}_{(s,s)} \end{pmatrix} \cdot$$

$$\cdot \begin{pmatrix} (JA)'_{(k,n)} \\ S'_{(s,n)} \end{pmatrix} var^{-1}_{(n,n)}(\nu)\nu_{(n)}.$$

Since

$$(B_{(q,k)}, 0_{(q,s)}) \begin{pmatrix} N^{11}_{(k,k)}, & N^{12}_{(k,s)} \\ N^{21}_{(s,k)}, & N^{22}_{(s,s)} \end{pmatrix} \begin{pmatrix} B'_{(k,q)} \\ 0_{(s,q)} \end{pmatrix} = B_{(q,k)} N^{11}_{(k,k)} B'_{(k,q)}$$

and

$$\begin{pmatrix} N^{11}_{(k,k)}, & N^{12}_{(k,s)} \\ N^{21}_{(s,k)}, & N^{22}_{(s,s)} \end{pmatrix} \begin{pmatrix} B'_{(k,q)} \\ 0_{(s,q)} \end{pmatrix} = \begin{pmatrix} N^{11}_{(k,k)} \\ N^{21}_{(s,k)} \end{pmatrix} B'_{(k,q)},$$

we can write (without denoting the dimensions of matrices and vectors)

$$\begin{pmatrix} \hat{\Theta}(\nu) \\ \hat{\vartheta}(\nu) \end{pmatrix} = \begin{pmatrix} \Theta_0 \\ \vartheta_0 \end{pmatrix} + \begin{pmatrix} N^{11}, & N^{12} \\ N^{21}, & N^{22} \end{pmatrix} \begin{pmatrix} A'J' \\ S' \end{pmatrix} var^{-1}(\nu)\nu -$$

$$- \begin{pmatrix} N^{11} \\ N^{21} \end{pmatrix} B'(BN^{11}B')^{-1} \left[b + (N^{11}, N^{12}) \begin{pmatrix} A'J' \\ S' \end{pmatrix} var^{-1}(\nu)\nu \right] =$$

$$= \begin{pmatrix} \Theta_0 \\ \vartheta_0 \end{pmatrix} + \begin{pmatrix} (N^{11}A'J' + N^{12}S')var^{-1}(\nu)\nu - \\ (N^{21}A'J' + N^{22}S')var^{-1}(\nu)\nu - \end{pmatrix}$$

$$\begin{pmatrix} -N^{11}B'(BN^{11}B')^{-1}[b + B(N^{11}A'J' + N^{12}S')var^{-1}(\nu)\nu] \\ -N^{21}B'(BN^{11}B')^{-1}[b + B(N^{11}A'J' + N^{12}S')var^{-1}(\nu)\nu] \end{pmatrix}. \tag{2.5.11}$$

The relation (2.5.11) expresses the linear unbiased and jointly efficient estimator of the parameter hypervector within Model 5.

The next step is to derive the form of its covariance matrix:

$$var \begin{pmatrix} \hat{\Theta}(\nu) \\ \hat{\vartheta}(\nu) \end{pmatrix} = var \left[\begin{pmatrix} N^{11}, & N^{12} \\ N^{21}, & N^{22} \end{pmatrix} \begin{pmatrix} A'J' \\ S' \end{pmatrix} var^{-1}(\nu)\nu \right] +$$

$$+var \left[\begin{pmatrix} N^{11} \\ N^{21} \end{pmatrix} B'(BN^{11}B')^{-1}B(N^{11}, N^{12}) \begin{pmatrix} A'J' \\ S' \end{pmatrix} var^{-1}(\nu)\nu \right] -$$

$$-cov \left[\begin{pmatrix} N^{11}, & N^{12} \\ N^{21}, & N^{22} \end{pmatrix} \begin{pmatrix} A'J' \\ S' \end{pmatrix} var^{-1}(\nu)\nu, \begin{pmatrix} N^{11} \\ N^{21} \end{pmatrix} B' \cdot \right.$$

$$\left. \cdot (BN^{11}B')^{-1}B(N^{11}, N^{12}) \begin{pmatrix} A'J' \\ S' \end{pmatrix} var^{-1}(\nu)\nu \right] -$$

$$-cov \left[\begin{pmatrix} N^{11} \\ N^{21} \end{pmatrix} B'(BN^{11}B')^{-1}B(N^{11}, N^{12}) \begin{pmatrix} A'J' \\ S' \end{pmatrix} var^{-1}(\nu)\nu, \right.$$

$$\left. \begin{pmatrix} N^{11}, & N^{12} \\ N^{21}, & N^{22} \end{pmatrix} \begin{pmatrix} A'J' \\ S' \end{pmatrix} var^{-1}(\nu)\nu \right] =$$

$$= \begin{pmatrix} N^{11}, & N^{12} \\ N^{21}, & N^{22} \end{pmatrix} \begin{pmatrix} A'J' \\ S' \end{pmatrix} var^{-1}(\nu)(JA, S) \begin{pmatrix} N^{11}, & N^{12} \\ N^{21}, & N^{22} \end{pmatrix} +$$

$$+\begin{pmatrix} N^{11} \\ N^{21} \end{pmatrix} B'(BN^{11}B')^{-1}B(N^{11},N^{12}) \begin{pmatrix} A'J' \\ S' \end{pmatrix} var^{-1}(\nu)(JA,S)\cdot$$

$$\cdot\begin{pmatrix} N^{11} \\ N^{21} \end{pmatrix} B'(BN^{11}B')^{-1}B(N^{11},N^{12}) - \begin{pmatrix} N^{11}, & N^{12} \\ N^{21}, & N^{22} \end{pmatrix} \begin{pmatrix} A'J' \\ S' \end{pmatrix}\cdot$$

$$\cdot var^{-1}(\nu)(JA,S) \begin{pmatrix} N^{11} \\ N^{21} \end{pmatrix} B'(BN^{11}B')^{-1}B(N^{11},N^{12}) -$$

$$-\begin{pmatrix} N^{11} \\ N^{21} \end{pmatrix} B'(BN^{11}B')^{-1}B(N^{11},N^{12}) \begin{pmatrix} A'J' \\ S' \end{pmatrix} var^{-1}(\nu)\cdot$$

$$\cdot(JA,S)\begin{pmatrix} N^{11}, & N^{12} \\ N^{21}, & N^{22} \end{pmatrix},$$

which, because of

$$\begin{pmatrix} N^{11}, & N^{12} \\ N^{21}, & N^{22} \end{pmatrix} \begin{pmatrix} A'J' \\ S' \end{pmatrix} var^{-1}(\nu)(JA,S) = I$$

and

$$(N^{11},N^{12}) \begin{pmatrix} A'J' \\ S' \end{pmatrix} var^{-1}(\nu)(JA,S) \begin{pmatrix} N^{11} \\ N^{21} \end{pmatrix} = N^{11},$$

finally yields

$$var\begin{pmatrix} \hat{\Theta}(\nu) \\ \hat{\vartheta}(\nu) \end{pmatrix} = \begin{pmatrix} N^{11}, & N^{12} \\ N^{21}, & N^{22} \end{pmatrix} -$$

$$-\begin{pmatrix} N^{11} \\ N^{21} \end{pmatrix} B'(BN^{11}B')^{-1}B(N^{11},N^{12}) =$$

$$\begin{pmatrix} N^{11} - N^{11}B'(BN^{11}B')^{-1}BN^{11}, & N^{12} - N^{11}B'(BN^{11}B')^{-1}BN^{12} \\ N^{21} - N^{21}B'(BN^{11}B')^{-1}BN^{11}, & N^{22} - N^{21}B'(BN^{11}B')^{-1}BN^{12} \end{pmatrix}. \tag{2.5.12}$$

Model 4 (an incomplete direct measurement of a vector parameter with a system of conditions)

Model 6 (an incomplete indirect measurement of a vector parameter with a system of conditions)

The relations for the linear unbiased and jointly efficient estimator of the unknown parameter hypervector $(\Theta_1', \Theta_2', \vartheta')'$ in the model

$$
\begin{aligned}
&\zeta = J\Theta_{10} + s(\vartheta_0) + J\Delta\Theta_1 + S\Delta\vartheta + \varphi_0 + \varepsilon,\ \nu = \zeta - J\Theta_{10} - s(\vartheta_0),\\
&\Theta = (\Theta_1', \Theta_2')',\ (B, C)(\Delta\Theta_1', \Delta\Theta_2')' + b = 0,\\
&E[\nu] = J\Delta\Theta_1 + S\Delta\vartheta,\ E[\varphi_0] = 0,\ E[\varepsilon] = 0,\\
&var(\zeta) = var(\nu) = var(\varphi_0) + var(\varepsilon),\ cov(\varphi_0, \varepsilon) = 0,\\
&R[(J, S)] = p + s \le n,\ R(S) = s \le n,\ R(J) = p \le n,\\
&R[(B, C)] = q \le k,\ R(C) = k - p \le q,\ R[var(\nu)] = n
\end{aligned}
$$

is a special case of the correponding relations in the model

$$
\begin{aligned}
&\zeta = Jf(\Theta_{10}) + s(\vartheta_0) + JA\Delta\Theta_1 + S\Delta\vartheta + \varphi_0 + \varepsilon,\\
&\nu = \zeta - Jf(\Theta_{10}) - s(\vartheta_0),\\
&B\Delta\Theta_1 + C\Delta\Theta_2 + b = 0,\\
&E[\nu] = JA\Delta\Theta_1 + S\Delta\vartheta,\ E[\varphi_0] = 0,\ E[\varepsilon] = 0,\\
&var(\zeta) = var(\nu) = var(\varphi_0) + var(\varepsilon),\ cov(\varphi_0, \varepsilon) = 0,\\
&R[(JA, S)] = p + s \le n,\ R(A) = l \le p,\ R(J) = p \le n,\\
&R[(B, C)] = q \le k,\ R(C) = k - l,\ R[var(\nu)] = n
\end{aligned}
$$

for $A = I$ and $l = p$.

For the sake of completeness, the expressions for the linear unbiased and jointly efficient estimator of the unknown parameter hypervector $(\Theta_1', \Theta_2', \vartheta')'$ will be derived in detail using the least squares method[6]; moreover, this procedure enables us to compare different approaches to deriving the resulting expressions.

In accordance with the least squares method, a point $(\Delta\Theta_1', \Delta\vartheta')'$ of

[6]The reader who wants to derive expressions for the linear unbiased and jointly efficient estimator of the hypervector $(\Theta_1', \Theta_2', \vartheta')'$ using the same procedure as in the previous cases has to realize that the condition $b + B\Delta\Theta_1 + C\Delta\Theta_2 = 0$ binds the values of the vector $(\Theta_1', \Theta_2')'$, it does not concern the values of the vector ϑ.

the linear manifold $\mathcal{T} : \{(\Delta\Theta_1', \Delta\Theta_2')' : b + B\Delta\Theta_1 + C\Delta\Theta_2 = 0\}$ is sought such that the quadratic form

$$Q(\Delta\Theta_1, \Delta\vartheta, u) = (JA\Delta\Theta_1 + S\Delta\vartheta - u)' var^{-1}(\nu)(JA\Delta\Theta_1 + S\Delta\vartheta - u) \tag{2.5.13}$$

assumes its minimum here for every realization u of the random variable ν.

The auxiliary Lagrange function is $\Phi(\Delta\Theta_1, \Delta\Theta_2, \Delta\vartheta, u) = (JA\Delta\Theta_1 + S\Delta\vartheta - u)' var^{-1}(\nu)(JA\Delta\Theta_1 + S\Delta\vartheta - u) - 2\lambda'(b + B\Delta\Theta_1 + C\Delta\Theta_2)$. As $\partial\Phi(\Delta\Theta_1, \Delta\Theta_2, \Delta\vartheta, u)/\partial\Delta\Theta_1 = 2A'J' var^{-1}(\nu)(JA\Delta\Theta_1 + S\Delta\vartheta - u) - 2B'\lambda$, $\partial\Phi(\Delta\Theta_1, \Delta\Theta_2, \Delta\vartheta, u)/\partial\Delta\Theta_2 = -2C'\lambda$ and $\partial\Phi(\Delta\Theta_1, \Delta\Theta_2, \Delta\vartheta, u)/\partial\Delta\vartheta = 2S' var^{-1}(\nu)(JA\Delta\Theta_1 + S\Delta\vartheta - u)$, the quadratic form (2.5.13) acquires the extremum at the point $(\widehat{\Delta\Theta}_1'(u), \widehat{\Delta\vartheta}'(u))' \in \mathcal{T}$, where $(\widehat{\Delta\Theta}_1'(u), \widehat{\Delta\vartheta}'(u))'$ is a solution of the equation system

$$A'J' var^{-1}(\nu)JA\Delta\Theta_1 + A'J' var^{-1}(\nu)S\Delta\vartheta = A'J' var^{-1}(\nu)u + B'\lambda,$$

$$S' var^{-1}(\nu)JA\Delta\Theta_1 + S' var^{-1}(\nu)\Delta\vartheta = S' var^{-1}(\nu)u,$$

which implies that

$$\begin{pmatrix} \widehat{\Delta\Theta}_1(u) \\ \widehat{\Delta\vartheta}(u) \end{pmatrix} = \begin{pmatrix} A'J' var^{-1}(\nu)JA, & A'J' var^{-1}(\nu)S \\ S' var^{-1}(\nu)JA, & S' var^{-1}(\nu)S \end{pmatrix}^{-1} \cdot \begin{pmatrix} A'J' var^{-1}(\nu)u + B'\lambda \\ S' var^{-1}(\nu)u \end{pmatrix}.$$

Using the above mentioned notation

$$\begin{pmatrix} A'J' var^{-1}(\nu)JA, & A'J' var^{-1}(\nu)S \\ S' var^{-1}(\nu)JA, & S' var^{-1}(\nu)S \end{pmatrix}^{-1} = \begin{pmatrix} N^{11}, & N^{12} \\ N^{21}, & N^{22} \end{pmatrix},$$

where N^{11}, N^{12}, N^{21} and N^{22} are defined by (2.5.8), we can write

$$\widehat{\Delta\Theta}_1(u) = N^{11}(A'J' var^{-1}(\nu)u + B'\lambda) + N^{12}S' var^{-1}(\nu)u,$$

$$\widehat{\Delta\vartheta}(u) = N^{21}(A'J' var^{-1}(\nu)u + B'\lambda) + N^{22}S' var^{-1}(\nu)u.$$

Inserting the expression for $\widehat{\Delta\Theta}_1(u)$ into the condition $b + B\Delta\Theta_1 + C\Delta\Theta_2 = 0$ we get, together with the annulled first partial derivative of the Lagrange function with respect to $\Delta\Theta_2$, the equation system

$$BN^{11}B'\lambda + C\widehat{\Delta\Theta}_2(u) = -[b + BN^{11}A'J' var^{-1}(\nu)u + BN^{12}S' var^{-1}(\nu)u],$$

$$C'\lambda = 0$$

for determining $\widehat{\Delta\Theta}_2(u)$ and λ.

The hypermatrix

$$\begin{pmatrix} BN^{11}B', & C \\ C', & 0 \end{pmatrix},$$

as a consequence of the regularity assumptions within Model 6, is regular even if its submatrix $BN^{11}B'$ generally need not be regular (the $q \times q$-dimensional zero matrix is not regular). This means that the inverse of the mentioned block matrix exists, it cannot be expressed in an explicit way using the rules for the inverse of a regular block matrix. Thus, let

$$\begin{pmatrix} BN^{11}B', & C \\ C', & 0 \end{pmatrix}^{-1} = \begin{pmatrix} P_{11}, & P_{12} \\ P_{21}, & P_{22} \end{pmatrix}$$

(see the procedure applied on p. 140, the relation (2.4.13)); then

$$\begin{pmatrix} \lambda \\ \widehat{\Delta\Theta}_2(u) \end{pmatrix} =$$

$$= \begin{pmatrix} P_{11}, & P_{12} \\ P_{21}, & P_{22} \end{pmatrix} \begin{pmatrix} -[b + BN^{11}A'J'var^{-1}(\nu)u + BN^{12}S'var^{-1}(\nu)u] \\ 0 \end{pmatrix},$$

which implies

$$\lambda = -P_{11}[b + BN^{11}A'J'var^{-1}(\nu)u + BN^{12}S'var^{-1}(\nu)u]$$

and

$$\widehat{\Delta\Theta}_2(u) = -P_{21}[b + BN^{11}A'J'var^{-1}(\nu)u + BN^{12}S'var^{-1}(\nu)u],$$

thus

$$\widehat{\Delta\Theta}_1(u) =$$
$$(N^{11}A'J' + N^{12}S')var^{-1}(\nu)u - N^{11}B'P_{11}[b + BN^{11}A'J'var^{-1}(\nu)u +$$
$$BN^{12}S'var^{-1}(\nu)u],$$

$$\widehat{\Delta\vartheta}(u) =$$
$$(N^{21}A'J' + N^{22}S')var^{-1}(\nu)u - N^{21}B'P_{11}[b + BN^{11}A'J'var^{-1}(\nu)u +$$
$$BN^{12}S'var^{-1}(\nu)u]$$

and

$$\widehat{\Delta\Theta}_2(u) = -P_{21}[b + BN^{11}A'J'var^{-1}(\nu)u + BN^{12}S'var^{-1}(\nu)u].$$

We still have to convince ourselves that the point $(\widehat{\Delta\Theta}_1'(u), \widehat{\Delta\vartheta}'(u))'$ is the minimum point of the quadratic form (2.5.13). The idea of proving it is the same as in the case of the corresponding simple model; for this reason it is not given here.

The LS-estimator≡the linear unbiased and jointly efficient estimator of the hypervector $(\Theta_1', \Theta_2', \vartheta')'$ in Model 6 is

$$\begin{pmatrix} \hat{\Theta}_1(\nu) \\ \hat{\vartheta}(\nu) \\ \hat{\Theta}_2(\nu) \end{pmatrix} = \begin{pmatrix} \Theta_{10} \\ \vartheta_0 \\ \Theta_{20} \end{pmatrix} + \begin{pmatrix} N^{11}, & N^{12} \\ N^{21}, & N^{22} \\ 0, & 0 \end{pmatrix} \begin{pmatrix} A'J' \\ S' \end{pmatrix} var^{-1}(\nu)\nu -$$

$$- \begin{pmatrix} N^{11}B'P_{11} \\ N^{21}B'P_{11} \\ P_{21} \end{pmatrix} \left[b + B(N^{11}, N^{12}) \begin{pmatrix} A'J' \\ S' \end{pmatrix} var^{-1}(\nu)\nu \right] = \begin{pmatrix} \Theta_{10} \\ \vartheta_0 \\ \Theta_{20} \end{pmatrix} +$$

$$+ \begin{pmatrix} (N^{11}A'J' + N^{12}S')var^{-1}(\nu)\nu - N^{11}B'P_{11}[b + B(N^{11}A'J' + N^{12}S')var^{-1}(\nu)\nu] \\ (N^{21}A'J' + N^{22}S')var^{-1}(\nu)\nu - N^{21}B'P_{11}[b + B(N^{11}A'J' + N^{12}S')var^{-1}(\nu)\nu] \\ -P_{21}[b + B(N^{11}A'J' + N^{12}S')var^{-1}(\nu)\nu] \end{pmatrix} \tag{2.5.14}$$

and it is characterized by the covariance matrix of the form

$$var \begin{pmatrix} \hat{\Theta}_1(\nu) \\ \hat{\vartheta}(\nu) \\ \hat{\Theta}_2(\nu) \end{pmatrix} =$$

$$= \begin{pmatrix} N^{11} - N^{11}B'P_{11}BN^{11}, & N^{12} - N^{11}B'P_{11}BN^{12}, & -N^{11}B'P_{12} \\ N^{21} - N^{21}B'P_{11}BN^{11}, & N^{22} - N^{21}B'P_{11}BN^{12}, & -N^{21}B'P_{12} \\ -P_{21}BN^{11}, & -P_{21}BN^{12}, & -P_{22} \end{pmatrix}. \tag{2.5.15}$$

The auxiliary relations $P_{11}BN^{11}B'P_{11} = P_{11}$, $P_{21}BN^{11}B'P_{12} = -P_{22}$ and $P_{21}BN^{11}B'P_{11} = 0$, being an analogy of the expressions from pp. 142–143, have been applied here.

The next step consists in comparing the structure of corresponding relations in the simple and mixed models on the one hand and the mutual relation of the structures of expressions within the Models 1 to 6 on the other.

If the perturbing component φ in a mixed model is absent, i.e., $s(\cdot) \equiv 0$ and $\varphi_0 \equiv 0$ as well, the relationships within the mixed model are identical with the corresponding relationshisp in its simple form. For example, in a mixed form of Model 6

$$\widehat{\Delta\Theta}_1(\nu) = (N^{11}A'J' + N^{12}S')var^{-1}(\nu)\nu -$$

$$-N^{11}B'P_{11}[b + B(N^{11}A'J' + N^{12}S')var^{-1}(\nu)\nu],$$

$$\widehat{\Delta\Theta}_2(\nu) = -P_{21}[b + B(N^{11}A'J' + N^{12}S')var^{-1}(\nu)\nu]$$

(see the first and the third component of the vector (2.5.14)). If $S \equiv 0$ and $\varphi_0 \equiv 0$, then $\nu = \eta$, $var(\nu) = var(\eta) = var(\varepsilon) = \Sigma$, $N^{11} = (A'J'\Sigma^{-1}JA)^{-1}$, $N^{12} \equiv 0$ and $N^{22} \equiv 0$; as a consequence of this, $BN^{11}B' = B(A'J'\Sigma^{-1}JA)^{-1}B'$ and the hypermatrix $\begin{pmatrix} BN^{11}B', & C \\ C' & 0 \end{pmatrix}$ (p. 160) identifies with the hypermatrix $\begin{pmatrix} B(A'J'\Sigma^{-1}JA)^{-1}B', & C \\ C' & 0 \end{pmatrix}$ (p. 139) and consequently also the matrices $\begin{pmatrix} P_{11}, & P_{12} \\ P_{21}, & P_{22} \end{pmatrix}$ and $\begin{pmatrix} Q_{11}, & Q_{12} \\ Q_{21}, & Q_{22} \end{pmatrix}$ become identical. In this case, evidently, $\hat{\vartheta}(\nu) \equiv 0$. Taking into account all these facts, we obtain

$$\widehat{\Delta\Theta}_1(\eta) = (A'J'\Sigma^{-1}JA)^{-1}A'J'\Sigma^{-1}\eta -$$

$$-(A'J'\Sigma^{-1}JA)^{-1}B'Q_{11}[b + B(A'J'\Sigma^{-1}JA)^{-1}A'J'\Sigma^{-1}\eta],$$

$$\widehat{\Delta\Theta}_2(\eta) = -Q_{21}[b + B(A'J'\Sigma^{-1}JA)^{-1}A'J'\Sigma^{-1}\eta],$$

which are relations determining the LS-estimator of the unknown parameters in the simple form of Model 6 (compare them with (2.4.18)). In an analogous way these simplifications are reflected in the corresponding covariance matrices.

Let us take into account the mixed form of Model 6. If $C \equiv 0$, the hypermatrix $\begin{pmatrix} BN^{11}B', & C \\ C' & 0 \end{pmatrix}$ attains the form $\begin{pmatrix} BN^{11}B', & 0 \\ 0, & 0 \end{pmatrix}$, whose block $BN^{11}B'$ is regular as a consequence of the assumptions of regularity, thus $P_{11} = (BN^{11}B')^{-1}$ and $P_{21} \equiv 0$ and

$$\widehat{\Delta\Theta}(\nu) = (N^{11}A'J' + N^{12}S')var^{-1}(\nu)\nu -$$

$$-N^{11}B'(BN^{11}B')^{-1}[b+B(N^{11}A'J'+N^{12}S')var^{-1}(\nu)\nu],$$

$$\widehat{\Delta\vartheta}(\nu)=(N^{21}A'J'+N^{22}S')var^{-1}(\nu)\nu-$$

$$-N^{21}B'(BN^{11}B')^{-1}[b+B(N^{11}A'J'+N^{12}S')var^{-1}(\nu)\nu],$$

which are the relations determining the LS-estimator of the vector parameters $\Delta\Theta$ and $\Delta\vartheta$ in the mixed linear Model 5 (compare with the relationships (2.5.11) and (2.5.12)).

Furthermore, if $B\equiv 0$ and $b\equiv 0$ (the components of the unknown parameter vector Θ are not bound by a condition), we immediately obtain the relations

$$\widehat{\Delta\Theta}(\nu)=(N^{11}A'J'+N^{12}S')var^{-1}(\nu)\nu,$$

$$\widehat{\Delta\vartheta}(\nu)=(N^{21}A'J'+N^{22}S')var^{-1}(\nu)\nu,$$

expressing the LS-estimators of the parameters $\Delta\Theta$ and $\Delta\vartheta$ in the mixed form of Model 2 (compare these with relationships (2.5.11) and (2.5.12)). The corresponding covariance matrices can be simplified in an analogous way.

2.5.2 Linear regression in estimation theory

The problem formulated on page 145, namely the problem of determining the optimum values of the useful signal at the observation points and outside of this set (evidently under the condition that the formulated problem is physically reasonable, i.e., the useful signal is defined outside the set of observation points) and its dispersion is solved in this section.

Determining the optimum value of the useful signal at the observation points, we speak of a filtration of the observed vector; determining the optimum value of the signal outside this domain, we speak of a prediction of the observed vector. The formulated problem is solvable under the condition that the vector $cov(\zeta,\zeta_{up})$ of the covariances between the components of the observation vector ζ and the filtered or predicted useful signal ζ_{up} is known. The problem how to utilize, optimumly, information involved in this covariance vector for approximating the random variable ζ_{up} by the observation vector ζ was solved in Section 1.3.3 under the condition that the mean values $E[\zeta]$ and $E[\zeta_{up}]$ are known. The same problem will be solved now in its statistical form, when neither the mean value of the observation vector ζ nor the mean value of the random variable ζ_{up} are known, the optimum estimators of these mean values can be obtained within individual models of measurement as linear combinations of LS-estimators of the unknown parameters, evidently under the condition of known functions $f(\cdot)$, $s(\cdot)$ and $g(\cdot)$ which express the relations between the theoretical values of the directly observable quantities and the theoretical values of the unknown parameters or the relations between the unknown parameters.

The solution of the formulated problem requires us to apply the following symbols (connected with relations (2.5.1) and (2.5.1′))

$$\zeta_u = \zeta - \varepsilon = Jf(\Theta_0) + s(\vartheta_0) + JA\Delta\Theta + S\Delta\vartheta + \varphi_0,$$

$$\nu_u = \zeta_u - Jf(\Theta_0) - s(\vartheta_0),$$

$$\zeta_{up} = f_p(\Theta_0) + s_p(\vartheta_0) + a_p'\Delta\Theta + s_p'\Delta\vartheta + \varphi_{0p},$$

where $f_p(\cdot)$ and $s_p(\cdot)$ are known functions implied by the theory about the measured object, $a_p = \partial f_p(\Theta_0)/\partial\Theta$, $s_p = \partial s_p(\vartheta_0)/\partial\vartheta$, the index u denotes the useful component of the vector ζ or ν, the index p means the predicted or filtered random variables or deterministic values necessary for determining them.

The random variable $\hat{\zeta}$ satisfying the natural requirement of the minimum dispersion

$$D[\zeta_{up} - \hat{\zeta}_{up}(\zeta)] = min \tag{2.5.16}$$

of the residuum $\zeta_{up} - \hat{\zeta}_{up}(\zeta)$ which is simultaneously unbiased (i.e., $E[\hat{\zeta}_{up}(\zeta)] = E[\zeta_{up}]$, see p. 69) is considered to be the optimum approximation of the random variable ζ_{up}. If the unbiasedness condition of the sought approximation $\hat{\zeta}(\zeta)$ satisfying the optimality criterion (2.5.16) is fulfilled, this criterion becomes identical with the optimality criterion of the form (1.3.8).

Now, we are able to solve the formulated problem. The relation between the unknown vector parameters Θ and ϑ on the one hand and the mean value of the observation vector differs within various linear models of measurement, therefore the resulting expressions are given for individual models. The universality of Model 2 enables us to derive the necessary relations for it alone.

Thus, the problem is to determine a linear and unbiased approximation of the random variable $\zeta_{up} = \nu_{up} + f_p(\Theta_0) + s_p(\vartheta_0)$ within the model

$$\zeta = Jf(\Theta_0) + s(\vartheta_0) + JA\Delta\Theta + S\Delta\vartheta + \varphi_0 + \varepsilon,\ \nu = \zeta - Jf(\Theta_0) - s(\vartheta_0),$$

$$E[\nu] = JA\Delta\Theta + S\Delta\vartheta,\ E[\varphi_0] = 0,\ E[\varepsilon] = 0,$$

$$var(\nu) = var(\zeta) = var(\varepsilon) + var(\varphi_0),\ cov(\varepsilon, \varphi_0) = 0,$$

$$R(A) = k \leq p,\ R(S) = s \leq n,\ R(J) = p \leq n,\ R[(JA, S)] = k + s \leq n,$$
$$R[var(\nu)] = n.$$

The random vector $\nu = \zeta - Jf(\Theta_0) - s(\vartheta_0)$ and the random variable $\nu_{up} = \zeta_{up} - f_p(\Theta_0) - s_p(\vartheta_0)$, $E[\nu] = JA\Delta\Theta + S\Delta\vartheta$, $E[\nu_{up}] = a_p'\Delta\Theta + s_p'\Delta\vartheta$, $var(\nu) = var(\zeta)$ are taken into account here (the reasons for it are the same as when estimating the unknown parameters) under the assumption that $cov(\zeta, \zeta_{up}) = cov(\nu, \nu_{up})$ is known. Since $\zeta_{up}(\zeta) = \nu_{up}(\nu) + f_p(\Theta_0) + s_p(\vartheta_0)$ and $\hat{\zeta}_{up}(\zeta) - \zeta_{up} = \hat{\nu}_{up}(\nu) - \nu_{up}$, the problem is transformed into the problem of determining an unbiased approximation $\hat{\nu}_{up}(\nu)$ satisfying the requirement $D[\nu_{up} - \hat{\nu}_{up}(\nu)] = min$.

The linear approximation $L_p'\nu + l_p$ ($L_p \in \mathcal{R}^n$, $l_p \in \mathcal{R}^1$) of the random variable ν_{up} is unbiased if and only if $E[L_p'\nu + l_p] = E[\nu_{up}]$, which means that the equality

$$L_p'(JA, S)\begin{pmatrix}\Delta\Theta \\ \Delta\vartheta\end{pmatrix} + l_p = (a_p', s_p')\begin{pmatrix}\Delta\Theta \\ \Delta\vartheta\end{pmatrix}$$

has to be fulfilled for any vector $(\Delta\Theta', \Delta\vartheta)' \in \mathcal{R}^{k+s}$, which implies $l_p \equiv 0$; the set of the solutions of our problem has been restricted to the set of linear combinations of the form $\{L_p'\nu : L_p'(JA, S) = (a_p', s_p')\}$. The approximation belonging to this class, satisfying simultaneously the condition

$$D[\nu_{up} - L_p'\nu] = min$$

is said to be optimum.

Thus the minimum point of this function with the subsidiary condition has to be determined. The Lagrange auxiliary function is of the form

$$\Phi(L_p) = D[\nu_{up}] - 2L_p'cov(\nu, \nu_{up}) + L_p'var(\nu)L_p - 2[L_p'(JA, S) - (a_p', s_p')]\lambda,$$

where λ is a $(k+s)$-dimensional Lagrange multiplier. The stationary point of this function is the point determined by the solution of the equation

$$-cov(\nu, \nu_{up}) + var(\nu)L_p - (JA, S)\lambda = 0,$$

i.e., the point

$$L_{p0} = var^{-1}(\nu)cov(\nu, \nu_{up}) + var^{-1}(\nu)(JA, S)\lambda; \qquad (2.5.17)$$

the $(k+s)$-dimensional multiplier λ is a solution of the equation

$$cov(\nu_{up},\nu)var^{-1}(\nu)(JA,S)+\lambda'(JA,S)'var^{-1}(\nu)(JA,S)-(a'_p,s'_p)=0$$

obtained by inserting the expression (2.5.17) into the condition $L'_p(JA,S)=(a'_p,s'_p)$. It implies

$$\lambda'=[(a'_p,s'_p)-cov(\nu_{up},\nu)var^{-1}(\nu)(JA,S)][(JA,S)'var^{-1}(\nu)(JA,S)]^{-1},$$

and finally

$$L'_{p0}=cov(\nu_{up},\nu)var^{-1}(\nu)\{I-(JA,S)[(JA,S)'var^{-1}(\nu)(JA,S)]^{-1}(JA,S)\cdot$$
$$\cdot var^{-1}(\nu)\}+(a'_p,s'_p)[(JA,S)'var^{-1}(\nu)(JA,S)]^{-1}(JA,S)var^{-1}(\nu).$$

This stationary point is the minimum point of the Lagrange function (the reader can convince himself of it using the proper method based on the definition of the positive definiteness of the matrix $var(\nu)$, see also p. 133).

The optimum (in the sense of the criterion $D[\nu_{up}-\hat{\nu}_{up}(\nu)]=min$) linear unbiased approximation of the random variable ν_{up} is of the form

$$\hat{\nu}_{up}(\nu)=cov(\nu_{up},\nu)var^{-1}(\nu)\{\nu-(JA,S)[(JA,S)'var^{-1}(\nu)(JA,S)]^{-1}\cdot$$
$$\cdot(JA,S)'var^{-1}(\nu)\nu\}+$$
$$+(a'_p,s'_p)[(JA,S)'var^{-1}(\nu)(JA,S)]^{-1}(JA,S)'var^{-1}(\nu)\nu.$$

Comparing this expression with (2.5.2), we conclude that it can be recorded in the form

$$\hat{\nu}_{up}(\nu)=cov(\nu_{up},\nu)var^{-1}(\nu)\left[\nu-(JA,S)\begin{pmatrix}\widehat{\Delta\Theta}(\nu)\\ \widehat{\Delta\vartheta}(\nu)\end{pmatrix}\right]+$$
$$+(a'_p,s'_p)\begin{pmatrix}\widehat{\Delta\Theta}(\nu)\\ \widehat{\Delta\vartheta}(\nu)\end{pmatrix}. \tag{2.5.18}$$

Here $(JA,S)\begin{pmatrix}\widehat{\Delta\Theta}(\nu)\\ \widehat{\Delta\vartheta}(\nu)\end{pmatrix}=JA\widehat{\Delta\Theta}(\nu)+S\widehat{\Delta\vartheta}(\nu)=\hat{E}[\nu]$ represents the unbiased and jointly efficient estimator of the mean value of the observation vector (we know that $E[\nu]=JA\Delta\Theta+S\Delta\vartheta$) and, analogously,

$(a'_p, s'_p) \begin{pmatrix} \widehat{\Delta\Theta}(\nu) \\ \widehat{\Delta\vartheta}(\nu) \end{pmatrix} = a'_p \widehat{\Delta\Theta}(\nu) + s'_p \widehat{\Delta\vartheta}(\nu) = \hat{E}[\nu_{up}] = \hat{E}[\nu_p]$ is the unbiased and jointly efficient estimator of the mean value of the predicted random variable ν_p and it, as a consequence of the fact that the mean value of the error components is assumed to be zero, is identical with the estimator of the mean value of the predicted useful component. Using these facts for rewriting (2.5.18), we obtain the relation

$$\hat{\nu}_{up}(\nu) = \hat{E}[\nu_{up}] + cov(\nu_{up}, \nu) var^{-1}(\nu)(\nu - \hat{E}[\nu]) \tag{2.5.19}$$

which is a statistical analogy of the relation (1.3.9).

Substituting now for $\hat{E}[\nu_{up}]$ and $\hat{E}[\nu]$ in (2.5.19) the expressions corresponding to the individual models, we obtain the explicit formulas for the optimum unbiased linear approximation of the predicted useful component of the observation vector within these linear models of measurement.

Before writing these expressions, we derive the dispersion of the random variable $\hat{\nu}_{up}(\nu)$. In accordance with the definition

$$D[\hat{\nu}_{up}(\nu)] = D[\hat{E}[\nu_{up}]] + D[cov(\nu_{up}, \nu) var^{-1}(\nu)(\nu - \hat{E}[\nu])] +$$

$$+ 2cov\{\hat{E}[\nu_{up}],\ cov(\nu_{up}\nu) var^{-1}(\nu)(\nu - \hat{E}[\nu])\}$$

(it has to be realized that an estimator of the mean value is a random variable, thus it is characterized by its dispersion, which implies furthermore, that $D[\nu - E[\nu]] \neq D[\nu - \hat{E}[\nu]]$). Since

$$D[cov(\nu_{up}, \nu) var^{-1}(\nu)(\nu - \hat{E}[\nu])] = D[cov(\nu_{up}, \nu) var^{-1}(\nu)\nu] +$$

$$+ D[cov(\nu_{up}, \nu) var^{-1}(\nu)\hat{E}[\nu]] -$$

$$- 2cov(\nu_{up}, \nu) var^{-1}(\nu) cov(\nu, \hat{E}[\nu]) var^{-1}(\nu) cov(\nu, \nu_{up})$$

and

$$cov\{\hat{E}[\nu_{up}], cov(\nu_{up}, \nu) var^{-1}(\nu)(\nu - \hat{E}[\nu])\} =$$

$$= cov(\nu_{up}, \nu) var^{-1}(\nu) cov(\nu, \hat{E}[\nu_{up}]) - cov(\nu_{up}, \nu) var^{-1}(\nu) cov(\hat{E}[\nu], \hat{E}[\nu_{up}]),$$

the next problem is to determine $var(\hat{E}[\nu])$, $cov(\nu, \hat{E}[\nu])$, $cov(\nu, \hat{E}[\nu_{up}])$ and $cov(\hat{E}[\nu], \hat{E}[\nu_{up}])$. Within Model 2

$$var(\hat{E}[\nu]) = var\left[(JA, S)\begin{pmatrix}\widehat{\Delta\Theta}(\nu)\\ \widehat{\Delta\vartheta}(\nu)\end{pmatrix}\right] = (JA, S)\begin{pmatrix}\widehat{\Delta\Theta}(\nu)\\ \widehat{\Delta\vartheta}(\nu)\end{pmatrix}\begin{pmatrix}A'J'\\ S'\end{pmatrix} =$$

$$= (JA, S)\begin{pmatrix}N^{11}, & N^{12}\\ N^{21}, & N^{22}\end{pmatrix}\begin{pmatrix}A'J'\\ S'\end{pmatrix} = cov(\nu, \hat{E}[\nu]),$$

$$cov\left[\nu, (JA, S)\begin{pmatrix}\widehat{\Delta\Theta}(\nu)\\ \widehat{\Delta\vartheta}(\nu)\end{pmatrix}\right] =$$

$$= var(\nu)var^{-1}(\nu)(JA, S)\begin{pmatrix}N^{11}, & N^{12}\\ N^{21}, & N^{22}\end{pmatrix}\begin{pmatrix}A'J'\\ S'\end{pmatrix} =$$

$$= (JA, S)\begin{pmatrix}N^{11}, & N^{12}\\ N^{21}, & N^{22}\end{pmatrix}\begin{pmatrix}A'J'\\ S'\end{pmatrix},$$

further, taking into account that $\hat{E}[\nu_{up}] = (a_p', s_p')\begin{pmatrix}\widehat{\Delta\Theta}(\nu)\\ \widehat{\Delta\vartheta}(\nu)\end{pmatrix}$, we get

$$cov(\nu, \hat{E}[\nu_{up}]) = (JA, S)\begin{pmatrix}N^{11}, & N^{12}\\ N^{21}, & N^{22}\end{pmatrix}\begin{pmatrix}a_p\\ s_p\end{pmatrix}$$

and

$$cov(\hat{E}[\nu], \hat{E}[\nu_{up}]) = (JA, S)\begin{pmatrix}N^{11}, & N^{12}\\ N^{21}, & N^{22}\end{pmatrix}\begin{pmatrix}a_p\\ s_p\end{pmatrix}$$

(the relations (2.5.9) and (2.5.10) have been applied here). These partial results show that the random variables $\hat{E}[\nu]$ and $(\nu - \hat{E}[\nu])$ are uncorrelated because of $cov(\hat{E}[\nu_{up}], \nu - \hat{E}[\nu]) = cov(\hat{E}[\nu_{up}], \nu) - cov(\hat{E}[\nu_{up}], \hat{E}[\nu]) = 0$ and $var(\nu - \hat{E}[\nu]) = var(\nu) - var(\hat{E}[\nu])$, since $cov(\nu, \hat{E}[\nu]) = var(\hat{E}[\nu])$ (realize, that $var(\nu - E[\nu]) = var(\nu)$, as $E[\nu]$ is the theoretical mean value of the random vector ν, not its estimator) and as a consequence of this the sought dispersion of the random variable $\hat{\nu}_{up}(\nu)$ is

$$D[\hat{\nu}_{up}(\nu)] = D[\hat{E}[\nu_{up}]] + cov(\nu_{up}, \nu)var^{-1}(\nu)cov(\nu, \nu_{up}) -$$

$$- cov(\nu_{up}, \nu)var^{-1}(\nu)var(\hat{E}[\nu])var^{-1}(\nu)cov(\nu, \nu_{up}); \tag{2.5.20}$$

this expression holds for each of the fundamental linear models of measurement.

Now, we are fully prepared for writing explicit formulas of **the optimum unbiased linear prediction of the useful component of the observation vector and its dispersion** within individual linear models of measurement.

Model 2 and its special case **Model 1**

$$\hat{E}[\nu] = (JA, S)\begin{pmatrix} N^{11}, & N^{12} \\ N^{21}, & N^{22} \end{pmatrix}\begin{pmatrix} A'J' \\ S' \end{pmatrix} var^{-1}(\nu)\nu,$$

$$var(\hat{E}[\nu]) = (JA, S)\begin{pmatrix} N^{11}, & N^{12} \\ N^{21}, & N^{22} \end{pmatrix}\begin{pmatrix} A'J' \\ S' \end{pmatrix},$$

$$\hat{E}[\nu_{up}] = (a_p', s_p')\begin{pmatrix} N^{11}, & N^{12} \\ N^{21}, & N^{22} \end{pmatrix}\begin{pmatrix} A'J' \\ S' \end{pmatrix} var^{-1}(\nu)\nu,$$

$$D[\hat{E}[\nu_{up}]] = (a_p', s_p')\begin{pmatrix} N^{11}, & N^{12} \\ N^{21}, & N^{22} \end{pmatrix}\begin{pmatrix} a_p \\ s_p \end{pmatrix},$$

$$\hat{\nu}_{up}(\nu) = \{[(a_p', s_p') - cov(\nu_{up}, \nu)var^{-1}(\nu)(JA, S)]\begin{pmatrix} N^{11}, & N^{12} \\ N^{21}, & N^{22} \end{pmatrix}\cdot$$

$$\cdot\begin{pmatrix} A'J' \\ S' \end{pmatrix} + cov(\nu_{up}, \nu)\}var^{-1}(\nu)\nu,$$

$$D[\hat{\nu}_{up}(\nu)] = (a_p', s_p')\begin{pmatrix} N^{11}, & N^{12} \\ N^{21}, & N^{22} \end{pmatrix}\begin{pmatrix} a_p \\ s_p \end{pmatrix} + cov(\nu_{up}, \nu)var^{-1}(\nu)\cdot$$

$$\cdot cov(\nu, \nu_{up}) - cov(\nu_{up}, \nu)var^{-1}(\nu)(JA, S)\begin{pmatrix} N^{11}, & N^{12} \\ N^{21}, & N^{22} \end{pmatrix}\begin{pmatrix} A'J' \\ S' \end{pmatrix}\cdot$$

$$\cdot var^{-1}(\nu)cov(\nu, \nu_{up}).$$

For $A = 1,\ J = i = (1, 1, \ldots, 1)'$ and $a_p = 1$ we obtain the expressions within Model 1.

Model 5 and its special case **Model 3**

$$\hat{E}[\nu] = (JA, S)\left\{\begin{pmatrix} N^{11}, & N^{12} \\ N^{21}, & N^{22} \end{pmatrix}\begin{pmatrix} A'J' \\ S' \end{pmatrix} var^{-1}(\nu)\nu - \right.$$

$$\left. - \begin{pmatrix} N^{11} \\ N^{21} \end{pmatrix} B'(BN^{11}B')^{-1}\left[b + B(N^{11}, N^{12})\begin{pmatrix} A'J' \\ S' \end{pmatrix} var^{-1}(\nu)\nu\right]\right\},$$

$$var(\hat{E}[\nu]) = (JA, S)\left[\begin{pmatrix} N^{11}, & N^{12} \\ N^{21}, & N^{22} \end{pmatrix} - \right.$$

$$\left. - \begin{pmatrix} N^{11} \\ N^{21} \end{pmatrix} B'(BN^{11}B')^{-1}B(N^{11}, N^{12})\right]\begin{pmatrix} A'J' \\ S' \end{pmatrix},$$

$$\hat{E}[\nu_{up}] = (a_p', s_p')\left\{\begin{pmatrix} N^{11}, & N^{12} \\ N^{21}, & N^{22} \end{pmatrix}\begin{pmatrix} A'J' \\ S' \end{pmatrix} var^{-1}(\nu)\nu - \right.$$

$$\left. - \begin{pmatrix} N^{11} \\ N^{21} \end{pmatrix} B'(BN^{11}B')^{-1}\left[b + B(N^{11}, N^{12})\begin{pmatrix} A'J' \\ S' \end{pmatrix} var^{-1}(\nu)\nu\right]\right\},$$

$$D[\hat{E}[\nu_{up}]] = (a_p', s_p')\left[\begin{pmatrix} N^{11}, & N^{12} \\ N^{21}, & N^{22} \end{pmatrix} - \right.$$

$$\left. - \begin{pmatrix} N^{11} \\ N^{21} \end{pmatrix} B'(BN^{11}B')^{-1}B(N^{11}, N^{12})\right]\begin{pmatrix} a_p \\ s_p \end{pmatrix},$$

$$\hat{\nu}_{up}(\nu) = \left\{[(a_p', s_p') - cov(\nu_{up}, \nu)var^{-1}(\nu)(JA, S)]\begin{pmatrix} N^{11}, & N^{12} \\ N^{21}, & N^{22} \end{pmatrix} \cdot \right.$$

$$\left. \begin{pmatrix} A'J' \\ S' \end{pmatrix} + cov(\nu_{up}, \nu)\right\} var^{-1}(\nu)\nu - [(a_p', s_p') - cov(\nu_{up}, \nu)var^{-1}(JA, S)]$$

$$\cdot \begin{pmatrix} N^{11} \\ N^{21} \end{pmatrix} B'(BN^{11}B')^{-1}\left[b + B(N^{11}, N^{12})\begin{pmatrix} A'J' \\ S' \end{pmatrix} var^{-1}(\nu)\nu\right],$$

$$D[\hat{\nu}_{up}(\nu)] = (a_p', s_p')\left[\begin{pmatrix} N^{11}, & N^{12} \\ N^{21}, & N^{22} \end{pmatrix} - \begin{pmatrix} N^{11} \\ N^{21} \end{pmatrix} B'(BN^{11}B')^{-1} \cdot \right.$$

$$\left. \cdot B(N^{11}, N^{12})\right]\begin{pmatrix} a_p \\ s_p \end{pmatrix} + cov(\nu_{up}, \nu)var^{-1}(\nu)cov(\nu, \nu_{up}) -$$

$$-cov(\nu_{up},\nu)var^{-1}(\nu)(JA,S)\left[\begin{pmatrix} N^{11}, & N^{12} \\ N^{21}, & N^{22} \end{pmatrix} - \right.$$

$$\left. - \begin{pmatrix} N^{11} \\ N^{21} \end{pmatrix} B'(BN^{11}B')^{-1}B(N^{11},N^{12})\right]\begin{pmatrix} A'J' \\ S' \end{pmatrix} var^{-1}(\nu)cov(\nu,\nu_{up}).$$

The relations for Model 3 are special cases of these relations for $A = I$ and $a_p = i' = (1,1,\ldots,1)$.

Model 6 and its special case **Model 4**

$$\hat{E}[\nu] = (JA,S)\left\{\begin{pmatrix} N^{11}, & N^{12} \\ N^{21}, & N^{22} \end{pmatrix}\begin{pmatrix} A'J' \\ S' \end{pmatrix} var^{-1}(\nu)\nu - \right.$$

$$\left. - \begin{pmatrix} N^{11} \\ N^{21} \end{pmatrix} B'P_{11}\left[b + B(N^{11},N^{12})\begin{pmatrix} A'J' \\ S' \end{pmatrix} var^{-1}(\nu)\nu\right]\right\},$$

$$var(\hat{E}[\nu]) = (JA,S)\left[\begin{pmatrix} N^{11}, & N^{12} \\ N^{21}, & N^{22} \end{pmatrix} - \right.$$

$$\left. - \begin{pmatrix} N^{11} \\ N^{21} \end{pmatrix} B'P_{11}B(N^{11},N^{12})\right]\begin{pmatrix} A'J' \\ S' \end{pmatrix},$$

$$\hat{E}[\nu_{up}] = (a'_p,s'_p)\left\{\begin{pmatrix} N^{11}, & N^{12} \\ N^{21}, & N^{22} \end{pmatrix}\begin{pmatrix} A'J' \\ S' \end{pmatrix} var^{-1}(\nu)\nu - \right.$$

$$\left. - \begin{pmatrix} N^{11} \\ N^{21} \end{pmatrix} B'P_{11}\left[b + B(N^{11},N^{12})\begin{pmatrix} A'J' \\ S' \end{pmatrix} var^{-1}(\nu)\nu\right]\right\},$$

$$D[\hat{E}[\nu_{up}]] = (a'_p,s'_p)\left[\begin{pmatrix} N^{11}, & N^{12} \\ N^{21}, & N^{22} \end{pmatrix} - \right.$$

$$\left. - \begin{pmatrix} N^{11} \\ N^{21} \end{pmatrix} B'P_{11}B(N^{11},N^{12})\right]\begin{pmatrix} a_p \\ s_p \end{pmatrix},$$

$$\hat{\nu}_{up}(\nu) = \left\{[(a'_p,s'_p) - cov(\nu_{up},\nu)var^{-1}(\nu)(JA,S)]\begin{pmatrix} N^{11}, & N^{12} \\ N^{21}, & N^{22} \end{pmatrix} \cdot \right.$$

$$\left. \cdot \begin{pmatrix} A'J' \\ S' \end{pmatrix} + cov(\nu_{up},\nu)\right\} var^{-1}(\nu)\nu -$$

$$-[(a'_p, s'_p) - cov(\nu_{up}, \nu)var^{-1}(\nu)(JA, S)]$$

$$\cdot\begin{pmatrix} N^{11} \\ N^{21} \end{pmatrix} B'P_{11}\left[b + B(N^{11}, N^{12})\begin{pmatrix} A'J' \\ S' \end{pmatrix} var^{-1}(\nu)\nu\right],$$

$$D[\hat{\nu}_{up}(\nu)] = (a'_p, s'_p)\left[\begin{pmatrix} N^{11}, & N^{12} \\ N^{21}, & N^{22} \end{pmatrix} - \begin{pmatrix} N^{11} \\ N^{21} \end{pmatrix} B'P_{11}B(N^{11}, N^{12})\right]\cdot$$

$$\cdot\begin{pmatrix} a_p \\ s_p \end{pmatrix} + cov(\nu_{up}, \nu)var^{-1}(\nu)cov(\nu, \nu_{up}) - cov(\nu_{up}, \nu)var^{-1}(\nu)(JA, S)\cdot$$

$$\cdot\left[\begin{pmatrix} N^{11}, & N^{12} \\ N^{21}, & N^{22} \end{pmatrix} - \begin{pmatrix} N^{11} \\ N^{21} \end{pmatrix} B'P_{11}B(N^{11}, N^{12})\right]\cdot$$

$$\cdot\begin{pmatrix} A'J' \\ S' \end{pmatrix} var^{-1}(\nu)cov(\nu, \nu_{up}).$$

The relations for Model 4 are special cases of these relations for $A = I$ and $a_p = i' = (1, 1, \ldots, 1)$.

Analyse now the relation

$$\hat{\nu}_{up}(\nu) = cov(\nu_{up}, \nu)var^{-1}(\nu)\nu - [cov(\nu_{up}, \nu)var^{-1}(\nu)(JA, S) - (a'_p, s'_p)]\cdot$$

$$\cdot[(JA, S)'var^{-1}(\nu)(JA, S)]^{-1}(JA, S)'var^{-1}(\nu)\nu$$

for the optimum linear prediction of the useful part of the observation vector in Model 2 in the case of $p = i \in \{1, 2, \ldots, n\}$ and $(a'_i, s'_i) = \{JA, S\}_{i\cdot}$ (the ith row of the design matrix (JA, S)). The reader can easily convince himself that

$$\{(JA, S)\}_{i\cdot} = cov(\nu_i, \nu)var^{-1}(\nu)(JA, S); \tag{2.5.21}$$

it suffices to realize that

$$cov(\nu_i, \nu)var^{-1}(\nu) = (cov(\nu_i, \nu_1), cov(\nu_i, \nu_2), \ldots, cov(\nu_i, \nu_n))$$

$$\begin{pmatrix} D[\nu_1], & cov(\nu_1, \nu_2), & \ldots, & cov(\nu_1, \nu_i), & \ldots, & cov(\nu_1, \nu_n) \\ cov(\nu_2, \nu_1), & D[\nu_2], & \ldots, & cov(\nu_2, \nu_i), & \ldots, & cov(\nu_2, \nu_n) \\ \ldots & \ldots & \ldots & \ldots & \ldots & \ldots \\ cov(\nu_i, \nu_1), & cov(\nu_i, \nu_2), & \ldots, & D[\nu_i], & \ldots, & cov(\nu_i, \nu_n) \\ \ldots & \ldots & \ldots & \ldots & \ldots & \ldots \\ cov(\nu_n, \nu_1), & cov(\nu_n, \nu_2), & \ldots, & cov(\nu_n, \nu_i), & \ldots, & D[\nu_n] \end{pmatrix}^{-1} =$$

$$= (0, 0, \ldots, 1, \ldots, 0)$$

(the n-dimensional random vector consisting of zeros with an exception of a unit in the ith place), which directly implies (2.5.21). The expression for $\hat{\nu}_{ui}(\nu)$ rewritten using this yields

$$\hat{\nu}_{ui}(\nu) = cov(\nu_{ui}, \nu)var^{-1}(\nu)\nu + [cov(\nu_i, \nu) - cov(\nu_{ui}, \nu)]var^{-1}(\nu)\cdot$$
$$\cdot(JA, S)[(JA, S)'var^{-1}(\nu)(JA, S)]^{-1}(JA, S)'var^{-1}(\nu)\nu. \tag{2.5.22}$$

Furthermore, $cov(\nu_i, \nu) - cov(\nu_{ui}, \nu) = cov(\nu_i - \nu_{ui}, \nu) = cov(\varepsilon_i, \varphi_0 + \varepsilon)$ (as $\nu_i - \nu_{ui} = \varphi_{0i} + \varepsilon_i - \varphi_{0i} = \varepsilon_i$ and $\nu = (JA, S)(\Delta\Theta', \Delta\vartheta')' + \varphi_0 + \varepsilon$). If, in addition to it, $cov(\varepsilon, \varphi_0) = 0$, then $cov(\nu_i, \nu) - cov(\nu_{ui}, \nu) = cov(\varepsilon_i, \varepsilon)$. The error component of the observation vector possesses the property that the correlation $\varrho(\varepsilon_i, \varepsilon_{i+m})$, i, $i + m \in \{1, 2, \ldots, n\}$ strongly decreases with increasing m (the case $\varrho(\varepsilon_i, \varepsilon_{i+m}) = 0$, $m \neq 0$ is frequent), which implies that the term

$$[cov(\nu_i, \nu) - cov(\nu_{ui}, \nu)]var^{-1}(\nu)(JA, S)[(JA, S)'var^{-1}(\nu)(JA, S)]^{-1}\cdot$$
$$\cdot(JA, S)'var^{-1}(\nu)\nu$$

in (2.5.22) may be characterized as a correction term which ensures that the approximation of the form (2.5.22) is unbiased (the same conclusion results from the following consideration: if the problem is to determine the optimum linear approximation of the observation vector, the correction term is zero; for the sake of completeness we must realize that in this case $\hat{\nu}_i(\nu) = \nu_i$). Neglecting this term, we obtain a linear approximation of the random variable of the form

$$\tilde{\nu}_{ui}(\nu) = cov(\nu_{ui}, \nu)var^{-1}(\nu)\nu,$$

satisfying the criterion $D[\tilde{\nu}_{ui}(\nu) - \nu_i] = min$ which is not unbiased. Its bias defined as the difference between the mean values of the random variables $\hat{\nu}_{ui}(\nu)$ and $\tilde{\nu}_{ui}(\nu)$ is

$$b(\tilde{\nu}_{ui}(\nu)) = \{(JA, S)\}_{i\cdot} \begin{pmatrix} \Delta\Theta \\ \Delta\vartheta \end{pmatrix} - cov(\nu_{ui}, \nu)var^{-1}(\nu)(JA, S) \begin{pmatrix} \Delta\Theta \\ \Delta\vartheta \end{pmatrix} =$$

$$= [cov(\nu_i, \nu) - cov(\nu_{ui}\nu)]var^{-1}(JA, S) \cdot \begin{pmatrix} \Delta\Theta \\ \Delta\vartheta \end{pmatrix}$$

and, as a consequence of the mentioned properties of the error component ε, it acquires neglectible values.

The results of this analysis can be generalized for any $p \notin \{1, 2, \ldots, n\}$. A practical consequence is a conclusion that the relation

$$\tilde{\nu}_{up}(\nu) = cov(\nu_{up}, \nu) var^{-1}(\nu)\nu,$$

whose dispersion is

$$D[\tilde{\nu}_{up}(\nu)] = cov(\nu_{up}, \nu) var^{-1} cov(\nu, \nu_{up}),$$

can be used instead of the rigorous relations (2.5.22), namely if the vector (a'_p, s'_p) is not known with a sufficient degree of accuracy (for further details see [22]).

2.6 The optimum elimination of systematic influences

A systematic influence that affects measurement results can be of either a deterministic or stochastic nature. It is self-evident that the measurement results can be influenced by both deterministic and stochastic effects simultaneously. A drift in the equipment used for measurement can serve as an example of a systematic deterministic influence, the influence of vibrations of a building in which a laboratory measurement is being performed is an example of a systematic stochastic influence.

The essential difference between the systematic influences of these types consists in the fact that as far as a deterministic systematic influence is concerned, it can be modelled by a function of an unknown vector parameter ϑ (if we take into account a drift in the equipment used, the function $s(\cdot)$ is, e.g., a polynomial of a certain degree, where the components of the vector ϑ are its coefficients), and a stochastic systematic influence can be modelled by an n-dimensional random variable φ_0, $E[\varphi_0] = 0$ with a known variance matrix $var(\varphi)$.

The mixed linear model given at the beginning of Section 2.5 represents an effective tool for describing the influence of systematic affects on the measurement results; analogously, the results derived in Section 2.5.1 can be used as an effective tool for obtaining immediately the algorithms for estimating unknown parameters with, optimumly, eliminated systematic affects.

In accordance with Section 2.2, let X be a p-dimensional vector of theoretical values of directly observable quantities of the measured object, ξ the n-dimensional random vector modelling the measurement of the directly observable components of the vector X, Θ is the theoretical value of a k-dimensional vector of unknown parameters characterizing the measured object whose LS-estimators are being determined on the basis of the measured results and $X = f(\vartheta_0) + A\Delta\Theta$ is the theoretical model expressing the relations between the theoretical value of a directly observable quantity X and the theoretical value of the unknown vector parameter $\Theta = \Theta_0 + \Delta\Theta$, $B\Delta\Theta + b = 0$, see Models 3 and 5, or $(B,C)(\Delta\Theta_1', \Delta\Theta_2')' + b = 0$, see Models 4 and 6, are the conditions binding the values of components of the parameter vector Θ.

If the measurement is not affected by any systematic influence, the theoretical model $X = f(\Theta_0) + A\Delta\Theta$ is assigned a stochastic model

$$\xi = Jf(\Theta_0) + JA\Delta\Theta + \varepsilon,\ B\Delta\Theta + b = 0,\ E[\xi] = Jf(\Theta_0) + JA\Delta\Theta,$$

$$E[\varepsilon] = 0,\ var(\xi) = var(\varepsilon),$$

where the regularity of the model is ensured by the conditions $R(A_{(p,k)}) = k \leq p$, $R(J_{(n,p)}) = p \leq n$, $R(JA) = k \leq n$, $R(B_{(q,k)}) = q \leq p$.

If the measurement of the vector X is affected by a systematic deterministic influence modelled by a function $s(\cdot)$, $\vartheta \in \mathcal{R}^s$ (if this function is not linear, the usual procedure consisting in its development into the Taylor series and in neglecting the terms of degrees higher than one, thus $s(\vartheta) \doteq s(\vartheta_0) + [\partial s(\vartheta_0)/\partial\vartheta']\Delta\vartheta = s(\vartheta_0) + S\Delta\vartheta$), then instead of a realization of the random variable ξ we get a realization of a random variable $\xi + s(\vartheta_0) + S\Delta\vartheta = Jf(\Theta_0) + JA\Delta\Theta + s(\vartheta_0) + S\Delta\vartheta + \varepsilon$, $E[\xi + s(\vartheta_0) + S\Delta\vartheta] = Jf(\Theta_0) + JA\Delta\Theta + s(\vartheta_0) + S\Delta\vartheta$, $var[\xi + s(\vartheta_0) + S\Delta\vartheta] = var(\xi) = var(\varepsilon)$; if this measurement is affected by a stochastic effect φ_0, $E[\varphi_0] = 0$, whose covariance matrix $var(\varphi_o)$ is known; instead of a realization of the random variable ξ we obtain a realization of a random variable $\xi + \varphi_0 = Jf(\vartheta_0) + JA\Delta\Theta + \varepsilon + \varphi_0$, $E[\xi + \varphi_0] = Jf(\Theta_0) + JA\Delta\Theta$, characterized by the variance matrix $var(\xi + \varphi_0) = var(\varepsilon) + var(\varphi_0)$ under the condition that $cov(\varepsilon, \varphi_0) = 0$; if the measurement considered is influenced by both deterministic and a stochastic effects, instead of a realization of the random variable ξ we get a realization of a random variable $\zeta = \xi + s(\vartheta_0) + S\Delta\vartheta + \varphi_0 = Jf(\Theta_0) + JA\Delta\Theta + s(\vartheta_0) + S\Delta\vartheta + \varepsilon + \varphi_o$, $E[\zeta] = Jf(\Theta_0) + s(\vartheta_0) + JA\Delta\Theta + S\Delta\vartheta$, $var(\zeta) = var(\varphi_0)$, if $cov(\varepsilon, \varphi_0) = 0$.

The random variable ζ is described by the mixed model (2.5.1′) in which now, when this model concerns the measurement influenced both by systematic deterministic and stochastic effects, the unknown parameter Θ characterizing the measured object is interpreted as a useful part, the vector parameter ϑ of the deterministic systematic influence and the random vector φ_0 modelling the stochastic systematic influence represent the noise part of the observation vector.

The problem is to determine the LS-estimator of the useful parameter Θ and its covariance matrix. The solution within individual models of measurement is given by the relations:

Model 1

$$\hat{\Theta}(\nu) = \Theta_0 + \{i'var^{-1}(\nu)i - i'var^{-1}(\nu)S[S'var^{-1}(\nu)S]^{-1}S'var^{-1}(\nu)i\}^{-1}\cdot$$
$$\cdot\{i' - i'var^{-1}(\nu)S[S'var^{-1}(\nu)S]^{-1}S'\}var^{-1}(\nu)\nu,$$
$$D[\hat{\Theta}(\nu)] = \{i'var^{-1}(\nu)i - i'var^{-1}(\nu)S[S'var^{-1}(\nu)S]^{-1}S'var^{-1}(\nu)i\}^{-1}; \tag{2.6.1}$$

Model 2

$$\hat{\Theta}(\nu) = \Theta_0 + \{A'J'var^{-1}(\nu)JA - A'J'var^{-1}(\nu)S[S'var^{-1}(\nu)S]^{-1}\cdot$$
$$\cdot S'var^{-1}(\nu)JA\}^{-1}\{A'J' - A'J'var^{-1}(\nu)S[S'var^{-1}(\nu)S]^{-1}S'\}var^{-1}(\nu)\nu,$$
$$var[\hat{\Theta}(\nu)] = \{A'J'var^{-1}(\nu)JA - A'J'var^{-1}(\nu)S[S'var^{-1}(\nu)S]^{-1}\cdot$$
$$\cdot S'var^{-1}(\nu)JA\}^{-1}; \tag{2.6.2}$$

Model 3

$$\hat{\Theta}(\nu) = \Theta_0 + (N^{11}J' + N^{12}S')var^{-1}(\nu)\nu - N^{11}B'(BN^{11}B')^{-1}[b+$$
$$+B(N^{11}J' + N^{12}S')var^{-1}(\nu)\nu],$$
$$var[\hat{\Theta}(\nu)] = N^{11} - N^{11}B'(BN^{11}B')^{-1}BN^{11},$$

where (2.6.3)

$$N^{11} = \{J'var^{-1}(\nu)J - J'var^{-1}(\nu)S[S'var^{-1}(\nu)S]^{-1}S'var^{-1}(\nu)J\}^{-1}$$

and

$$N^{12} = -N^{11}J'var^{-1}(\nu)S[S'var^{-1}(\nu)S]^{-1};$$

Model 4

$$\begin{pmatrix} \hat{\Theta}_1(\nu) \\ \hat{\Theta}_2(\nu) \end{pmatrix} = \begin{pmatrix} \Theta_{10} \\ \Theta_{20} \end{pmatrix} + \begin{pmatrix} N^{11}, & N^{12} \\ 0, & 0 \end{pmatrix} \begin{pmatrix} J' \\ S' \end{pmatrix} var^{-1}(\nu)\nu -$$

$$- \begin{pmatrix} N^{11}B'P_{11} \\ P_{21} \end{pmatrix} \left[b + B(N^{11}, N^{12}) \begin{pmatrix} J' \\ S' \end{pmatrix} var^{-1}(\nu)\nu \right] = \begin{pmatrix} \Theta_{10} \\ \Theta_{20}) \end{pmatrix} +$$

$$+ \begin{pmatrix} (N^{11}J' + N^{12}S')var^{-1}(\nu)\nu - N^{11}B'P_{11}[b + B(N^{11}J' + N^{12}S')var^{-1}(\nu)\nu] \\ -P_{21}[b + B(N^{11}J' + N^{12}S')var^{-1}(\nu)\nu] \end{pmatrix},$$

$$var \begin{pmatrix} \hat{\Theta}_1(\nu) \\ \hat{\Theta}_2(\nu) \end{pmatrix} = \begin{pmatrix} N^{11} - N^{11}B'P_{11}BN^{11}, & -N^{11}B'P_{12} \\ -P_{21}BN^{11}, & -P_{22} \end{pmatrix},$$

where

$$N^{11} = \{J'var^{-1}(\nu)J - J'var^{-1}(\nu)S[S'var^{-1}(\nu)S]^{-1}S'var^{-1}(\nu)J\}^{-1},$$

$$N^{12} = -N^{11}J'var^{-1}(\nu)S[S'var^{-1}(\nu)S]^{-1}$$

and

$$\begin{pmatrix} P_{11}, & P_{12} \\ P_{21}, & P_{22} \end{pmatrix} = \begin{pmatrix} BN^{11}B', & C \\ C', & 0 \end{pmatrix}^{-1};$$

Model 5

$$\hat{\Theta}(\nu) = \Theta_0 + (N^{11}A'J' + N^{12}S')var^{-1}(\nu)\nu - N^{11}B'(BN^{11}B')^{-1}[b+$$
$$B(N^{11}A'J' + N^{12}S')var^{-1}(\nu)\nu],$$

$$var[\hat{\Theta}(\nu)] = N^{11} - N^{11}B'(BN^{11}B')^{-1}BN^{11},$$

where (2.6.4)

$$N^{11} = \{A'J'var^{-1}(\nu)JA - A'J'var^{-1}(\nu)S[S'var^{-1}(\nu)S]^{-1}\cdot$$
$$\cdot S'var^{-1}(\nu)JA\}^{-1}$$

and

$$N^{12} = -N^{11}A'J'var^{-1}(\nu)S[S'var^{-1}(\nu)S]^{-1};$$

Model 6

$$\begin{pmatrix} \hat{\Theta}_1(\nu) \\ \hat{\Theta}_2(\nu) \end{pmatrix} = \begin{pmatrix} \Theta_{10} \\ \Theta_{20} \end{pmatrix} + \begin{pmatrix} N^{11}, & N^{12} \\ 0, & 0 \end{pmatrix} \begin{pmatrix} A'J' \\ S' \end{pmatrix} var^{-1}(\nu)\nu -$$

$$- \begin{pmatrix} N^{11}B'P_{11} \\ P_{21} \end{pmatrix} \left[b + B(N^{11}, N^{12}) \begin{pmatrix} A'J' \\ S' \end{pmatrix} var^{-1}(\nu)\nu \right] = \begin{pmatrix} \Theta_{10} \\ \Theta_{20} \end{pmatrix} +$$

$$\begin{pmatrix} (N^{11}A'J' + N^{12}S')var^{-1}(\nu)\nu - N^{11}B'P_{11}[b + B(N^{11}A'J' + N^{12}S')var^{-1}(\nu)\nu] \\ -P_{21}[b + B(N^{11}A'J' + N^{12}S')var^{-1}(\nu)\nu] \end{pmatrix},$$

$$var \begin{pmatrix} \hat{\Theta}_1(\nu) \\ \hat{\Theta}_2(\nu) \end{pmatrix} = \begin{pmatrix} N^{11} - N^{11}B'P_{11}BN^{11}, & -N^{11}B'P_{12} \\ -P_{21}BN^{11}, & -P_{22} \end{pmatrix},$$

where (2.6.6)

$$N^{11} = \{A'J'var^{-1}(\nu)JA - A'J'var^{-1}(\nu)S[S'var^{-1}(\nu)S]^{-1}.$$
$$\cdot S'var^{-1}(\nu)JA\}^{-1},$$
$$N^{12} = -N^{11}A'J'var^{-1}(\nu)S[S'var^{-1}(\nu)S]^{-1}$$

and

$$\begin{pmatrix} P_{11}, & P_{12} \\ P_{21}, & P_{22} \end{pmatrix} = \begin{pmatrix} BN^{11}B', & C \\ C', & 0 \end{pmatrix}^{-1},$$

which have been derived in Section 2.5.1 (the same symbols were used there). These relations express the linear unbiased and jointly efficient estimators of the unknown useful parameters such that the perturbing systematic effect of both types is eliminated optimumly (i.e., in the sense of the least squares method).

If the measurement is affected by a systematic stochastic affect only, all the terms containing the matrix S are avoided and the matrix $var(\nu)$ acquires the form $var(\nu) = var(\zeta) = var(\varepsilon + \varphi_0)$ $(= var(\varepsilon) + var(\varphi_0)$, if $cov(\varepsilon, \varphi_0) = 0$, i.e., if the error component ε of the observation vector and the random vector modelling the stochastic systematic affect are uncorrelated). If the measurement is influenced by a systematic deterministic effect only and the systematic stochastic effect is absent, then $var(\nu) = var(\varepsilon) = \Sigma$. If the measurement results are not influenced by any systematic effects, the expressions for the LS-estimator of the unknown parameter Θ are identical with the corresponding expressions within the simple forms of individual models of measurement (compare also with the relations derived in Section 2.4).

An advantage of this statistically rigorous approach to the optimum elimination of systematic influences consists in it not being necessary to enumerate the optimum estimator of the noise parameter ϑ.

Intuitively, the procedure of producing a priori elimination of systematic effects directly from the measured data and the processing of the clean data using well-known procedures for determining the LS-estimator of the unknown parameter Θ seems to be more logical.

A procedure of producing a priori elimination of the systematic influences of both types is modelled by a random variable

$$\nu^{(el)} = \nu - S\Delta\vartheta - \varphi_0, \tag{2.6.7}$$

where $\Delta\vartheta$ is the theoretical value of the parameter of the systematic deterministic affect (rigorously speaking, if $s(\vartheta_0) \neq 0$, certain elimination of the systematic deterministic influence is performed within the transformation of the random vector ζ to the random vector $\nu = \zeta - Jf(\Theta_0) - s(\vartheta_0)$; this operation causes a change of the mean value, $E[\nu] = E[\zeta] - Jf(\Theta_0) - s(\vartheta_0)$, it causes no change in the variance matrix, $var(\nu) = var(\zeta)$, as neither $Jf(\Theta_0)$ nor $s(\vartheta_0)$ are random variables; the case $s(\vartheta_0) \equiv 0$ occurs very frequently in practical situations). The theoretical value of the vector parameter ϑ is unknown, similarly the vector φ_0 is not known in its separate form. Unique and simultaneously full information on systematic influences affecting the measurement is obtained directly in the measurement results, i.e., in a realization of the random variable ν. Each a priori elimination of systematic affects must be based on it alone. Applying the results obtained in Sections 2.5.1 and 2.5.2, we get the optimum approximation $\hat{\nu}^{(el)}(\nu)$

$$\hat{\nu}^{(el)}(\nu) = \nu - S\widehat{\Delta\vartheta}(\nu) - \hat{\varphi}_0(\nu),$$

of the random variable $\nu^{(el)}$ defined by (2.6.7); here $\widehat{\Delta\vartheta}(\nu)$ is the LS-estimator of the parameter $\Delta\vartheta$ (in Section 2.5.1 its explicit forms for individual measurement models are derived), $\hat{\varphi}_0(\nu)$ is the optimum linear approximation of the n-dimensional random vector modelling the systematic stochastic effect. An explicit expression for $\hat{\varphi}_0(\nu)$ can be obtained by a modification of (2.5.19). As the optimum linear approximation of the components of the random vector φ_0 at the observation points is sought, the index p attains the values $p = 1, 2, \dots, n$. For any $i \notin \{1, 2, \dots, n\}$ the relationship (2.5.19) is of the form

$$\hat{\nu}_{ui}(\nu) = \hat{E}[\nu_{ui}] + cov(\nu_{ui}, \nu)var^{-1}(\nu - \hat{E}[\nu])$$

and expresses the optimum linear approximation of the ith component of the vector ν_u (the same symbols as in Section 2.5.2 are used here); thus the random variable

$$\hat{\nu}_{ui}(\nu) - \hat{E}[\nu_{ui}] = cov(\nu_{ui}, \nu)var^{-1}(\nu - \hat{E}[\nu])$$

is the optimum linear approximation of the ith component of the random vector φ_0; therefore the sought vector $\hat{\varphi}_0(\nu)$, which is the optimum linear approximation of the vector φ_0 modelling the systematic random influence, is of the form

$$\hat{\varphi}_0(\nu) = \hat{\nu}_u(\nu) - \hat{E}[\nu_u] = cov(\nu_u, \nu)var^{-1}(\nu)(\nu - \hat{E}[\nu]),$$

where $cov(\nu_u, \nu)$ is the $n \times n$-dimensional matrix of covariances between the components of the observation vector and its useful component ν_u, $\hat{E}[\nu]$ is an estimator of the mean value of the observation vector (for their explicit forms within individual models of measurement, see Section 2.5.2).

The mean value, the variance matrix of the random vector $\hat{\varphi}_0(\nu)$, and its covariance with the observation vector ν are

$$E[\hat{\varphi}_0(\nu)] = 0$$

($\hat{E}[\nu]$ is an unbiased estimator of the mean value of the observation vector ν, thus $E[\hat{E}[\nu]] = E[\nu]$),

$$var[\hat{\varphi}_0(\nu)] = cov(\nu_u, \nu)var^{-1}(\nu)var(\nu - \hat{E}[\nu])var^{-1}(\nu)cov(\nu, \nu_u) =$$

$$= cov(\nu_u, \nu)var^{-1}(\nu)cov(\nu, \nu_u) -$$

$$- cov(\nu_u, \nu)var^{-1}(\nu)var(\hat{E}[\nu])var^{-1}(\nu)cov(\nu, \nu_u)$$

(since $var(\nu - \hat{E}[\nu]) = var(\nu) - var(\hat{E}[\nu])$, see p. 168) and

$$cov[\nu, \hat{\varphi}_0(\nu)] = cov(\nu, \nu - \hat{E}[\nu])var^{-1}(\nu)cov(\nu, \nu_u) =$$

$$= cov(\nu, \nu_u) - var(\hat{E}[\nu])var^{-1}(\nu)cov(\nu, \nu_u)$$

(because of $cov(\nu, \hat{E}[\nu]) = var(\hat{E}[\nu])$, see p. 168), respectively.

The mean value and the covariance matrix of the random vector $\hat{\nu}^{(el)}(\nu) = \nu - S\widehat{\Delta\vartheta}(\nu) - \hat{\varphi}_0(\nu)$ are

$$E[\hat{\nu}^{(el)}(\nu)] = E[\nu] - S\Delta\vartheta$$

(realize that $\widehat{\Delta\vartheta}(\nu)$ is an unbiased estimator of the parameter $\Delta\vartheta$ and

$$var[\hat{\nu}^{(el)}(\nu)] = var(\nu) + S var[\widehat{\Delta\vartheta}(\nu)]S' + var[\hat{\varphi}_0(\nu)] -$$

$$-cov[\nu, \widehat{\Delta\vartheta}(\nu)]S' - S cov[\widehat{\Delta\vartheta}(\nu), \nu] - cov[\nu, \hat{\varphi}_0(\nu)] - cov[\hat{\varphi}_0(\nu), \nu] +$$

$$+S cov[\widehat{\Delta\vartheta}(\nu), \hat{\varphi}_0(\nu)] + cov[\hat{\varphi}_0(\nu), \widehat{\Delta\vartheta}(\nu)]S',$$

respectively.

This expression for the variance of the random vector $\hat{\nu}^{(el)}(\nu)$ can be simplified as consequence of the fact that the LS-estimator of the parameter $\Delta\vartheta$ and the LS-approximation of the random component φ_0 are uncorrelated. This assertion will be proved, as usual, for Model 2:

$$cov[\widehat{\Delta\vartheta}(\nu), \hat{\varphi}_0(\nu)] =$$

$$= cov\left[(N^{21}, N^{22})\begin{pmatrix} A'J' \\ S' \end{pmatrix} var^{-1}(\nu)\nu, cov(\nu_u, \nu)var^{-1}(\nu)\nu - \right.$$

$$\left. - cov(\nu_u, \nu)var^{-1}(\nu)(JA, S)\begin{pmatrix} N^{11}, & N^{12} \\ N^{21}, & N^{22} \end{pmatrix}\begin{pmatrix} A'J' \\ S' \end{pmatrix} var^{-1}(\nu)\nu\right] =$$

$$= \left[(N^{21}, N^{22})\begin{pmatrix} A'J' \\ S' \end{pmatrix} var^{-1}(\nu) - (N^{21}, N^{22})\begin{pmatrix} A'J' \\ S' \end{pmatrix} var^{-1}(\nu)\cdot\right.$$

$$\left.\cdot(JA, S)\begin{pmatrix} N^{11}, & N^{12} \\ N^{21}, & N^{22} \end{pmatrix}\begin{pmatrix} A'J' \\ S' \end{pmatrix} var^{-1}(\nu)\right] cov(\nu, \nu_u) = 0.$$

Thus the final expression for the variance of the random vector $\hat{\nu}^{(el)}(\nu)$ reads

$$var[\hat{\nu}^{(el)}(\nu)] = var(\nu) + S var[\widehat{\Delta\vartheta}(\nu)]S' + var[\hat{\varphi}_0(\nu)] -$$

$$-cov[\nu, \widehat{\Delta\vartheta}(\nu)]S' - S cov[\widehat{\Delta\vartheta}(\nu), \nu] - cov[(\nu, \hat{\varphi}_0(\nu)] - cov[\hat{\varphi}_0(\nu), \nu]. \tag{2.6.8}$$

Now we shall prove that the $n \times n$-dimensional variance matrix of the form (2.6.8) must be singular, i.e., that $R(var[\hat{\nu}^{(el)}(\nu)]) < n$. The assertion will be proved for Model 2 again: here

$$\hat{\nu}^{(el)}(\nu) = \nu - S\widehat{\Delta\vartheta}(\nu) - \hat{\varphi}_0(\nu) =$$

$$= \left[I - S(N^{21}A'J' + N^{22}S')var^{-1}(\nu) - cov(\nu_u, \nu)var^{-1}(\nu) + \right.$$

$$+ cov(\nu_u,\nu)var^{-1}(\nu)(JA,S)\begin{pmatrix} N^{11}, & N^{12} \\ N^{21}, & N^{22}\end{pmatrix}\begin{pmatrix} A'J' \\ S'\end{pmatrix} var^{-1}(\nu)\Big]\,\nu =$$

$$= (I - E_0)\nu,$$

where E_0 is an elimination matrix of the form

$$E_0 = S(N^{21}A'J' + N^{22}S')var^{-1}(\nu) - cov(\nu_u,\nu)var^{-1}(\nu) -$$

$$-cov(\nu_u,\nu)var^{-1}(\nu)(JA,S)\begin{pmatrix} N^{11}, & N^{12} \\ N^{21}, & N^{22}\end{pmatrix}\begin{pmatrix} A'J' \\ S'\end{pmatrix} var^{-1}(\nu).$$

We shall show that $E_0S = S$ and $E_0JA = 0$ (there exists a whole class of matrices with these properties, not each of them ensures the optimum elimination of systematic effects). The facts that

$$S(N^{21}A'J' + N^{22}S')var^{-1}(\nu)S =$$

$$= S(0,I)\begin{pmatrix} N^{11}, & N^{12} \\ N^{21}, & N^{22}\end{pmatrix}\begin{pmatrix} A'J' \\ S'\end{pmatrix} var^{-1}(\nu)(JA,S)\begin{pmatrix} 0 \\ I\end{pmatrix} =$$

$$= S(0,I)\begin{pmatrix} 0 \\ I\end{pmatrix} = S$$

and

$$(JA,S)\begin{pmatrix} N^{11}, & N^{12} \\ N^{21}, & N^{22}\end{pmatrix}\begin{pmatrix} A'J' \\ S'\end{pmatrix} var^{-1}(\nu)S =$$

$$= (JA,S)\begin{pmatrix} N^{11}, & N^{12} \\ N^{21}, & N^{22}\end{pmatrix}\begin{pmatrix} A'J' \\ S'\end{pmatrix} var^{-1}(\nu)(JA,S)\begin{pmatrix} 0 \\ I\end{pmatrix} =$$

$$= (JA,S)\begin{pmatrix} 0 \\ I\end{pmatrix} = S$$

immediately imply that $E_0S = S$. Analogously, since

$$S(N^{21}A'J' + N^{22}S')var^{-1}(\nu)JA =$$

$$= S(0,I)\begin{pmatrix} N^{11}, & N^{12} \\ N^{21}, & N^{22}\end{pmatrix}\begin{pmatrix} A'J' \\ S'\end{pmatrix} var^{-1}(\nu)(JA,S)\begin{pmatrix} I \\ 0\end{pmatrix} =$$

$$= S(0,I)\begin{pmatrix} I \\ 0\end{pmatrix} = 0$$

and

$$(JA, S)\begin{pmatrix} N^{11}, & N^{12} \\ N^{21}, & N^{22} \end{pmatrix}\begin{pmatrix} A'J' \\ S' \end{pmatrix} var^{-1}(\nu)JA =$$

$$= (JA, S)\begin{pmatrix} N^{11}, & N^{12} \\ N^{21}, & N^{22} \end{pmatrix}\begin{pmatrix} A'J' \\ S' \end{pmatrix} var^{-1}(\nu)(JA, S)\begin{pmatrix} I \\ 0 \end{pmatrix} = JA$$

we obtain $E_0JA = 0$.

The fact $E_0S = S$ implies that the rank of the $m \times n$-dimensional transformation matrix $(I - E_0)$ must be smaller than n. By definition, namely, the rank of the matrix states the number of its linearly independent columns or rows. As a consequence of the fact that $(I - E_0)S = 0$ $(S \not\equiv 0)$, the rows of the matrix $(I - E_0)$ are linearly dependent[7].

The covariance matrix $var[\hat{\nu}^{(el)}(\nu)]$ can be expressed by means of the transformation matrix $(I - E_0)$ in the following way

$$var[\hat{\nu}^{(el)}(\nu)] = var[(I - E_0)\nu] = (I - E_0)var(\nu)(I - E_0)';$$

its rank, as a consequence of the fact that $R(I-E_0) < n$, is $R(var[\hat{\nu}^{(el)}(\nu)]) = R[(I - E_0)var(\nu)(I - E_0)'] < n$, however $R[var(\nu)] = n$.

This implies a very important conclusion: For determining the LS-estimator of the parameter Θ from the data after eliminating systematic affects (of the deterministic or the stochastic type) the relations derived in Section 2.4 cannot be used because of the assumption of the regularity of the covariance matrix of the observation vector y. The unique statistically rigorous way consists in applying the universal model ([21], p. 186), in which the fact that the regularity conditions are not fulfilled is taken into account; instead of inverses of regular matrices the special kinds of generalized inverses of non-regular matrices are applied within it; using them we obtain

$$\widehat{\Delta\Theta}(\hat{\nu}^{(el)}(\nu)) = [(A'J')^-_{m\{var[\hat{\nu}^{(el)}(\nu)]\}}]'\hat{\nu}^{(el)}(\nu),$$

$$var[\widehat{\Delta\Theta}(\hat{\nu}^{(el)}(\nu))] = [(A'J')^-_{m\{var[\hat{\nu}^{(el)}(\nu)]\}}]'var[\hat{\nu}^{(el)}(\nu)](A'J')^-_{m\{var[\hat{\nu}^{(el)}(\nu)]\}}, \tag{2.6.9}$$

where $(A'J')^-_{m\{var[\hat{\nu}^{(el)}(\nu)]\}}$ is the minimum $var[\hat{\nu}^{(el)}(\nu)]$-seminorm generalized inverse of the matrix $A'J'$ (a generalized inverse of a matrix A, which

---

[7] The following theorem can be proved: A homogeneous linear equation $x_1\alpha_1 + x_2\alpha_2 + \ldots + x_n\alpha_n = 0$, where $\alpha_1, \alpha_2, \ldots, \alpha_n$ are given n-dimensional vectors and $x_1, x_2 \ldots, x_n \in \mathcal{R}^1$ are unknown, possesses a non-trivial solution $(x_1, x_2, \ldots, x_n)' \neq 0$ if and only if the vectors $\alpha_1, \alpha_2, \ldots, \alpha_n$ are linearly dependent.

need not be either square or regular, is a matrix A^- satisfying the condition $AA^-A = A$; let N be any positive semidefinite matrix and the matrix A possess the full rank in columns, then $(A')^-_{m(N)}$ is a generalized inverse of the matrix A ensuring the joint efficiency of an unbiased estimator of the parameter Θ, see pp. 240–242; for further detail concerning the generalized inverses of matrices see, e.g., [21], pp. 35 and 187).

Applying the theorem of the equivalence of estimating models ([21], p. 286) we can prove that the estimator (2.6.9) is equivalent with the estimator (2.6.2) of the form

$$\widehat{\Delta\Theta}(\nu) =$$

$$= \{A'J'var^{-1}(\nu)JA - A'J'var^{-1}(\nu)S[S'var^{-1}(\nu)S]^{-1}S'var^{-1}(\nu)JA\}^{-1}\cdot$$

$$\cdot\{A'J'var^{-1}(\nu) - A'J'var^{-1}(\nu)S[S'var^{-1}(\nu)S]^{-1}S'var^{-1}(\nu)\}\nu,$$

$$var[\widehat{\Delta\Theta}(\nu)] =$$

$$= \{A'J'var^{-1}(\nu)JA - A'J'var^{-1}(\nu)S[S'var^{-1}(\nu)S]^{-1}S'var^{-1}(\nu)JA\}^{-1}.$$

At the end the **fundamental assertion on the elimination of systematic affects** has to be stated: In Model 2 there does not exist any better (in the sense of the minimum least squares principle) estimator of the parameter ϑ based on the input data after the eliminated systematic affect than the estimators of the form (2.6.2) or (2.6.9), which are equivalent.

The proof of this assertion, which goes beyond the framework of this publication, can be found in [21], p. 286.

The fundamental assertion of the elimination of systematic affects ensures that the best estimators of the parameter Θ are obtained within individual models of measurement by means of the relations (2.6.1) to (2.6.6).

The special structure of the model for processing data influenced by a systematic deterministic and a systematic stochastic effect after their elimination, when the input random vector is of the form

$$\hat{\nu}^{(el)}(\nu) = \nu - S\widehat{\Delta\vartheta}(\nu) - \hat{\varphi}_0(\nu)$$

and is characterized by the singular covariance matrix

$$var[\hat{\nu}^{(el)}(\nu)] = var(\nu) + Svar[\widehat{\Delta\vartheta}(\nu)]S' + var[\hat{\varphi}_0(\nu)] - cov[\nu, \widehat{\Delta\vartheta}(\nu)]S' -$$

$$-Scov[\widehat{\Delta\vartheta}(\nu), \nu] - cov[\nu, \hat{\varphi}_0(\nu)] - cov[\hat{\varphi}_0(\nu), \nu],$$

enables us to write simple explicit expressions for the LS-estimators of the useful parameter Θ and its covariance matrix equivalent to the corresponding expressions within the calculation of the LS-estimators based on the original data, i.e., on the observation vector ν.

This special structure of the model for processing the inputs with a priori eliminated systematic influences consists in the fact that, although $var[\hat{\nu}^{(el)}(\nu)] = (I - E_0)var(\nu)(I - E_0)'$ is a singular matrix, the matrix $var(\nu)$ is regular and $(I - E_0)JA = JA$, where the matrix JA is assumed to possess the full rank in columns. As a consequence of these special properties, it can be proved that the matrix $A'J'var^-[\hat{\nu}^{(el)}(\nu)]JA$ is independent of the concrete choice of the generalized inverse of the matrix $var[\hat{\nu}^{(el)}(\nu)]$ and moreover is regular; this implies the existence of its inverse $\{A'J'var^-[\hat{\nu}^{(el)}(\nu)]JA\}^{-1}$, thus the estimator

$$\hat{\Theta}(\hat{\nu}^{(el)}(\nu)) = \Theta_0 + \{A'J'var^-[\hat{\nu}^{(el)}(\nu)]JA\}^{-1}A'J'var^-[\hat{\nu}^{(el)}(\nu)]\hat{\nu}^{(el)}(\nu),$$

$$var[\hat{\Theta}(\hat{\nu}^{(el)}(\nu))] = \{A'J'var^-[\hat{\nu}^{(el)}(\nu)]JA\}^{-1}$$

is equivalent with the estimator of the form (2.6.2) within Model 2 of the measurement (this assertion is a simple consequence of the theorem on the equivalence of the estimators for our case $JA = (I - E_0)JA$, see [21], p. 268.

This assertion enables us to write the following review of explicit relations within individual models of measurement:

Model 1

$$\hat{\Theta}[\hat{\nu}^{(el)}(\nu)] = \Theta_0 + \{i'var^-[\hat{\nu}^{(el)}(\nu)]i\}^{-1}i'var^-[\hat{\nu}^{(el)}(\nu)]\hat{\nu}^{(el)}(\nu),$$

$$D[\hat{\Theta}[\hat{\nu}^{(el)}(\nu)]] = \{i'var^-[\hat{\nu}^{(el)}(\nu)]i\}^{-1},$$

where (2.6.10)

$$\hat{\nu}^{(el)}(\nu) = \nu - S(N^{21}, N^{22})\begin{pmatrix} i' \\ S' \end{pmatrix} var^{-1}(\nu)\nu - cov(\nu_u, \nu)var^{-1}(\nu)\nu +$$

$$+cov(\nu_u, \nu)var^{-1}(\nu)(i, S)\begin{pmatrix} N^{11}, & N^{12} \\ N^{21}, & N^{22} \end{pmatrix}\begin{pmatrix} i' \\ S' \end{pmatrix} var^{-1}(\nu)\nu,$$

$$var[\hat{\nu}^{(el)}(\nu)] = var(\nu) - SN^{22}S' - iN^{12}S' - SN^{21}i' + cov(\nu_u, \nu)var^{-1}(\nu)\cdot$$

$$\cdot cov(\nu, \nu_u) - cov(\nu_u, \nu)var^{-1}(\nu)(i, S)\begin{pmatrix} N^{11}, & N^{12} \\ N^{21}, & N^{22} \end{pmatrix}\begin{pmatrix} i' \\ S' \end{pmatrix} var^{-1}(\nu)\cdot$$

$$\cdot cov(\nu,\nu_u) - cov(\nu,\nu_u) +$$

$$+(i,S)\begin{pmatrix} N^{11}, & N^{12} \\ N^{21}, & N^{22} \end{pmatrix}\begin{pmatrix} i' \\ S' \end{pmatrix} var^{-1}(\nu)cov(\nu,\nu_u) - cov(\nu_u,\nu) +$$

$$+cov(\nu_u,\nu)var^{-1}(\nu)(i,S)\begin{pmatrix} N^{11}, & N^{12} \\ N^{21}, & N^{22} \end{pmatrix}\begin{pmatrix} i' \\ S' \end{pmatrix},$$

N^{11}, N^{12}, N^{21} and N^{22} are defined by (2.5.8) for $JA = i$.

The relations (2.6.10) and (2.6.1) are equivalent.

Model 2

$$\hat{\Theta}[\hat{\nu}^{(el)}(\nu)] = \Theta_0 + \{A'J'var^{-}[\hat{\nu}^{(el)}(\nu)]JA\}^{-1}A'J'var^{-}[\hat{\nu}^{(el)}(\nu)]\hat{\nu}^{(el)}(\nu),$$

$$var[\hat{\Theta}[\hat{\nu}^{(el)}(\nu)]] = \{A'J'var^{-}[\hat{\nu}^{(el)}(\nu)]JA\}^{-1},$$

where (2.6.11)

$$\hat{\nu}^{(el)}(\nu) = \nu - S(N^{21},N^{22})\begin{pmatrix} A'J' \\ S' \end{pmatrix} var^{-1}(\nu)\nu - cov(\nu_u,\nu)var^{-1}(\nu)\nu +$$

$$+cov(\nu_u,\nu)var^{-1}(\nu)(JA,S)\begin{pmatrix} N^{11}, & N^{12} \\ N^{21}, & N^{22} \end{pmatrix}\begin{pmatrix} A'J' \\ S' \end{pmatrix} var^{-1}(\nu)\nu,$$

$$var[\hat{\nu}^{(el)}(\nu)] = var(\nu) - SN^{22}S' - JAN^{12}S' - SN^{21}A'J' +$$

$$+cov(\nu_u,\nu)var^{-1}(\nu)cov(\nu,\nu_u) - cov(\nu_u,\nu)var^{-1}(\nu)\cdot$$

$$\cdot(JA,S)\begin{pmatrix} N^{11}, & N^{12} \\ N^{21}, & N^{22} \end{pmatrix}\begin{pmatrix} A'J' \\ S' \end{pmatrix} var^{-1}(\nu)cov(\nu,\nu_u) -$$

$$-cov(\nu,\nu_u) + (JA,S)\begin{pmatrix} N^{11}, & N^{12} \\ N^{21}, & N^{22} \end{pmatrix}\begin{pmatrix} A'J' \\ S' \end{pmatrix} var^{-1}(\nu)cov(\nu,\nu_u) -$$

$$-cov(\nu_u,\nu) + cov(\nu_u,\nu)var^{-1}(\nu)(JA,S)\begin{pmatrix} N^{11}, & N^{12} \\ N^{21}, & N^{22} \end{pmatrix}\begin{pmatrix} A'J' \\ S' \end{pmatrix},$$

N^{11}, N^{12}, N^{21} and N^{22} are defined by (2.5.8).

The relations (2.6.11) and (2.6.2) are equivalent.

Model 5 and its special case **Model 3**

$$\hat{\Theta}[\hat{\nu}^{(el)}(\nu)] = \Theta_0 + \{A'J'var^-[\hat{\nu}^{(el)}(\nu)]JA\}^{-1} - \{A'J'var^-[\hat{\nu}^{(el)}(\nu)]JA\}^{-1}$$
$$\cdot B'(B\{A'J'var^-[\hat{\nu}^{(el)}(\nu)]JA\}^{-1}B')^{-1}(b + B\{A'J'var^-[\hat{\nu}^{(el)}(\nu)]JA\}^{-1}\cdot$$
$$\cdot A'J'var^-[\hat{\nu}^{(el)}(\nu)]\hat{\nu}^{(el)}(\nu)),$$
$$var\{\hat{\Theta}[\hat{\nu}^{(el)}(\nu)]\} =$$
$$= \{A'J'var^-[\hat{\nu}^{(el)}(\nu)]JA\}^{-1} - \{A'J'var^-[\hat{\nu}^{(el)}(\nu)]JA\}^{-1}$$
$$B'(B\{A'J'var^-[\hat{\nu}^{(el)}(\nu)]JA\}^{-1}B')^{-1}B\{A'J'var^-[\hat{\nu}^{(el)}(\nu)]JA\}^{-1},$$

where (2.6.12)

$$\hat{\nu}^{(el)}(\nu) = \nu - S\Bigg\{(N^{21}, N^{22})\begin{pmatrix} A'J' \\ S' \end{pmatrix} var^{-1}(\nu)\nu -$$
$$- N^{21}B'(BN^{11}B')^{-1}\left[b + B(N^{11}, N^{12})\begin{pmatrix} A'J' \\ S' \end{pmatrix} var^{-1}(\nu)\nu\right]\Bigg\} -$$
$$-cov(\nu_u, \nu)var^{-1}(\nu)\nu + cov(\nu_u, \nu)var^{-1}(\nu)\cdot$$
$$\cdot(JA, S)\Bigg\{\begin{pmatrix} N^{11}, & N^{12} \\ N^{21}, & N^{22} \end{pmatrix}\begin{pmatrix} A'J' \\ S' \end{pmatrix} var^{-1}(\nu)\nu -$$
$$- \begin{pmatrix} N^{11} \\ N^{21} \end{pmatrix} B'(BN^{11}B')^{-1}\left[b + B(N^{11}, N^{12})\begin{pmatrix} A'J' \\ S' \end{pmatrix} var^{-1}(\nu)\nu\right]\Bigg\},$$
$$var[\hat{\nu}^{(el)}(\nu)] = var(\nu) - SN^{22}S' - JAN^{12}S' - SN^{21}A'J' +$$
$$+SN^{21}B'(BN^{11}B')^{-1}BN^{12}S' + SN^{21}B'(BN^{11}B')^{-1}BN^{11}A'J' +$$
$$+JAN^{11}B'(BN^{11}B')^{-1}BN^{12}S' + cov(\nu_u, \nu)var^{-1}(\nu)cov(\nu, \nu_u) -$$
$$-cov(\nu_u, \nu)var^{-1}(\nu)(JA, S)\left[\begin{pmatrix} N^{11}, & N^{12} \\ N^{21}, & N^{22} \end{pmatrix} -\right.$$
$$\left.- \begin{pmatrix} N^{11} \\ N^{21} \end{pmatrix} B'(BN^{11}B')^{-1}B(N^{11}, N^{12})\right]\begin{pmatrix} A'J' \\ S' \end{pmatrix} var^{-1}(\nu)cov(\nu, \nu_u) -$$
$$-cov(\nu, \nu_u) + (JA, S)\left[\begin{pmatrix} N^{11}, & N^{12} \\ N^{21}, & N^{22} \end{pmatrix} -\right.$$
$$\left.- \begin{pmatrix} N^{11} \\ N^{21} \end{pmatrix} B'(BN^{11}B')^{-1}B(N^{11}, N^{12})\right]\begin{pmatrix} A'J' \\ S' \end{pmatrix} var^{-1}(\nu)cov(\nu, \nu_u) -$$

$$-cov(\nu_u,\nu)+cov(\nu_u,\nu)var^{-1}(\nu)(JA,S)\left[\begin{pmatrix} N^{11}, & N^{12} \\ N^{21}, & N^{22}\end{pmatrix} -\right.$$

$$\left.-\begin{pmatrix} N^{11} \\ N^{21}\end{pmatrix} B'(BN^{11}B')^{-1}B(N^{11},N^{12})\right]\begin{pmatrix} A'J' \\ S'\end{pmatrix};$$

N^{11}, N^{12}, N^{21} and N^{22} possess the same meaning as in Model 2.

The relations for Model 3 are special cases of the given relations for $A = I$.

The relations (2.6.12) and (2.6.5) are equivalent within Model 5, for $A = I$ the relations equivalent with (2.6.3) within Model 3 are obtained.

Model 6 and its special case **Model 4**

$$\begin{pmatrix} \hat{\Theta}_1[\hat{\nu}^{(el)}(\nu)] \\ \hat{\Theta}_2[\hat{\nu}^{(el)}(\nu)]\end{pmatrix} = \begin{pmatrix} \Theta_{10} \\ \Theta_{20}\end{pmatrix} +$$

$$+\begin{pmatrix} \{A'J'var^{-}[\hat{\nu}^{(el)}(\nu)]JA\}^{-1} \\ 0\end{pmatrix} A'J'var^{-}[\hat{\nu}^{(el)}(\nu)]\hat{\nu}^{(el)}(\nu)-$$

$$-\begin{pmatrix} \{A'J'var^{-}[\hat{\nu}^{(el)}(\nu)]JA\}^{-1} \\ -Q_{21}\end{pmatrix} (b+B\{A'J'var^{-}[\hat{\nu}^{(el)}(\nu)]JA\}^{-1}\cdot$$

$$\cdot A'J'var^{-}[\hat{\nu}^{(el)}(\nu)]\hat{\nu}^{(el)}(\nu)),$$

$$var\begin{pmatrix} \hat{\Theta}_1[\hat{\nu}^{(el)}(\nu)] \\ \hat{\Theta}_2[\hat{\nu}^{(el)}(\nu)]\end{pmatrix}$$

$$= \begin{pmatrix} var[\hat{\Theta}_1(\hat{\nu}^{(el)}(\nu))], & cov[\hat{\Theta}_1(\hat{\nu}^{(el)}(\nu)),\hat{\Theta}_2(\hat{\nu}^{(el)}(\nu))] \\ cov[\hat{\Theta}_2(\hat{\nu}^{(el)}(\nu)),\hat{\Theta}_1(\hat{\nu}^{(el)}(\nu))], & var[\hat{\Theta}_2(\hat{\nu}^{(el)}(\nu))]\end{pmatrix},$$

where (2.6.13)

$$var[\hat{\Theta}_1(\hat{\nu}^{(el)}(\nu))] = \{A'J'var^{-}[\hat{\nu}^{(el)}(\nu)]JA\}^{-1} - \{A'J'var^{-}[\hat{\nu}^{(el)}(\nu)]JA\}^{-1}$$

$$\cdot B'Q_{11}B\{A'J'var^{-}[\hat{\nu}^{(el)}(\nu)]JA\}^{-1},$$

$$cov[\hat{\Theta}_1(\hat{\nu}^{(el)}(\nu)),\hat{\Theta}_2(\hat{\nu}^{(el)}(\nu))] = -\{A'J'var^{-}[\hat{\nu}^{(el)}(\nu)]JA\}^{-1}B'Q_{12} =$$

$$= \{cov[\hat{\Theta}_2(\hat{\nu}^{(el)}(\nu)),\hat{\Theta}_1(\hat{\nu}^{(el)}(\nu))\}',$$

$$var[\hat{\Theta}_2(\hat{\nu}^{(el)}(\nu))] = -Q_{22};$$

here

$$\hat{\nu}^{(el)}(\nu) = \nu - S\Big\{(N^{21}, N^{22})\begin{pmatrix} A'J' \\ S' \end{pmatrix} var^{-1}(\nu)\nu -$$

$$- N^{21}B'P_{11}\left[b + B(N^{11}, N^{12})\begin{pmatrix} A'J' \\ S' \end{pmatrix} var^{-1}(\nu)\nu\right]\Big\} -$$

$$-cov(\nu_u, \nu)var^{-1}(\nu)\nu + cov(\nu_u, \nu)var^{-1}(\nu)(JA, S)$$

$$\cdot\begin{pmatrix} N^{11}, & N^{12} \\ N^{21}, & N^{22} \end{pmatrix}\begin{pmatrix} A'J' \\ S' \end{pmatrix} var^{-1}(\nu)\nu - cov(\nu_u, \nu)var^{-1}(\nu)(JA, S)$$

$$\cdot\begin{pmatrix} N^{11} \\ N^{21} \end{pmatrix} B'P_{11}\left[b + B(N^{11}, N^{12})\begin{pmatrix} A'J' \\ S' \end{pmatrix} var^{-1}(\nu)\nu\right],$$

$$var[\hat{\nu}^{(el)}(\nu)] = var(\nu) - SN^{22}S' - SN^{21}A'J' - JAN^{12}S' +$$

$$+SN^{21}B'P_{11}BN^{12}S' + SN^{21}B'P_{11}BN^{11}A'J' +$$

$$+JAN^{11}B'P_{11}BN^{12}S' + cov(\nu_u, \nu)var^{-1}(\nu)cov(\nu, \nu_u) -$$

$$-cov(\nu_u, \nu)var^{-1}(\nu)(JA, S)\left[\begin{pmatrix} N^{11}, & N^{12} \\ N^{21}, & N^{22} \end{pmatrix} -\right.$$

$$\left.- \begin{pmatrix} N^{11} \\ N^{21} \end{pmatrix} B'P_{11}B(N^{11}, N^{12})\right]\begin{pmatrix} A'J' \\ S' \end{pmatrix} - cov(\nu, \nu_u) + (JA, S)\cdot$$

$$\cdot\left[\begin{pmatrix} N^{11}, & N^{12} \\ N^{21}, & N^{22} \end{pmatrix} - \begin{pmatrix} N^{11} \\ N^{21} \end{pmatrix} B'P_{11}B(N^{11}, N^{12})\right]\begin{pmatrix} A'J' \\ S' \end{pmatrix}\cdot$$

$$var^{-1}(\nu)cov(\nu, \nu_u) - cov(\nu_u, \nu) + cov(\nu_u, \nu)var^{-1}(\nu)(JA, S)\cdot$$

$$\cdot\left[\begin{pmatrix} N^{11}, & N^{12} \\ N^{21}, & N^{22} \end{pmatrix} - \begin{pmatrix} N^{11} \\ N^{21} \end{pmatrix} B'P_{11}B(N^{11}, N^{21})\right]\begin{pmatrix} A'J' \\ S' \end{pmatrix},$$

N^{11}, N^{12}, N^{21} and N^{22} are determined by the relations (2.5.8),

$$\begin{pmatrix} P_{11}, & P_{12} \\ P_{21}, & P_{22} \end{pmatrix} = \begin{pmatrix} BN^{11}B', & C \\ C', & 0 \end{pmatrix}^{-1}$$

and $$\begin{pmatrix} Q_{11}, & Q_{12} \\ Q_{21}, & Q_{22} \end{pmatrix} = \begin{pmatrix} B\{A'J'var^{-}[\hat{\nu}^{(el)}(\nu)]JA\}^{-1}B', & C \\ C', & 0 \end{pmatrix}^{-1}.$$

The relations for Model 4 are special cases of these relations for $A = I$.

The relations (2.6.13) and (2.6.6) are equivalent within Model 6, substituting $A = I$ in (2.6.13), we obtain relations equivalent with relations (2.6.3) holding within Model 3.

2.7 Estimators of the unit dispersion

The covariance matrix of the observation vector in all procedures for estimating unknown vector parameters and predicting or filtering was assumed to be known. This assumption is not fullfilled in general.

However, in the simplest case, which occurs very frequently, the covariance matrix of the observation vector is of the form $\sigma^2 H$, where H is a known symmetric and positive definite matrix and σ^2 is an unknown coefficient (the parameter σ^2 of the covariance matrix is a parameter of the second order in contradistinction to the unknown parameters Θ or $(\Theta', \vartheta')'$ being parameters of the first order). The reader can easily make sure that in this case the unknown coefficient σ^2 does not occur in any expression used for estimating the first order parameters (see relations (2.4.1), (2.4.4), (2.4.8), (2.4.14), (2.4.16) and (2.4.18) in the case of the simple linear models and (2.5.9), (2.5.11) and (2.5.14) in the case of the mixed ones); knowledge about it is necessary for determining the variances of the estimators of the first order parameters.

The aim of this part is to show the optimum way of estimating the unknown parameter σ^2, called a unit dispersion.

2.7.1 Estimators of the unit dispersion in the simple models of measurement

A unit dispersion is estimated by the expression

$$\hat{\sigma}^2 = v'H^{-1}v/(n + q - k), \tag{2.7.1}$$

where n is the dimension of the observation vector ξ, k the dimension of the unknown vector parameter Θ, q determines the number of conditions to be fulfilled by the components of the unknown vector parameter Θ, and v is an n-dimensional random vector called a vector of residuals. Its form and fundamental statistical characteristics within individual models of measurement are given in the following survey (the same symbols are used as in Section 2.4):

Model 1

$$v = \eta - i\widehat{\Delta\Theta}(\eta) = [I - i(i'H^{-1}i)^{-1}i'H^{-1}]\eta,$$
$$E[v] = 0,\ var(v) = \sigma^2[H - i(i'H^{-1}i)^{-1}i'],\ R[var(v)] = n - 1,$$

see the relation (2.4.1) (in our case $k = 1$ and $q = 0$);

Model 2

$$v = \eta - JA\widehat{\Delta\Theta}(\eta) = [I - JA(A'J'H^{-1}JA)^{-1}A'J'H^{-1}]\eta,$$
$$E[v] = 0,\ var(v) = \sigma^2[H - JA(A'J'H^{-1}JA)^{-1}A'J'],$$
$$R[var(v)] = n - k,$$

see the relation (2.4.4) (in our case $q = 0$);

Model 3

$$v = \eta - J\widehat{\Delta\Theta}(\eta) = \Big(I - J(J'H^{-1}J)^{-1}\cdot$$
$$\cdot\{I - B'[B(J'H^{-1}J)^{-1}B']^{-1}B(J'H^{-1}J)^{-1}\}J'H^{-1}\Big)\eta +$$
$$+J(J'H^{-1}J)^{-1}B'[B(J'H^{-1}J)^{-1}B']^{-1}b,$$
$$E[v] = 0,\ var(v) = \sigma^2\{H - J(J'H^{-1}J)^{-1}J' +$$
$$+J(J'H^{-1}J)^{-1}B'[B(J'H^{-1}J)^{-1}B']^{-1}B(J'H^{-1}J)^{-1}J'\},$$
$$R[var(v)] = n + q - k,$$

see the relation (2.4.8);

Model 4

$$v = \eta - J\widehat{\Delta\Theta}_1(\eta) =$$
$$= \{I - J(J'H^{-1}J)^{-1}[I - B'Q_{11}B(J'H^{-1}J)^{-1}]J'H^{-1}\}\eta +$$
$$+J(J'H^{-1}J)^{-1}B'Q_{11}b,$$
$$E[v] = 0,\ var(v) =$$

$$= \sigma^2[H - J(J'H^{-1}J)^{-1}J' + J(J'H^{-1}J)^{-1}B'Q_{11}B(J'H^{-1}J)^{-1}J'],$$
$$R[var(v)] = n + q - k,$$

see the first vector of the relation (2.4.14);

Model 5

$$v = \eta - JA\widehat{\Delta\Theta}(\eta) =$$
$$= \Big(I - JA(A'J'H^{-1}JA)^{-1}\{I - B'[B(A'J'H^{-1}JA)^{-1}B']^{-1}B\cdot$$
$$\cdot(A'J'H^{-1}JA)^{-1}\}A'J'H^{-1}\Big)\eta +$$
$$+JA(A'J'H^{-1}JA)^{-1}B'[B(A'J'H^{-1}JA)^{-1}B']^{-1}b,$$
$$E[v] = 0,\ var(v) = \sigma^2\{H - JA(A'J'H^{-1}JA)^{-1}A'J' +$$
$$+JA(A'J'H^{-1}JA)^{-1}B'[B(A'J'H^{-1}JA)^{-1}B']^{-1}B(A'J'H^{-1}JA)^{-1}A'J'\},$$
$$R[var(v)] = n + q - k,$$

see the relation (2.4.16);

Model 6

$$v = \eta - JA\widehat{\Delta\Theta}_1(\eta) =$$
$$= \{I - JA(A'J'H^{-1}JA)^{-1}[I - B'Q_{11}B(A'J'H^{-1}JA)^{-1}]A'J'H^{-1}\}\eta +$$
$$+JA(A'J'H^{-1}JA)^{-1}B'Q_{11}b,$$
$$E[v] = 0,\ var(v) = \sigma^2[H - JA(A'J'H^{-1}JA)^{-1}A'J' + JA(A'J'H^{-1}JA)^{-1}\cdot$$
$$\cdot B'Q_{11}B(A'J'H^{-1}JA)^{-1}A'J']$$
$$R[var(v)] = n + q - k,$$

see the first vector in the relation (2.4.18).

We now investigate, in greater detail, the properties of the estimator of the unit dispersion of the form (2.7.1). We shall prove that this estimator of the unit dispersion is invariant, which means that this estimator of the parameter of the second order does not depend on the actual value of the unknown parameter of the first order Θ. As usual, the assertion will be proved for Model 2. In this case

$$v = [I - JA(A'J'H^{-1}JA)^{-1}A'J'H^{-1}]\eta$$

and

$$v'H^{-1}v =$$
$$= \eta'[I - H^{-1}JA(A'J'H^{-1}JA)^{-1}A'J']H^{-1}[I - JA(A'J'H^{-1}JA)^{-1}\cdot$$
$$\cdot A'J'H^{-1}]\eta = \eta[H^{-1} - H^{-1}JA(A'J'H^{-1}JA)^{-1}A'J'H^{-1}]\eta =$$
$$= (\eta - JA\Delta\Theta)'[H^{-1} - H^{-1}JA(A'J'H^{-1}JA)^{-1}A'J'H^{-1}](\eta - JA\Delta\Theta),$$

since $A'J'[H^{-1}-H^{-1}JA(A'J'H^{-1}JA)^{-1}A'J'H^{-1}] = A'J'H^{-1}-A'J'H^{-1} = 0$ and analogously $[H^{-1} - H^{-1}JA(A'J'H^{-1}JA)^{-1}A'J'H^{-1}]JA = 0$.

Further, it will be proved that an invariant estimator of the unit dispersion of the form (2.7.1) is unbiased. The proof is performed for Model 2 again (in this case $q = 0$):

$$E[\hat{\sigma}^2] = E[v'H^{-1}v]/(n-k) = E[Tr(v'H^{-1}v)]/(n-k) =$$
$$= E[Tr(H^{-1}vv')]/(n-k) = Tr(H^{-1}E[vv'])/(n-k) =$$
$$= Tr[H^{-1}var(v)]/(n-k) =$$
$$= Tr\{\sigma^2 H^{-1}[H - JA(A'J'H^{-1}JA)^{-1}A'J']\}/(n-k) =$$
$$= \sigma^2 Tr[I_{(n,n)} - H^{-1}JA(A'J'H^{-1}JA)^{-1}A'J']/(n-k) =$$
$$= \sigma^2\{Tr(I_{(n,n)}) - Tr[H^{-1}JA(A'J'H^{-1}JA)^{-1}A'J']\}/(n-k) =$$
$$= \sigma^2\{Tr(I_{(n,n)}) - Tr[(A'J'H^{-1}JA)^{-1}A'J'H^{-1}JA]\}/(n-k) =$$
$$= \sigma^2[Tr(I_{(n,n)}) - Tr(I_{(k,k)})]/(n-k) = \sigma^2.$$

The following simple rules were applied for deriving these sequences of the relations:

Let V be an $n \times n$-dimensional matrix. Its trace $Tr(V)$ is defined by the relation $Tr(V) = \sum_{i=1}^{n}\{V\}_{i,i}$, where $\{V\}_{i,i}$, $i = 1, 2, \ldots, n$ are diagonal elements of the matrix V. It can be easily proved that the trace of any square matrix $A_{(m,n)}B_{(n,m)}$ is $Tr(A_{(m,n)}B_{(n,m)}) = Tr(B_{(n,m)}A_{(m,n)})$; furthermore, $Tr(aA + bB) = aTr(A) + bTr(B)$ for any matrices A and B of the same dimension and any real constants $a, b \in \mathcal{R}^1$, $Tr(a) = a$, where $a \in \mathcal{R}^1$ and $E[Tr(vv')] = Tr(E[vv'])$.

Finally it is necessary to determine the dispersion of an invariant and unbiased estimator of the unit dispersion of the form (2.7.1):

$$D[\hat{\sigma}^2] = D[v'H^{-1}v/(n+k-q)] = D[v'H^{-1}v]/(n+q-k)^2,$$

which, because of $E[v'H^{-1}v] = \sigma^2(n+q-k)$, yields

$$D[v'H^{-1}v] = E[\{v'H^{-1}v - \sigma^2(n+q-k)\}^2] =$$
$$= E[v'H^{-1}vv'H^{-1}v] - 2\sigma^2(n+q-k)E[v'H^{-1}v] + \sigma^4(n+q-k)^2 =$$
$$= E[v'H^{-1}vv'H^{-1}v] - \sigma^4(n+q-k)^2.$$

Since

$$E[v'H^{-1}vv'H^{-1}v] = E[Tr(v'H^{-1}v)Tr(v'H^{-1}v)] =$$
$$= E[Tr(H^{-1}vv')Tr(H^{-1}vv')] = E[Tr(H^{-1}vv' \otimes H^{-1}vv')] =$$
$$= E[Tr\{(H^{-1} \otimes H^{-1})(vv' \otimes vv')\}] =$$
$$= Tr\{(H^{-1} \otimes H^{-1})E[vv' \otimes vv']\} = Tr[(H^{-1} \otimes H^{-1})\Psi_v],$$

where $\Psi_v = E[vv' \otimes vv']$ is an $n^2 \times n^2$-dimensional matrix of the 4th central moments of the random variable v, we obtain

$$D[v'H^{-1}v/(n+q-k)] = D[\hat{\sigma}^2] = Tr[(H^{-1} \otimes H^{-1})\Psi_v]/(n+q-k)^2 - \sigma^4 \tag{2.7.2}$$

(the number $(v'H^{-1}v)^2$ has been expressed in the form $(v'H^{-1}v)^2 = v'H^{-1} \cdot vv'H^{-1}v$; the fact that $v'H^{-1}v = Tr(v'H^{-1}v) = Tr(H^{-1}vv')$ and the relations $Tr(A \otimes B) = Tr(A)Tr(B)$ and $(AB) \otimes (CD) = (A \otimes C)(B \otimes D)$ have been applied here; the Kronecker multiplication of two matrices $A_{(n,m)}$ and $B_{(j,k)}$ is defined in the following way

$$A_{(n,m)} \otimes B_{(j,k)} = \begin{pmatrix} a_{11}B_{(j,k)}, & a_{12}B_{(j,k)}, & \dots, & a_{1m}B_{(j,k)} \\ a_{21}B_{(j,k)}, & a_{22}B_{(j,k)}, & \dots, & a_{2m}B_{(j,k)} \\ \dots & \dots & \dots & \dots \\ a_{n1}B_{(j,k)}, & a_{n2}B_{(j,k)}, & \dots, & a_{nm}B_{(j,k)} \end{pmatrix},$$

the result is an $(n \times j) \times (m \times k)$-dimensional matrix).

The last step consists in investigating the relations between the matrices Ψ_v and Ψ_ξ (or Ψ_η) of the 4th central moments of the n-dimensional vector of residuals v and the n-dimensional observation vector ξ (or η). $\Psi_\xi = \Psi_\eta$, as $\xi - E[\xi] = \eta - E[\eta]$; further, as the vector of residuals can be expressed in the form $v = P_i(\eta - E[\eta])$, where the transformation matrices within individual linear simple models of measurement are (the index i denotes the number of the model considered)

$$P_1 = I - i(i'H^{-1}i)^{-1}i'H^{-1},$$

$$P_2 = I - JA(A'J'H^{-1}JA)^{-1}A'J'H^{-1},$$
$$P_3 = I - J(J'H^{-1}J)^{-1}\{I - B'[B(J'H^{-1}J)^{-1}B']^{-1}B(J'H^{-1}J)^{-1}\}J'H^{-1},$$
$$P_4 = I - J(J'H^{-1}J)^{-1}[I - B'Q_{11}B(J'H^{-1}J)^{-1}]J'H^{-1},$$
$$P_5 = I - JA(A'J'H^{-1}JA)^{-1}\{I - B'[B(A'J'H^{-1}JA)^{-1}B']^{-1}\cdot$$
$$\cdot B(A'J'H^{-1}JA)^{-1}\}A'J'H^{-1},$$
$$P_6 = I - JA(A'J'H^{-1}JA)^{-1}[I - B'Q_{11}B(A'J'H^{-1}JA)^{-1}]A'J'H^{-1},$$

we can write

$$\Psi_v = E[vv' \otimes vv'] = E[P_i(\eta - E[\eta])(\eta - E[\eta])'P_i' \otimes P_i(\eta - E[\eta])\cdot$$
$$\cdot(\eta - E[\eta])'P_i'] =$$
$$= E[(P_i \otimes P_i)(\eta - E[\eta])(\eta - E[\eta])' \otimes (\eta - E[\eta])(\eta - E[\eta])'(P_i' \otimes P_i')] =$$
$$= (P_i \otimes P_i)E[(\eta - E[\eta])(\eta - E[\eta])' \otimes (\eta - E[\eta])(\eta - E[\eta])'](P_i' \otimes P_i')$$

and since, due to the definition, $E[(\eta - E[\eta])(\eta - E[\eta])' \otimes (\eta - E[\eta])(\eta - E[\eta])'] = \Psi_\eta$, we have proved that

$$\Psi_v = (P_i \otimes P_i)\Psi_\eta(P_i' \otimes P_i'), \; i = 1, 2, \ldots, 6. \tag{2.7.3}$$

Thus the dispersion of an invariant and unbiased estimator of the unit dispersion (2.7.1) is determined by the relations (2.7.2) and (2.7.3), evidently under the condition of the known matrix of the 4th central moments of the observation vector.

These last mentioned relations will be analysed in more detail in Section 2.9 in the case of a normally distributed observation vector.

2.7.2 Estimators of the unit dispersion in mixed linear models

If the covariance matrix of the observation vector ζ within linear mixed models is of the form $var(\zeta) = var(\nu) = \sigma^2 H$, where H is a known symmetric and positive definite matrix, the unknown unit dispersion σ^2 is estimated using the relation

$$\hat{\sigma}^2 = v'H^{-1}v/(n + q - k - s), \tag{2.7.4}$$

where n is the dimension of the observation vector ζ or ν, k the dimension of the unknown vector parameter Θ, s the dimension of the unknown vector parameter ϑ (the parameter of the mean value of the random variable φ, see Section 2.5) and q the number of conditions binding the values of the components of the unknown vector parameter Θ, v is the n-dimensional random vector of residuals which have in individual models the form:

Model 2 and its special case **Model 1**

$$v = \nu - (JA, S)\begin{pmatrix} \widehat{\Delta\Theta}(\nu) \\ \widehat{\Delta\vartheta}(\nu) \end{pmatrix} =$$

$$\left[I - (JA, S)\begin{pmatrix} M^{11}, & M^{12} \\ M^{21}, & M^{22} \end{pmatrix}\begin{pmatrix} A'J' \\ S' \end{pmatrix} H^{-1}\right]\nu,$$

see the relation (2.5.9), where

$$\begin{pmatrix} M^{11}, & M^{12} \\ M^{21}, & M^{22} \end{pmatrix} = \left[\begin{pmatrix} A'J' \\ S' \end{pmatrix} H^{-1}(JA, S)\right]^{-1}, \; M^{ij} = \sigma^{-2}N^{ij}, \; i,j = 1,2;$$

$$E[v] = 0, \; var(v) = \sigma^2\left[H - (JA, S)\begin{pmatrix} M^{11}, & M^{12} \\ M^{21}, & M^{22} \end{pmatrix}\begin{pmatrix} A'J' \\ S' \end{pmatrix}\right],$$

$$R[var(v)] = n - k - s;$$

within Model 1 $A = 1$, $J = i$ and $k = 1$.

Model 5 and its special case **Model 3**

$$v = \nu - (JA, S)\begin{pmatrix} \widehat{\Delta\Theta}(\nu) \\ \widehat{\Delta\vartheta}(\nu) \end{pmatrix} = \left\{I - (JA, S)\left[\begin{pmatrix} M^{11}, & M^{12} \\ M^{21}, & M^{22} \end{pmatrix} -\right.\right.$$

$$\left.\left. - \begin{pmatrix} M^{11} \\ M^{21} \end{pmatrix} B'(BM^{11}B')^{-1}B(M^{11}, M^{12})\right]\begin{pmatrix} A'J' \\ S' \end{pmatrix} H^{-1}\right\}\nu +$$

$$+(JA, S)\begin{pmatrix} M^{11} \\ M^{21} \end{pmatrix} B'(BM^{11}B')^{-1}b,$$

see the relation (2.5.11),

M^{11}, M^{12}, M^{21} and M^{22} have the same meaning as within Model 2;

$$E[v] = 0,\ var(v) = \sigma^2 \left\{ H - (JA, S) \left[\begin{pmatrix} M^{11}, & M^{12} \\ M^{21}, & M^{22} \end{pmatrix} - \right.\right.$$
$$\left.\left. - \begin{pmatrix} M^{11} \\ M^{21} \end{pmatrix} B'(BM^{11}B')^{-1}B(M^{11}, M^{12}) \right] \begin{pmatrix} A'J' \\ S' \end{pmatrix} \right\},$$
$$R[var(v)] = n + q - k - s;$$

within Model 3 $A = I$.

Model 6 and its special case **Model 4**

$$v = \nu - (JA, S) \begin{pmatrix} \widehat{\Delta\Theta}_1(\nu) \\ \widehat{\Delta\vartheta}(\nu) \end{pmatrix} = \left\{ I - (JA, S) \left[\begin{pmatrix} M^{11}, & M^{12} \\ M^{21}, & M^{22} \end{pmatrix} - \right.\right.$$
$$\left.\left. - \begin{pmatrix} M^{11} \\ M^{21} \end{pmatrix} B'R_{11}B(M^{11}, M^{12}) \right] \begin{pmatrix} A'J' \\ S' \end{pmatrix} H^{-1} \right\} \nu +$$
$$+ (JA, S) \begin{pmatrix} M^{11} \\ M^{21} \end{pmatrix} B'R_{11}b,$$

see the relation (2.5.14),

M^{11}, M^{12}, M^{21} and M^{22} has the same meaning as within Model 2 and

$$\begin{pmatrix} BM^{11}B', & C \\ C', & 0 \end{pmatrix}^{-1} = \begin{pmatrix} R_{11}, & R_{12} \\ R_{21}, & R_{22} \end{pmatrix};$$

$$E[v] = 0,\ var(v) = \sigma^2 \left\{ H - (JA, S) \left[\begin{pmatrix} M^{11}, & M^{12} \\ M^{21}, & M^{22} \end{pmatrix} - \right.\right.$$
$$\left.\left. - \begin{pmatrix} M^{11} \\ M^{21} \end{pmatrix} B'R_{11}B(M^{11}, M^{12}) \right] \begin{pmatrix} A'J' \\ S' \end{pmatrix} \right\},$$
$$R[var(v)] = n + q - k - s;$$

within Model 4 $A = I$.

Within the mixed linear models the estimator (2.7.4) of the unit dispersion is its invariant and unbiased estimator. The proof of this assertion is the same as in the case of the simple form of the linear models of measurement, only the hypervector $(\Delta\Theta', \Delta\vartheta')'$ of the unknown parameters and the design hypermatrix (JA, S) have to be taken into account instead of the unknown parameter vector $\Delta\Theta$ and the design matrix JA, respectively.

The dispersion of the unit dispersion estimator (2.7.4) is

$$D[v'H^{-1}v/(n+q-k-s)] = Tr[(H^{-1}\otimes H^{-1})\Phi_v]/(n+q-k-s)^2 - \sigma^4, \tag{2.7.5}$$

where $\Phi_v = (Q_i \otimes Q_i)\Psi_\nu(Q_i' \otimes Q_i')$ is the matrix of the 4th central moments of the random vector of residuals in the corresponding linear mixed model, Ψ_ν is the matrix of the 4th central moments of the observation vector of the corresponding mixed model and the matrices Q_i have (every index denotes the number of the model considered) the following meaning:

$$Q_1 = I - (i, S)\begin{pmatrix} M^{11}, & M^{12} \\ M^{21}, & M^{22} \end{pmatrix}\begin{pmatrix} i' \\ S' \end{pmatrix} H^{-1},$$

$$Q_2 = I - (JA, S)\begin{pmatrix} M^{11}, & M^{12} \\ M^{21}, & M^{22} \end{pmatrix}\begin{pmatrix} A'J' \\ S' \end{pmatrix} H^{-1},$$

$$Q_3 = I - (J, S)\left[\begin{pmatrix} M^{11}, & M^{12} \\ M^{21}, & M^{22} \end{pmatrix} - \begin{pmatrix} M^{11} \\ M^{21} \end{pmatrix} B'(BM^{11}B')^{-1}\cdot\right.$$

$$\left.\cdot B(M^{11}, M^{12})\right]\begin{pmatrix} J' \\ S' \end{pmatrix} H^{-1},$$

$$Q_4 = I - (J, S)\left[\begin{pmatrix} M^{11}, & M^{12} \\ M^{21}, & M^{22} \end{pmatrix} - \begin{pmatrix} M^{11} \\ M^{21} \end{pmatrix} B'R_{11}B(M^{11}, M^{12})\right]\cdot$$

$$\cdot\begin{pmatrix} J' \\ S' \end{pmatrix} H^{-1},$$

$$Q_5 = I - (JA, S)\left[\begin{pmatrix} M^{11}, & M^{12} \\ M^{21}, & M^{22} \end{pmatrix} - \begin{pmatrix} M^{11} \\ M^{21} \end{pmatrix} B'(BM^{11}B')^{-1}\cdot\right.$$

$$\left.\cdot B(M^{11}, M^{12})\right]\begin{pmatrix} A'J' \\ S' \end{pmatrix} H^{-1},$$

$$Q_6 = I - (JA, S)\left[\begin{pmatrix} M^{11}, & M^{12} \\ M^{21}, & M^{22} \end{pmatrix} - \begin{pmatrix} M^{11} \\ M^{21} \end{pmatrix} B'R_{11}B(M^{11}, M^{12})\right]\cdot$$

$$\cdot\begin{pmatrix} A'J' \\ S' \end{pmatrix} H^{-1}.$$

Knowledge of the matrix Φ_v or Ψ_ν is necessary for determining the dispersion of the unit dispersion estimator in the general case.

If the covariance matrix of the observation vector ζ is of the form $var(\zeta) = \sum_{i=1}^{p} \sigma_i^2 H_i$, where H_i are known positive definite matrices and the ratio $\sigma_1^2 : \sigma_2^2 : \ldots : \sigma_p^2$ of the unknown unit dispersions is known, the relation (2.7.4) is used for estimating the parameter σ_1^2; in this case $H = H_1 + (\sigma_2^2/\sigma_1^2)H_2 + \ldots + (\sigma_p^2/\sigma_1^2)H_p$.

2.7.3 Estimators of the unit dispersion from data affected by a systematic deterministic influence

Let the measurement of the vector X be affected by a systematic deterministic influence modelled by a function $s(\vartheta) \doteq s(\vartheta_0) + [\partial s(\vartheta_0)/\partial\vartheta']\Delta\vartheta = s(\vartheta_0) + S\Delta\vartheta$, $\vartheta = \vartheta_0 + \Delta\vartheta \in \mathcal{R}^s$; the covariance matrix of the observation vector η is assumed in the form $var(\eta) = \sigma^2 H$ (no systematic random influence is assumed). Then the invariant unbiased estimator of the unit dispersion σ^2 is given by the relation (2.7.4); for determining it is necessary to determine the LS-estimator of the useless parameter of the systematic deterministic affect.

2.8 Estimators of the parameters of the covariance matrix $var(\zeta) = \sum_{i=1}^{p} \kappa_i V_i$ the estimation of an unknown covariance matrix of the observation vector in a replicated linear model of measurement

The previous sections dealt with the simplest structure of the covariance matrix of the observation vector, namely $var(\zeta) = \sigma^2 H$, where H is a known symmetric and positive definite matrix and σ^2 is an unknown parameter of the second order—the unit dispersion.

The covariance matrix of an observation vector need not have always this simple form. This can be demonstrated by the case of a mixed linear model. In this case $var(\zeta) = var(\varepsilon) + var(\varphi)$, where ε is an error component and φ the random component of the useful signal, and these two random vectors are assumed to be uncorrelated.

Let $var(\varepsilon) = \sigma_1^2 H_1$ and $var(\varphi_0) = \sigma_2^2 H_2$, where H_1 and H_2 are known symmetric and positive definite matrices and σ_1^2 and σ_2^2 $(\sigma_1^2 \neq \sigma_2^2)$ are unknown coefficients. Then the covariance matrix of the observation vector

ζ is of the form $var(\zeta) = var(\varepsilon) + var(\varphi_0) = \sigma_1^2 H_1 + \sigma_2^2 H_2$ and the problem of estimating is to determine the LS-estimator of the unknown parameter $(\Theta', \vartheta')'$ of the first order and, simultaneously, the invariant and unbiased estimator of the parameters σ_1^2 and σ_2^2 of the second order (in reality the estimators of the second order parameters have to been determined first, and using them we obtain the estimator of the first order parameters).

The mentioned covariance matrix of the observation vector ζ of the form $var(\zeta) = \sigma_1^2 H_1 + \sigma_2^2 H_2$ represents the simplest case of a covariance matrix $\sum_{i=1}^{p} \kappa_i V_i$, where κ_i are unknown coefficients of the second order and V_i are known symmetric matrices; in general, the matrices V_i need not be positive semidefinite and the coefficients κ_i, $i = 1, 2, \ldots, p$ non-negative, nevertheless the matrix $\sum_{i=1}^{p} \kappa_i V_i = var(\zeta)$ must be positive definite within regular models of measurement). The importance of this structure of the covariance matrix consists in the fact that by using it one can express any covariance matrix with a priori unknown elements. Let $\sigma_{i,j}$, $\sigma_{i,j} = \sigma_{j,i}$, $i, j = 1, 2, \ldots, n$ be the (i, j)th element of the covariance matrix $var(\eta)$ of the observation vector η. Then

$$var(\eta) = \sum_{i=1}^{n} \sum_{j=1}^{n} \sigma_{i,j} e_i e_j' = \sum_{i=1}^{n} \sigma_{i,i} e_i e_i' + \sum_{i=1}^{n} \sum_{j=i+1}^{n} \sigma_{i,j} (e_i e_j' + e_j e_i'),$$

where $e_1 = (1, 0, \ldots, 0)'$, $e_2 = (0, 1, \ldots, 0)', \ldots,$ $e_n = (0, 0, \ldots, 1)'$ are unit vectors of the n-dimensional linear vector space $\mathcal{R}^n$. In this case $p = n(n+1)/2$. If the covariance matrix $var(\eta)$ of the observation vector η is

$$var(\eta) = \begin{pmatrix} \sigma_1^2, & 0, & \ldots, & 0 \\ 0, & \sigma_2^2, & \ldots, & 0 \\ \ldots & \ldots & \ldots & \ldots \\ 0, & 0, & \ldots, & \sigma_n^2 \end{pmatrix}$$

(uncorrelated measurement with different dispersions), then

$$var(\eta) = \sum_{i=1}^{n} \sigma_i^2 e_i e_i'.$$

The problem of determining, in some sense, the optimum invariant and unbiased estimators of the unknown parameters κ_i, $i = 1, 2, \ldots, p$, of the second order and, by means of them, LS-estimators of the unknown parameters $(\Theta', \vartheta')'$ of the first order goes beyond the framework of this textbook. The aim of the next lines is to initiate the reader to the problem, to make him familiar with the fundamental notions and procedures used in this sphere of estimating.

Consider Model 2 in its simple form

$$\eta = JA\Delta\Theta + \varepsilon, \ E[\varepsilon] = 0, \ var(\eta) = E[\varepsilon\varepsilon'] = \textstyle\sum_{i=1}^{p} \kappa_i V_i,$$

where κ_i are unknown parameters of the second order and V_i known symmetric matrices of the type $n \times n$; $\Psi = E[(\varepsilon\varepsilon') \otimes (\varepsilon\varepsilon')]$ is the matrix of the 4th central moments of the observation vector η.

In this case, a (κ, Ψ)-locally best invariant and unbiased estimator of an unbiasedly and invariantly estimable linear function $f'\kappa$, $\kappa \in \mathcal{K}$, of the second order parameters is

$$\widehat{f'\kappa} = \eta' T \eta,$$

where

$$T = MSM, \ M = I - JA(A'J'JA)^{-1}A'J'$$

and the matrix S is a solution of the equation system

$$(M \otimes M)\tilde{\Psi}(M \otimes M)vec(S) = (M \otimes M)\tilde{V}(f'\kappa + \lambda),$$

$$\tilde{V}(M \otimes M)vec(S) = f. \tag{2.8.1}$$

Here $vec(S)$ is an n^2-dimensional column vector, which arises by ordering the columns of the $n \times n$-dimensional matrix S into one column, i.e.,

$$[vec(S_{(n,n)})]' = \{vec[(s_{1(n)}, s_{2(n)}, \ldots, s_{n(n)})]\}' = (s'_{1(n)}, s'_{2(n)}, \ldots, s'_{n(n)}),$$

$\tilde{\Psi}$ is an $n^2 \times n^2$-dimensional matrix of the form

$$\tilde{\Psi} = [vec(\Psi_{11}), vec(\Psi_{12}), \ldots, vec(\Psi_{1n}), vec(\Psi_{21}),$$

$$vec(\Psi_{22}), \ldots, vec(\Psi_{2n}), \ldots, vec(\Psi_{n1}), vec(\Psi_{n2}), \ldots, vec(\Psi_{nm})],$$

where $\Psi_{ij} = E[\varepsilon_i\varepsilon_j \otimes \varepsilon\varepsilon']$ is the $n \times n$-dimensional (i,j)th block of the matrix Ψ, $i,j = 1,2,\ldots,n$, $\tilde{V}$ is an $n^2 \times p$-dimensional matrix of the form

$$\tilde{V} = [vec(V_1), vec(V_2), \ldots, vec(V_p)]$$

and $\lambda \in \mathcal{R}^p$ is an undetermined Lagrange vector multiplier ([19], p. 190).

An estimator of a linear function of the second order parameters is said to be invariant, if it is independent of the value of the first order parameter, which means that, for any k-dimensional vector δ, the matrix T has to satisfy the relation $\eta' T \eta = (\eta - LA\delta)' T (\eta - JA\delta)$, which occurs if and only if $TJA = 0$.

An unbiased estimator $\widehat{f'\kappa}$ of an unbiasedly and invariantly estimable linear function $f'\kappa$ of the second order parameters is an estimator fulfilling for each $\kappa \in \mathcal{K}$ the condition $E[\widehat{f'\kappa}] = f'\kappa$, where κ is the actual value of the vector parameter of the second order.

A (κ, Ψ)-locally best unbiased and invariant estimator of an unbiasedly and invariantly estimable linear function of the second order parameters is that estimator belonging to the class of all unbiased and invariant estimators of this function whose dispersion at the point (κ, Ψ) is minimum.

The problem of an invariant or invariant and unbiased estimability of a linear function of the unknown parameter of the second order is connected with the existence of a solution of the system of equations expressing the requirement (in our case invariance or simultaneous invariance and unbiasedness) of being satisfied by the resulting estimator.

In the case of the unknown unit dispersion in the covariance matrix $var(\eta) = \sigma^2 H$, the estimators of the first order parameters are invariant to it. This is not true if the covariance matrix is of the form $var(\eta) = \sum_{i=1}^{p} \kappa_i V_i$; the inverse of a matrix, which is a linear combination with unknown coefficients, is necessary for obtaining an explicit formula for this estimator. This is simultaneously the reason for seeking the estimators of the second order parameters directly from the observation vector.

Analyse now, on the basis of the assertion on a (κ, Ψ)-locally best estimator of a vector parameter of the second order, the properties of the unit dispersion estimator (2.7.1) in the model $\eta = JA\Delta\Theta + \varepsilon$, $E[\eta] = JA\Delta\Theta$, $var(\eta) = \sigma^2 I$. The estimator

$$\hat{\sigma}^2 = v'v/(n-k) = \eta'[I - JA(A'J'JA)^{-1}A'J']\eta/(n-k)$$

is invariant, as $[I - JA(A'J'JA)^{-1}A'J']JA = 0$, and unbiased because of $E[\eta'[I - JA(A'J'JA)^{-1}A'J']\eta/(n-k)] = E[Tr\{\eta'[I - JA(A'J'JA)^{-1}A'J']\eta/(n-k)\}] = \sigma^2[Tr(I_{(n,n)}) - Tr(I_{(k,k)})]/(n-k) = \sigma^2$ (see p. 193). The relation (2.7.2), expressing the dispersion of an estimator of the dispersion, clearly demonstrates its dependence on the actual value of the dispersion and on the matrix Ψ of the actual values of the 4th central moments of the observation vector:

$$D[\hat{\sigma}^2] =$$

$$= (n-k)^{-2} Tr\{[I - JA(A'J'JA)^{-1}A'J'] \otimes [I - JA(A'J'JA)^{-1}A'J']\Psi \cdot$$

$$\cdot [I - JA(A'J'JA)^{-1}A'J'] \otimes [I - JA(A'J'JA)^{-1}A'J']\} - \sigma^2,$$

as $I_{(n,n)} \otimes I_{(n,n)} = I_{(n^2,n^2)}$. To determine the estimator, we require further a priori information on the observation vector, namely knowledge of the actual values of the parameters σ^2 and Ψ. Such knowledge, of course, is necessary for deriving a (σ^2, Ψ)-locally best invariant and unbiased estimator of the unit dispersion. Assume, in accordance with the example 5.6.2 ([19], p. 191), that all the components of the observation vector η possess the same excess γ. For the sake of simplicity, let $n = 3$. Here

$$\Psi = \begin{pmatrix} (\gamma+3)\sigma^4, & 0, & 0, & 0, & \sigma^4, & 0, & 0, & 0, & \sigma^4 \\ 0, & \sigma^4, & 0, & \sigma^4, & 0, & 0, & 0, & 0, & 0 \\ 0, & 0, & \sigma^4, & 0, & 0, & 0, & \sigma^4, & 0, & 0 \\ 0, & \sigma^4, & 0, & \sigma^4, & 0, & 0, & 0, & 0, & 0 \\ \sigma^4, & 0, & 0, & 0, & (\gamma+3)\sigma^4, & 0, & 0, & 0, & \sigma^4 \\ 0, & 0, & 0, & 0, & 0, & \sigma^4, & 0, & \sigma^4, & 0 \\ 0, & 0, & \sigma^4, & 0, & 0, & 0, & \sigma^4, & 0, & 0 \\ 0, & 0, & 0, & 0, & 0, & \sigma^4, & 0, & \sigma^4, & 0 \\ \sigma^4, & 0, & 0, & 0, & \sigma^4, & 0, & 0, & 0, & (\gamma+3)\sigma^4 \end{pmatrix}$$

and $\tilde{\Psi} \equiv \Psi$. Furthermore, in this case $V = I_{(3,3)}$, which means that $\tilde{V} = (1,0,0,0,1,0,0,0,1)'$, $f = 1$ and $\kappa = \sigma^2$.

The matrices M and S are symmetric, therefore $vec(MSM) = (M \otimes M)vec(S)$ $(= vec(T))$. Inserting this into the system of equations (2.8.1) we obtain

$$(M \otimes M)\Psi vec(T) = (M \otimes M)vec(I)(\sigma^4 + \lambda),$$

$$[vec(I)]'vec(T) = 1,$$

where $M = I - JA(A'J'JA)^{-1}A'J'$ and λ is an undetermined Lagrange multiplier. The elements of the matrix T are a solution of the mentioned system of equations, the former ensuring the minimum of the dispersion of the invariant estimator of the unit dispersion at the point (σ^2, Ψ) and the latter its unbiasedness.

The matrix T within regular Model 2, under the given assumption of the form of the matrix Ψ, is

$$T = \frac{M}{i'[I + (\gamma/2)(M * M)]^{-1}diag(M)} - \frac{\gamma}{2} M\,Diag(T)M, \tag{2.8.2}$$

$$diag(T) = \frac{[I + (\gamma/2)(M * M)]^{-1}diag(M)}{i'[I + (\gamma/2)(M * M)]^{-1}diag(M)},$$

$$M = I - JA(A'J'JA)^{-1}A'J';$$

here γ is the excess of the components of the observation vector, $M * M$ means the Hadamard multiplication of matrices ($\{M * M\}_{i,j} = \{M\}_{i,j} \cdot \{M\}_{i,j}$), $Diag(T)$ is a matrix whose diagonal elements equal the diagonal elements of the matrix T and the other elements are zero and finally, $diag(T)$ is a vector whose elements equal the diagonal elements of the matrix T.

As in our case the matrix T of the form (2.8.2) does not depend on σ^2 however, it depends on γ by the help of which the matrix Ψ of the 4th central moments of the observation vector is defined, the estimator of the unit dispersion $\hat{\sigma} = \eta' T \eta$ is uniformly best with respect to σ^2 and locally best at the point Ψ.

In the simplest case of Model 1 of measurement: $\eta = i\Delta\Theta + \varepsilon$, $E[\eta] = i\Delta\Theta$, $var(\eta) = \sigma^2 I$ under the assumption that all the components of the observation vector have the same excess γ, the matrix T is

$$T = \frac{1}{n-1}\left(I - \frac{1}{n}ii'\right) = \frac{1}{n-1}\left(I - i(i'i)^{-1}i'\right),$$

which proves that within Model 1 under the mentioned assumptions the estimator of the unit dispersion

$$\hat{\sigma}^2 = \eta'[I - i(i'i)^{-1}i']\eta/(n-1) = v'v/(n-1),$$

being independent both of σ^2 and Ψ, is uniformly best, invariant and unbiased.

For the sake of complexity it has to be stated that, as a consequence of the fact that (κ, Ψ)-locally best estimators are taken into account, the corresponding estimate is much the better, the closer the chosen point (κ, Ψ) at which the estimate is determined is to its actual position. In the case of uniformly best estimators, the estimators and conseqently the estimates are independent of the actual values of the parameters κ and Ψ.

Finally, we treat the estimator of an unknown covariance matrix of an observation vector within a replicated model (experiment).

Let the covariance matrix of an observation vector be a priori completely unknown (no information on its structure is at our disposal), however there exists a possibility of realizing N times independently the n-dimensional observation vector ζ. Under the condition $N > n$, the estimator of the unknown covariance matrix $var(\zeta)$ of the observation vector ζ is

$$\widehat{var}(\zeta) = [1/(N-1)]\sum_{i=1}^{N}(\zeta_i - \overline{\zeta})(\zeta_i - \overline{\zeta})', \tag{2.8.3}$$

where $\overline{\xi} = (1/N)\sum_{i=1}^{N}\zeta_i$, $\zeta_i = (\zeta_{i1}, \zeta_{i2}, \ldots, \zeta_{in})'$, $i = 1, 2, \ldots, N$. The assumption $N > n$ must be fulfilled unconditionally, in the reverse case the estimator $\widehat{var}(\zeta)$ of a regular matrix $var(\zeta)$ becomes singular (cf. [1], pp. 121, 122).

The estimator $\widehat{var}(\zeta)$ (2.8.3) of the covariance matrix $var(\zeta)$ is its unbiased estimator. Realize that

$$E[\zeta_1] = E[\zeta_2] = \ldots = E[\zeta_N] = E[\zeta],$$

$$var(\zeta_1) = var(\zeta_2) = \ldots = var(\zeta_N) = var(\zeta),$$

$$E[\overline{\zeta}] = E\left((1/n)\textstyle\sum_{i=1}^{N}\zeta_i\right) = E[\zeta],$$

$$var(\zeta) = var\left((1/N)\textstyle\sum_{i=1}^{N}\zeta_i\right) = (1/N)^2 var\left(\textstyle\sum_{i=1}^{N}\xi_i\right) = (1/N)var(\zeta),$$

$$\textstyle\sum_{i=1}^{N}(\zeta_i - \overline{\zeta})(\zeta_i - \overline{\zeta})' = \sum_{i=1}^{N}\{(\zeta_i - E[\zeta]) - (\overline{\zeta} - E[\zeta])\}\cdot$$

$$\cdot\{(\zeta_i - E[\zeta]) - (\overline{\zeta} - E[\zeta])\}' =$$

$$\textstyle\sum_{i=1}^{N}(\zeta_i - E[\zeta])(\zeta_i - E[\zeta])' - \sum_{i=1}^{N}(\zeta_i - E[\zeta])(\overline{\zeta} - E[\zeta])' -$$

$$-\textstyle\sum_{i=1}^{N}(\overline{\zeta} - E[\zeta])(\zeta_i - E[\zeta])' + \sum_{i=1}^{N}(\overline{\zeta} - E[\zeta])(\overline{\zeta}_i - E[\zeta])',$$

$$\textstyle\sum_{i=1}^{N}(\zeta_i - E[\zeta])(\overline{\zeta} - E[\zeta])' = N(\overline{\zeta} - E[\zeta])(\overline{\zeta} - E[\zeta])'$$

and

$$\textstyle\sum_{i=1}^{N}(\overline{\zeta} - E[\zeta])(\zeta_i - E[\zeta])' = N(\overline{\zeta} - E[\zeta])(\overline{\zeta} - E[\zeta])',$$

thus

$$[1/(N-1)]E\left(\textstyle\sum_{i=1}^{N}(\zeta_i - \overline{\zeta})(\zeta_i - \overline{\zeta})'\right) =$$

$$= [1/(N-1)]\left(\textstyle\sum_{i=1}^{N} E[(\zeta_i - E[\zeta])(\zeta_i - E[\zeta])'] -\right.$$

$$\left. - NE[(\overline{\zeta} - E[\zeta])(\overline{\zeta} - E[\zeta])']\right) =$$

$$= [1/(N-1)][N var(\zeta) - var(\zeta)] = var(\zeta),$$

whereby it is proved that the estimator (2.8.1) of the covariance matrix is unbiased.

2.9 Normality of the observation vector

Normally distributed observation vectors occur very frequently. Before processing a measured data set, we are able to test the null hypothesis for its normality. If on basis of this testing, the measured data set can be considered to be normal, we have obtained very important information enabling us to get a better insight into both the measured data and the results of processing.

2.9.1 Properties of estimators of the first and second order parameters in the case of a normal observation vector

Let the observation vector possess a normal probability distribution. Then the estimator of the unknown vector parameter and the vector of residuals are normally distributed and the random vector $v'H^{-1}v$, by means of which the unit dispersion of the vector of random errors of the observation vector is estimated, has the central σ^2 chi-squares probability distribution with $n+q-k$ degrees of freedom (the simple form of the linear models of measurement are taken into account).

The assertions will be proved for the case of the simple form of Model 2. Let $\eta \sim N_n(JA\Delta\Theta, \sigma^2 H)$. Due to the theorem of a linear transformation of a normal random vector (see p. 78), the LQ estimator of the unknown vector parameter Θ of the form

$$\hat{\Theta}(\eta) = \Theta_0 + (A'J'H^{-1}JA)^{-1}A'J'H^{-1}\eta$$

has the normal probability distribution $N_k(\Theta, \sigma^2(A'J'H^{-1}JA)^{-1})$ and the n-dimensional vector of residuals

$$v = [I - JA(A'J'H^{-1}JA)^{-1}A'J'H^{-1}]\eta$$

the normal probability distribution $N_n(0, \sigma^2[H - JA(A'J'H^{-1}JA)^{-1}A'J'])$.

Let us investigate the properties of the random variable $v'H^{-1}v$. Since $v = [I - JA(A'J'H^{-1}JA)^{-1}A'J'H^{-1}]\eta = [I - JA(A'J'H^{-1}JA)^{-1}A'J'H^{-1}]\cdot(\eta - JA\Theta\Delta)$, we can write

$$v'H^{-1}v = (\eta - JA\Delta\Theta)'[H^{-1} - H^{-1}JA(A'J'H^{-1}JA)^{-1}A'J'H^{-1}]\cdot$$

$$\cdot(\eta - JA\Delta\Theta),$$

where $\eta - JA\Delta\Theta \sim N_n(0, \sigma^2 H)$. The matrix H is assumed to be regular, therefore it can be expressed in the form $H = H^{1/2}H^{1/2}$. By means of the random variable $\delta = H^{1/2}(\eta - JA\Delta\Theta)$ (δ has the normal probability distribution $N_n(0, \sigma^2)$), the quadratic form $v'H^{-1}v$ can be expressed as

$$v'H^{-1}v = \delta'[I - H^{-1/2}JA(A'J'H^{-1}JA)^{-1}A'J'H^{-1/2}]\delta. \tag{2.9.1}$$

The matrix $H^{-1/2}JA(A'J'H^{-1}JA)^{-1}A'J'H^{-1/2}$ is symmetric and idempotent (a matrix P is said to be idempotent if $P^2 = P$), which implies that it can be expressed in the form

$$H^{-1/2}JA(A'J'H^{-1}JA)^{-1}A'J'H^{-1/2} = Q'DQ, \tag{2.9.2}$$

where Q is an orthogonal matrix (i.e., $QQ' = I$) and the matrix D of the spectral numbers of the matrix $H^{-1/2}JA(A'J'H^{-1}JA)^{-1}A'J'H^{-1/2}$, being also idempotent ($D = QH^{-1/2}JA(A'J'H^{-1}JA)^{-1}A'J'H^{-1/2}Q'$ and $D^2 = D$), contains directly $R[H^{-1/2}JA(A'J'H^{-1}JA)^{-1}A'J'H^{-1/2}]$ units, the other elements of its diagonal are zeros, since only numbers 0 and 1 satisfy the condition $D = D^2$. As the matrix $H^{-1/2}JA(A'J'H^{-1}JA)^{-1}A'J'$ $H^{-1/2}$ is idempotent, its rank equals the trace, thus

$$R[H^{-1/2}JA(A'J'H^{-1}JA)^{-1}A'J'H^{-1/2}] =$$

$$= Tr[H^{-1/2}JA(A'J'H^{-1}JA)^{-1}A'J'H^{-1/2}] =$$
$$= Tr[(A'J'H^{-1}JA)^{-1}A'J'H^{-1}JA] = Tr(I_{(k,k)}) = k.$$

Using the relation (2.9.2), the relation (2.9.1) for the random variable $v'H^{-1}v$ can be written in the form

$$\delta'(I - Q'DQ)\delta = \delta'(Q'Q - Q'DQ)\delta = \Delta'(I - D)\Delta,$$

where the random variable $\Delta = Q\delta$ has the normal probability distribution $N_n(0, \sigma^2 I)$ and the rank of the matrix $I - D$ is $R(I - D) = n - k$. This means that $\Delta'(I - D)\Delta = \sum_{i=1}^{n-k} \Delta_i^2$, where the one-dimensional random variables Δ_i have the normal probability distribution $N(0, \sigma^2)$ (we must remember that the matrix $(I - D)$ of the type $n \times n$ is a diagonal matrix containing $n - k$ unit elements and k zero elements on the diagonal). If we realize now the way of generating a random variable with a central σ^2 chi-square probability distribution, we can conclude that the random variable $v'H^{-1}v$, under the condition of a normal observation vector, has the central

σ^2 chi-square probability distribution with $n-k$ degrees of freedom. Thus the mean value of the random variable $v'H^{-1}v$ is $E[v'H^{-1}v] = (n-k)\sigma^2$ and its dispersion equals $D[v'H^{-1}v] = 2(n-k)\sigma^4$, which means that, under the condition of normality of the observation vector, the estimator of the unit dispersion $\hat{\sigma}^2/(n-k)$ has the following properties: $E[\hat{\sigma}^2] = \sigma^2$ and $D[\hat{\sigma}^2] = 2\sigma^4/(n-k)$.

The assertion we have just proved that, under the condition of a normal observation vector, the random variable $v'H^{-1}v$ has the central σ^2 chi-square probability distribution with $n+q-k$ degrees of freedom (within the mixed linear models of measurement with $n+q-k-s$ degrees of freedom) implies a very important conclusion concerning the dispersion of the invariant and unbiased estimator of the unit dispersion $\hat{\sigma}^2 = v'H^{-1}v/(n+q-k)$ and $\hat{\sigma}^2 = v'H^{-1}v/(n+q-k-s)$, namely

$$D[\hat{\sigma}^2] = D[v'H^{-1}v/(n+q-k)] = 2\sigma^4/(n+q-k) \tag{2.9.3}$$

and

$$D[\hat{\sigma}^2] = D[v'H^{-1}v/(n+q-k-s)] = 2\sigma^4/(n+q-k-s)$$

in simple and mixed linear models, respectively (in Model 1 $q = 0$ and $k = 1$ and in Model 2 $q = 0$).

Evidently, the same relations will be obtained by taking into account the normality of the observation vector in the general relations for the dispersion of the estimator of the unit dispersion derived in Section 2.7. We shall demonstrate it by a simple example.

The problem is to determine the dispersion of the estimator of the unit dispersion according to the relation (2.7.2) in a simple form of Model 1 of measurement under the condition that $\eta \sim N_3(i\Theta, \sigma^2)$.

The first problem is to determine $Tr[(I^{-1} \otimes I^{-1})\Psi_v]$, where Ψ_v, according to (2.7.3), is $\Psi_v = (P_1 \otimes P_1)\Psi_\eta(P_1' \otimes P_1')$. In our case

$$P_1 = \begin{pmatrix} 1, & 0, & 0 \\ 0, & 1, & 0 \\ 0, & 0, & 1 \end{pmatrix} - (1/3)\begin{pmatrix} 1 \\ 1 \\ 1 \end{pmatrix}(1,1,1) = (1/3)\begin{pmatrix} 2, & -1, & -1 \\ -1, & 2, & -1 \\ -1, & -1, & 2 \end{pmatrix}$$

and since $P_1' = P_1$

$$P_1 \otimes P_1 = P_1' \otimes P_1' =$$

$$= (1/9)\begin{pmatrix} 4, & -2, & -2, & -2, & 1, & 1, & -2, & 1, & 1 \\ -2, & 4, & -2, & 1, & -2, & 1, & 1, & -2, & 1 \\ -2, & -2, & 4, & 1, & 1, & -2, & 1, & 1, & -2 \\ -2, & 1, & 1, & 4, & -2, & -2, & -2, & 1, & 1 \\ 1, & -2, & 1, & -2, & 4, & -2, & 1, & -2, & 1 \\ 1, & 1, & -2, & -2, & -2, & 4, & 1, & 1, & -2 \\ -2, & 1, & 1, & -2, & 1, & 1, & 4, & -2, & -2 \\ 1, & -2, & 1, & 1, & -2, & 1, & -2, & 4, & -2 \\ 1, & 1, & -2, & 1, & 1, & -2, & -2, & -2, & 4 \end{pmatrix}.$$

The 9×9-dimensional matrix of the central fourth moments of the three-dimensional random vector η is

$$\Psi_\eta = \begin{pmatrix} \Psi_{\eta 11}, & \Psi_{\eta 12}, & \Psi_{\eta 13} \\ \Psi_{\eta 21}, & \Psi_{\eta 22}, & \Psi_{\eta 23} \\ \Psi_{\eta 31}, & \Psi_{\eta 32}, & \Psi_{\eta 33} \end{pmatrix},$$

where

$$\Psi_{\eta 11} = \begin{pmatrix} E[\varepsilon_1^4], & E[\varepsilon_1^3\varepsilon_2], & E[\varepsilon_1^3\varepsilon_3] \\ E[\varepsilon_1^3\varepsilon_2], & E[\varepsilon_1^2\varepsilon_2^2], & E[\varepsilon_1^2\varepsilon_2\varepsilon_3] \\ E[\varepsilon_1^3\varepsilon_3], & E[\varepsilon_1^2\varepsilon_2\varepsilon_3], & E[\varepsilon_1^2\varepsilon_3^2] \end{pmatrix},$$

$$\Psi_{\eta 22} = \begin{pmatrix} E[\varepsilon_1^2\varepsilon_2^2], & E[\varepsilon_1\varepsilon_2^3], & E[\varepsilon_1\varepsilon_2^2\varepsilon_3] \\ E[\varepsilon_1\varepsilon_2^3], & E[\varepsilon_2^4], & E[\varepsilon_2^3\varepsilon_3] \\ E[\varepsilon_1\varepsilon_2^2\varepsilon_3], & E[\varepsilon_2^3\varepsilon_3], & E[\varepsilon_2^2\varepsilon_3^2] \end{pmatrix},$$

$$\Psi_{\eta 33} = \begin{pmatrix} E[\varepsilon_1^2\varepsilon_3^2], & E[\varepsilon_1\varepsilon_2\varepsilon_3^2], & E[\varepsilon_1\varepsilon_3^3] \\ E[\varepsilon_1\varepsilon_2\varepsilon_3^2], & E[\varepsilon_2^2\varepsilon_3^2], & E[\varepsilon_2\varepsilon_3^3] \\ E[\varepsilon_1\varepsilon_3^3], & E[\varepsilon_2\varepsilon_3^3], & E[\varepsilon_3^4] \end{pmatrix},$$

$$\Psi_{\eta 12} = \Psi_{\eta 21} = \begin{pmatrix} E[\varepsilon_1^3\varepsilon_2], & E[\varepsilon_1^2\varepsilon_2^2], & E[\varepsilon_1^2\varepsilon_2\varepsilon_3] \\ E[\varepsilon_1^2\varepsilon_2^2], & E[\varepsilon_1\varepsilon_2^3], & E[\varepsilon_1\varepsilon_2^2\varepsilon_3] \\ E[\varepsilon_1^2\varepsilon_2\varepsilon_3], & E[\varepsilon_1\varepsilon_2^2\varepsilon_3], & E[\varepsilon_1\varepsilon_2\varepsilon_3^2] \end{pmatrix},$$

$$\Psi_{\eta 13} = \Psi_{\eta 31} = \begin{pmatrix} E[\varepsilon_1^3\varepsilon_3], & E[\varepsilon_1^2\varepsilon_2\varepsilon_3], & E[\varepsilon_1^2\varepsilon_3^2] \\ E[\varepsilon_1^2\varepsilon_2\varepsilon_3], & E[\varepsilon_1\varepsilon_2^2\varepsilon_3], & E[\varepsilon_1\varepsilon_2\varepsilon_3^2] \\ E[\varepsilon_1^2\varepsilon_3^2], & E[\varepsilon_1\varepsilon_2\varepsilon_3^2], & E[\varepsilon_1\varepsilon_3^3] \end{pmatrix},$$

$$\Psi_{\eta 23} = \Psi_{\eta 32} = \begin{pmatrix} E[\varepsilon_1^2\varepsilon_2\varepsilon_3], & E[\varepsilon_1\varepsilon_2^2\varepsilon_3], & E[\varepsilon_1\varepsilon_2\varepsilon_3^2] \\ E[\varepsilon_1\varepsilon_2^2\varepsilon_3], & E[\varepsilon_2^3\varepsilon_3], & E[\varepsilon_2^2\varepsilon_3^2] \\ E[\varepsilon_1\varepsilon_2\varepsilon_3^2], & E[\varepsilon_2^2\varepsilon_3^2], & E[\varepsilon_2\varepsilon_3^3] \end{pmatrix}.$$

Due to the assumption $\eta \sim N_3(i\Theta, \sigma^2 I)$, the probability density of the random error $\varepsilon = (\varepsilon_1, \varepsilon_2, \varepsilon_3)'$ is

$$f(x_1, x_2, x_3) = \frac{1}{(2\pi)^{3/2}\sigma^3} \exp\left(-\frac{1}{2\sigma^2}\sum_{i=1}^{3} x_i^2\right),$$

the marginal densities are

$$f_{i,j}(x_i, x_j) = \frac{1}{(2\pi)^{3/2}\sigma^3} \exp\left(-\frac{1}{2\sigma^2}(x_i^2 + x_j^2)\right) \int_{\mathcal{R}^1} \exp\left(-\frac{1}{2\sigma^2}x_k^2\right) dx_k =$$

$$= \frac{1}{2\pi\sigma^2} \exp\left(-\frac{1}{2\sigma^2}(x_i^2 + x_j^2)\right), \quad i, j, k = 1, 2, 3, \; i \neq j \neq k,$$

and

$$f_i(x_i) = \frac{1}{(2\pi)^{1/2}\sigma} \exp\left(-\frac{x_i^2}{2\sigma^2}\right), \quad i = 1, 2, 3,$$

thus

$$E[\varepsilon_i^4] = \frac{1}{(2\pi)^{1/2}\sigma} \int_{\mathcal{R}^1} x_i^4 \exp\left(-\frac{x_i^2}{2\sigma^2}\right) dx_i = 3\sigma^4, \; i = 1, 2, 3,$$

(see p. 48),

$$E[\varepsilon_i^2\varepsilon_j^2] = \frac{1}{2\pi\sigma^2} \int_{\mathcal{R}^1} x_i^2 x_j^2 \exp\left(-\frac{1}{2\sigma^2}(x_i^2 + x_j^2)\right) dx_i dx_j =$$

$$= \frac{1}{(2\pi)^{1/2}\sigma} \int_{\mathcal{R}^1} x_i^2 \exp\left(-\frac{x_i^2}{2\sigma^2}\right) dx_i$$

$$\frac{1}{(2\pi)^{1/2}\sigma} \int_{\mathcal{R}^1} x_j^2 \exp\left(-\frac{x_j^2}{2\sigma^2}\right) dx_j = \sigma^4$$

and all the other 4th central moments, in which at least one power of the argument with respect to which it is integrated is odd, are zero, thus $E[\varepsilon_i^3\varepsilon_j] = 0$, $E[\varepsilon_i^2\varepsilon_j\varepsilon_k] = 0$, $i, j, k = 1, 2, 3$, $i \neq j \neq k$. Thus

$$\Psi_\eta = \sigma^4 \begin{pmatrix} 3, & 0, & 0, & 0, & 1, & 0, & 0, & 0, & 1 \\ 0, & 1, & 0, & 1, & 0, & 0, & 0, & 0, & 0 \\ 0, & 0, & 1, & 0, & 0, & 0, & 1, & 0, & 0 \\ 0, & 1, & 0, & 1, & 0, & 0, & 0, & 0, & 0 \\ 1, & 0, & 0, & 0, & 3, & 0, & 0, & 0, & 1 \\ 0, & 0, & 0, & 0, & 0, & 1, & 0, & 1, & 0 \\ 0, & 0, & 1, & 0, & 0, & 0, & 1, & 0, & 0 \\ 0, & 0, & 0, & 0, & 0, & 1, & 0, & 1, & 0 \\ 1, & 0, & 0, & 0, & 1, & 0, & 0, & 0, & 3 \end{pmatrix}$$

and since $I^{-1}_{(3,3)} \otimes I^{-1}_{(3,3)} = I_{(3,3)} \otimes I_{(3,3)} = I_{(9,9)}$, we obtain

$$(I^{-1} \otimes I^{-1})(P_1 \otimes P_1)\Psi_\eta(P_1' \otimes P_1') =$$

$$= (\sigma^4/81)\begin{pmatrix} 4, & -2, & -2, & -2, & 1, & 1, & -2, & 1, & 1 \\ -2, & 4, & -2, & 1, & -2, & 1, & 1, & -2, & 1 \\ -2, & -2, & 4, & 1, & 1, & -2, & 1, & 1, & -2 \\ -2, & 1, & 1, & 4, & -2, & -2, & -2, & 1, & 1 \\ 1, & -2, & 1, & -2, & 4, & -2, & 1, & -2, & 1 \\ 1, & 1, & -2, & -2, & -2, & 4, & 1, & 1, & -2 \\ -2, & 1, & 1, & -2, & 1, & 1, & 4, & -2, & -2 \\ 1, & -2, & 1, & 1, & -2, & 1, & -2, & 4, & -2 \\ 1, & 1, & -2, & 1, & 1, & -2, & -2, & -2, & 4 \end{pmatrix} \cdot$$

$$\cdot \begin{pmatrix} 3, & 0, & 0, & 0, & 1, & 0, & 0, & 0, & 1 \\ 0, & 1, & 0, & 1, & 0, & 0, & 0, & 0, & 0 \\ 0, & 0, & 1, & 0, & 0, & 0, & 1, & 0, & 0 \\ 0, & 1, & 0, & 1, & 0, & 0, & 0, & 0, & 0 \\ 1, & 0, & 0, & 0, & 3, & 0, & 0, & 0, & 1 \\ 0, & 0, & 0, & 0, & 0, & 1, & 0, & 1, & 0 \\ 0, & 0, & 1, & 0, & 0, & 0, & 1, & 0, & 0 \\ 0, & 0, & 0, & 0, & 0, & 1, & 0, & 1, & 0 \\ 1, & 0, & 0, & 0, & 1, & 0, & 0, & 0, & 3 \end{pmatrix} \cdot$$

$$\cdot \begin{pmatrix} 4, & -2, & -2, & -2, & 1, & 1, & -2, & 1, & 1 \\ -2, & 4, & -2, & 1, & -2, & 1, & 1, & -2, & 1 \\ -2, & -2, & 4, & 1, & 1, & -2, & 1, & 1, & -2 \\ -2, & 1, & 1, & 4, & -2, & -2, & -2, & 1, & 1 \\ 1, & -2, & 1, & -2, & 4, & -2, & 1, & -2, & 1 \\ 1, & 1, & -2, & -2, & -2, & 4, & 1, & 1, & -2 \\ -2, & 1, & 1, & -2, & 1, & 1, & 4, & -2, & -2 \\ 1, & -2, & 1, & 1, & -2, & 1, & -2, & 4, & -2 \\ 1, & 1, & -2, & 1, & 1, & -2, & -2, & -2, & 4 \end{pmatrix} =$$

$$= (\sigma^2/81)\begin{pmatrix} 14, & -4, & -4, & -4, & 8, & 2, & -4, & 2, & 8 \\ -7, & 5, & -1, & 5, & -7, & -1, & -1, & -1, & -1 \\ -7, & -1, & 5, & -1, & -1, & -1, & 5, & -1, & -7 \\ -7, & 5, & -1, & 5, & -7, & -1, & -1, & -1, & -1 \\ 8, & -4, & 2, & -1, & 14, & -4, & 2, & -4, & 8 \\ -1, & -1, & -1, & -1, & -7, & 5, & -1, & 5, & -7 \\ -7, & -1, & 5, & -1, & -1, & -1, & 5, & -1, & -7 \\ -1, & -1, & -1, & -1, & -7, & 5, & -1, & 5, & -7 \\ 8, & 2, & -4, & 2, & 8, & -4, & -4, & -4, & 14 \end{pmatrix}.$$

$$\cdot \begin{pmatrix} 4, & -2, & -2, & -2, & 1, & 1, & -2, & 1, & 1 \\ -2, & 4, & -2, & 1, & -2, & 1, & 1, & -2, & 1 \\ -2, & -2, & 4, & 1, & 1, & -2, & 1, & 1, & -2 \\ -2, & 1, & 1, & 4, & -2, & -2, & -2, & 1, & 1 \\ 1, & -2, & 1, & -2, & 4, & -2, & 1, & -2, & 1 \\ 1, & 1, & -2, & -2, & -2, & 4, & 1, & 1, & -2 \\ -2, & 1, & 1, & -2, & 1, & 1, & 4, & -2, & -2 \\ 1, & -2, & 1, & 1, & -2, & 1, & -2, & 4, & -2 \\ 1, & 1, & -2, & 1, & 1, & -2, & -2, & -2, & 4 \end{pmatrix}$$

and our problem is to determine the trace of the product of these matrices. This means that it suffices to determine the diagonal elements of the product only:

$$Tr[(I^{-1} \otimes I^{-1})(P_1 \otimes P_1)\Psi_\eta(P_1' \otimes P_1')] =$$

$$= (\sigma^4/81)(108 + 54 + 54 + 54 + 108 + 54 + 54 + 54 + 108) = 8\sigma^4.$$

Inserting this partial result into (2.7.2) we get

$$D[v'v/2] = \frac{Tr[(I^{-1} \otimes I^{-1})(P_1 \otimes P_1)\Psi_\eta(P_1' \otimes P_1')]}{2^2} - \sigma^4 =$$

$$= \frac{8}{4}\sigma^4 - \sigma^4 = \sigma^4,$$

which is a result identical with the result obtained by applying the relation (2.9.3).

It has been proved that the relation (2.7.1) within simple models of measurement and, analogously, the relation (2.7.4) in mixed linear models represent the invariant and unbiased estimators of the unit dispersion. For a normally distributed observation vector, these estimators can be proved ([21], p. 203) to be simultaneously efficient, i.e., in the class of all invariant and unbiased estimators of the unit dispersion there does not exist any estimator whose dispersion would be smaller than the dispersion of the unit dispersion estimator of the form (2.7.1) within the simple models and (2.7.4) within the mixed linear models of measurement (see the definition of joint efficiency on p. 111 applied to a one-dimensional random variable).

2.9.2 The confidence regions of the first and second order parameters in the case of a normal observation vector

The notion of a confidence region of the unknown estimated parameter will be explained for the simplest case of a one-dimensional parameter (in this case we speak of the confidence interval).

Each realization of an estimator $\hat{\Theta}(\xi)$ of a one-dimensional parameter Θ results in a unique number $\hat{\Theta}(x)$, where x is a realization of the n-dimensional observation vector ζ. Realizations of an estimator of the same parameter belonging to different realizations of the observation vector naturally differ more or less from one another, depending both on the probability distribution of the observation vector and the chosen estimating procedure. In connection with this the problem arises to determine the confidence with which this parameter has been estimated; the main reason for it consists in the fact that in all the considerations concerning the parameter Θ its actual value is substituted by a realization $\hat{\Theta}(x)$ of the estimator $\hat{\Theta}(\xi)$.

The conception of the reliability of an estimator of a one-dimensional unknown parameter in relation to the actual value of this parameter represents the confidence interval.

For a given sufficiently large probability $1-\alpha$ ($\alpha = 0.01$ or $\alpha = 0.05$), called the confidence level, for a chosen type of the estimator $\hat{\Theta}(\xi)$ of the unknown parameter Θ, a random interval $[\hat{\Theta}(\xi)-\Delta_{1\alpha}, \hat{\Theta}(\xi)+\Delta_{1\alpha}]$ included in the region of all the possible values of the unknown parameter such that it covers the actual value of Θ with the probability $1-\alpha$ is called the confidence interval; in other words, the actual value of the parameter Θ estimated by the random variable $\hat{\Theta}(\xi)$ lies with probability α outside the confidence interval $[\hat{\Theta}(\xi)-\Delta_{1\alpha}, \hat{\Theta}(\xi)+\Delta_{1\alpha}]$.

The definition of a confidence interval implies that for determining it the distribution function of the observation vector ξ, which induces the distribution function of the chosen estimator $\hat{\Theta}(\xi)$, has to be known.

The method of determining the confidence interval will be explained by the following example.

The problem is to determine the confidence interval of the mean value of the random variable $\xi \sim N(m, \sigma^2)$ on the confidence level $1-\alpha$; the estimator of the mean value is chosen in the form $\overline{\xi} = (ii')^{-1}i'\xi = (1/n)\sum_{i=1}^{n} \xi_i$ (an unbiased and efficient estimator of the mean value m), where ξ is an n-dimensional observation vector $\xi \sim N_n(m, \sigma^2 I)$; thus $\overline{\xi} \sim N(m, \sigma^2/n)$.

The problem formulated will be solved under the condition of both an a priori known unit dispersion σ^2 and of an a priori unknown σ^2 where the actual value is substituted by its unbiased invariant and efficient estimate.

a) Let σ^2 be a priori known. The random variable $\overline{\xi} - m$ possesses the normal probability distribution $N(0, \sigma^2/n)$ whose probability density is an even function, therefore it is logical to require that $\Delta_{1\alpha} = \Delta_{2\alpha} = \Delta_\alpha$. The problem is to determine the value Δ_α.

As the distribution function of the estimator $\overline{\xi}$ is known, it is possible to determine for any number $\Delta \in \mathcal{R}^1$ the probability that the random variable $\overline{\xi}$ realizes inside the interval $[m - \Delta, m + \Delta]$

$$P\{\omega : \overline{\xi}(\omega) \in [m - \Delta, m + \Delta]\} =$$

$$= F_{\overline{\xi}}(m + \Delta) - F_{\overline{\xi}}(m - \Delta) = 2F_{\overline{\xi}}(m + \Delta) - 1;$$

the fact that $F_{\overline{\xi}}(m - \Delta) = 1 - F_{\overline{\xi}}(m + \Delta)$, where $F_{\overline{\xi}}(\cdot)$, $x \in \mathcal{R}^1$, is the distribution function of a normally distributed random variable $\overline{\xi}$, has been taken into account.

On the basis of the known distribution function $F_{\overline{\xi}}(\cdot)$, $x \in \mathcal{R}^1$, of the random variable $\overline{\xi}$ a number Δ_α such that

$$P\{\omega : \overline{\xi}(\omega) \in [m - \Delta_\alpha, m + \Delta_\alpha]\} = 1 - \alpha \ (= 2F_{\overline{\xi}}(m + \Delta_\alpha) - 1)$$

can be determined; here $1 - \alpha$ is an a priori given confidence level. The sought number Δ_α is a solution of the equation

$$F_{\overline{\xi}}(m + \Delta_\alpha) = 1 - \alpha/2,$$

which means that

$$m + \Delta_\alpha = F_{\overline{\xi}}^{-1}(1 - \alpha/2) = F_{m,\sigma\sqrt{n}}^{-1}(1 - \alpha/2),$$

where $F_{m,\sigma\sqrt{n}}^{-1}(1 - \alpha/2)$ is the $(1 - \alpha/2)$-quantile of the normal probability distribution $N(m, \sigma^2/n)$. As a consequence of the well-known relation $F_{m,\sigma\sqrt{n}}^{-1}(1 - \alpha/2) = (\sigma/\sqrt{n})F_{0,1}^{-1}(1 - \alpha/2) + m$ between the quantiles of the normal probability distributions $N(0,1)$ and $N(m,\sigma^2/n)$ (see p. 49), we can write

$$\Delta_\alpha = (\sigma/\sqrt{n})F_{0,1}^{-1}(1-\alpha/2).$$

By this procedure, for a given m — the actual value of the mean value being estimated by the random variable $\bar{\xi}$ — the interval $[m-\Delta_\alpha, m+\Delta_\alpha]$ was determined, within which the estimator $\bar{\xi}$ is realized with the given probability $1-\alpha$. The center of this interval lies at the point m — the actual value of the unknown parameter being estimated. Shifting now the center of this interval to the random point $\bar{\xi}$, we get the sought random interval $[\bar{\xi}-\Delta_\alpha, \bar{\xi}+\Delta_\alpha]$ which, with probability $1-\alpha$, covers the actual value m of the unknown parameter estimated by the random variable $\bar{\xi}$. Thus the confidence interval of the mean value m estimated from a realization of an n-dimensional random variable $\xi \sim N_n(m,\sigma^2 I)$, using the estimator $\bar{\xi} = (i'i)^{-1}i'\xi$ and under the condition that σ^2 is a priori known, is of the form

$$\left[(1/n)\sum_{i=1}^{n}\xi_i - (\sigma/\sqrt{n})F_{0,1}^{-1}(1-\alpha/2),\right.$$

$$\left.(1/n)\sum_{i=1}^{n}\xi_i + (\sigma/\sqrt{n})F_{0,1}^{-1}(1-\alpha/2)\right].$$

The same procedure for deriving limits of the confidence region of a parameter estimated using a certain estimator from an observation vector with a known probability distribution can be used in all the other cases. Therefore, let us briefly repeat the essence of deriving the confidence interval of the mean value m of the random variable $\xi \sim N(m,\sigma^2)$ if it is estimated by a LS-estimator under the condition that the observation vector has a normal probability distribution $N_n(m,\sigma^2 I)$, where σ^2 is a priori known. Under the mentioned assumptions

$$\bar{\xi} = (1/n)\sum_{i=1}^{n}\xi_i \sim N(m,\sigma^2/n) \Rightarrow \frac{\bar{\xi}-m}{\sigma/\sqrt{n}} \sim N(0,1) \Rightarrow$$

$$\Rightarrow P\left\{\omega : \left|\frac{\bar{\xi}(\omega)-m}{\sigma/\sqrt{n}}\right| \leq F_{0,1}^{-1}(1-\alpha/2)\right\} = 1-\alpha,$$

where $\bar{x} = \bar{\xi}(\omega)$ is a realization of the random variable $\bar{\xi}$. However,

$$P\left\{\omega : \left|\frac{\bar{\xi}(\omega)-m}{\sigma/\sqrt{n}}\right| \leq F_{0,1}^{-1}(1-\alpha/2)\right\} =$$

$$= P\{\omega : -(\sigma/\sqrt{n})F_{0,1}^{-1}(1-\alpha/2) \le \bar{\xi}(\omega) - m \le (\sigma/\sqrt{n})F_{0,1}^{-1}(1-\alpha/2)\} =$$
$$= P\{\omega : m - (\sigma/\sqrt{n})F_{0,1}^{-1}(1-\alpha/2) \le \bar{\xi}(\omega) \le m + (\sigma/\sqrt{n})F_{0,1}^{-1}(1-\alpha/2)\},$$

which means that the estimator $\bar{\xi} = (1/n)\sum_{i=1}^{n} \xi_i$ is realized within the interval

$$[m - (\sigma/\sqrt{n})F_{0,1}^{-1}(1-\alpha/2), m + (\sigma/\sqrt{n})F_{0,1}^{-1}(1-\alpha/2)]$$

with probability $1-\alpha$; the confidence interval, i.e., a random interval which with probability $1-\alpha$ contains the actual value m of the mean value of the random variable ξ is

$$\left[(1/n)\sum_{i=1}^{n} \xi_i - (\sigma/\sqrt{n})F_{0,1}^{-1}(1-\alpha/2), \right.$$
$$\left. (1/n)\sum_{i=1}^{n} \xi_i + (\sigma/\sqrt{n})F_{0,1}^{-1}(1-\alpha/2)\right].$$

b) Let σ^2 be a priori unknown. In this case the random variable $(1/n)\sum_{i=1}^{n} \xi_i$ representing the LS-estimator of the mean value of the random variable $\xi \sim N(m, \sigma^2)$ on the basis of an n-dimensional observation vector $\xi \sim N_n(m, \sigma^2 I)$ has a normal probability distribution whose unit dispersion is not known, it has to be estimated from a realization of the observation vector ξ using the relation $\hat{\sigma}^2 = v'v/(n-1) = [1/(1-n)]\sum_{i=1}^{n}(\xi_i - \bar{\xi})^2$.

In order to determine the limits of the confidence interval in this case, we have to make use of the fact that under the assumptions mentioned the random variable

$$T = \sqrt{n}\frac{\bar{\xi} - m}{\left\{[1/(n-1)]\sum_{i=1}^{n}(\xi_i - \bar{\xi})^2\right\}^{1/2}}$$

possesses the Student's t probability distribution with $n-1$ degrees of freedom. The equality

$$P\{\omega : |T(\omega)| \le \Delta_\alpha\} = 1 - \alpha$$

for the given confidence level $1-\alpha$ is fulfilled just for the $(1-\alpha/2)$-quantile $t_{S_{n-1}}(1-\alpha/2)$ of the Student's t random variable with $n-1$ degrees of freedom. Since

$$P\{\omega : |T(\omega)| \le t_{S_{n-1}}(1-\alpha/2)\} =$$

$$= P\left\{\omega : \left|\sqrt{n}(\bar{\xi}(\omega) - m)\Big/\left([1/(n-1)]\sum_{i=1}^{n}(\xi_i(\omega) - \bar{\xi}(\omega))^2\right)^{1/2}\right| \leq \right.$$

$$\left. \leq t_{S_{n-1}}(1-\alpha/2)\right\} =$$

$$= P\left\{\omega : m - \left(\frac{1}{n(n-1)}\sum_{i=1}^{n}(\xi_i(\omega) - \bar{\xi}(\omega))^2\right)^{1/2} t_{S_{n-1}}(1-\alpha/2) \leq \right.$$

$$\left. \leq \bar{\xi}(\omega)x \leq m + \left(\frac{1}{n(n-1)}\sum_{i=1}^{n}(\xi_i(\omega) - \bar{\xi}(\omega))^2\right)^{1/2} t_{S_{n-1}}(1-\alpha/2\right\},$$

which, finally, implies that the confidence interval of the mean value of the random variable $\xi \sim N(m, \sigma^2)$, estimated from a realization of an n-dimensional observation vector $\xi \sim N_n(m, \sigma^2 I)$ with a priori unknown unit dispersion σ^2, is

$$\left[(1/n)\sum_{i=1}^{n}\xi_i - \left(\frac{1}{n(n-1)}\sum_{i=1}^{n}(\xi_i - \bar{\xi})^2\right)^{1/2} t_{S_{n-1}}(1-\alpha/2)\,,\right.$$

$$\left.(1/n)\sum_{i=1}^{n}\xi_i + \left(\frac{1}{n(n-1)}\sum_{i=1}^{n}(\xi_i - \bar{\xi})^2\right)^{1/2} t_{S_{n-1}}(1-\alpha/2)\right].$$

It has to be emphasized that in contradistinction to the alternative a) (the unit dispersion is assumed to be a priori known) in the case b) of the a priori unknown σ^2 the value

$$\Delta_\alpha = \left(\frac{1}{n(n-1)}\sum_{i=1}^{n}(\xi - \bar{\xi})^2\right)^{1/2} t_{S_{n-1}}(1-\alpha/2)$$

is random.

Let Θ be a k-dimensional unknown parameter; its confidence region is formed by a random set of points of the k-dimensional space $\mathcal{R}^k$, which is included into the set of all possible values of the vector parameter Θ and which, with the prescribed probability, contains the actual value of the unknown parameter.

In a simple case of a three-dimensional unknown vector Θ estimated from a realization of the n-dimensional observation vector $\xi \sim N_n(Jf(\Theta_0) + JA\Delta\Theta, \sigma^2 H)$, where H is a symmetric and positive definite matrix it will be shown that the confidence region of Θ has the form of an ellipsoid.

a) Let the unit dispersion σ^2 be a priori known. The LS-estimator of the parameter $\Theta = (\Theta_1, \Theta_2, \Theta_3)'$ in a simple Model 2 is $\hat{\Theta}(\eta) = \Theta_0 + (A'J'H^{-1}JA)^{-1}A'J'H^{-1}\eta \sim N_k(\Theta, \sigma^2(A'J'H^{-1}JA))$, $R[(A'J'H^{-1}JA)^{-1}] = k$ (see the relation (2.2.4) and the results of the section 2.9.1; $\eta = \xi - Jf(\Theta_0)$).

The fact that, according to the Pearson lemma (p. 84), the random variable $[\Theta - \hat{\Theta}(\eta)]'A'J'H^{-1}JA[\Theta - \hat{\Theta}(\eta)]$ possesses the central σ^2 chi-square probability distribution with 3 degrees of freedom, provides us with the basis for deriving the form of the confidence region for our case. In accordance with it,

$$P\{\omega : [\Theta - \hat{\Theta}(y)]'A'J'H^{-1}JA[\Theta - \hat{\Theta}(y)] \leq \sigma^2\chi_3^2(1-\alpha)\} = 1 - \alpha$$

($y = \eta(\omega)$ is a realization of the random vector $\eta = \xi - Jf(\Theta_0)$).

The equation

$$[\Theta - \hat{\Theta}(y)]'A'J'H^{-1}JA[\Theta - \hat{\Theta}(y)] = \sigma^2\chi_3^2(1-\alpha),$$

since the matrix $A'J'H^{-1}JA$ is positive definite, is the equation of the triaxial ellipsoid whose centre lies at the point $(\Theta_1, \Theta_2, \Theta_3)'$. The estimator $\hat{\Theta}(\eta)$ of the unknown parameter Θ is realized inside this ellipsoid with the probability $1-\alpha$. If we shift the centre of the ellipsoid to the random point $(\hat{\Theta}_1(\eta), \hat{\Theta}_2(\eta), \hat{\Theta}_3(\eta))'$, we get a random confidence ellipsoid

$$\mathcal{E} = \{u : (u - \hat{\Theta}(\eta))'A'J'H^{-1}JA(u - \hat{\Theta}(\eta)) \leq \sigma^2\chi_3^2(1-\alpha)\} \subset \mathcal{R}^3$$

containing with probability $1-\alpha$ the actual value of the parameter Θ estimated by the relation $\hat{\Theta}(\eta) = \Theta_0 + (A'J'H^{-1}JA)^{-1}A'J'H^{-1}\eta$ from the observation vector $\eta \sim N_n(JA\Delta\Theta, \sigma^2 H)$. Within Model 2 of an indirect measurement of a vector parameter the components of the vector Θ are not assumed to be bound by a condition, therefore the parametric space is $\mathcal{R}^k$ in general, in our case $\mathcal{R}^3$.

b) Let the unit dispersion of the observation vector be a priori unknown, it is estimated, in accordance with (2.7.4), by the relation

$$\hat{\sigma}^2(\eta) = v'(\eta)H^{-1}v(\eta)/(n-3)$$

(the arguments in this and the following relations are used in order to distinguish consequentially between the random variables and their realizations).

The random variable $\hat{\sigma}^2(\eta)$ possesses the central $[\sigma^2/(n-3)]$ chi-square probability distribution with $n-3$ degrees of freedom (remember that the observation vector η is assumed to have the normal probability distribution

$N_n(JA\Delta\Theta, \sigma^2 H)$ and that in our case $k = 3$). Due to the Pearson lemma, the random variable $[\Theta - \hat{\Theta}(\eta)]' A' J' H^{-1} J A [\Theta - \hat{\Theta}(\eta)]$ has the central σ^2chi-square probability distribution with 3 degrees of freedom.

We shall show that the random variables $v'(\eta) H^{-1} v(\eta)$ and $[\Theta - \hat{\Theta}(\eta)]'$ $A' J' H^{-1} J A [\Theta - \hat{\Theta}(\eta)]$ are stochastically independent. It suffices to prove the stochastic independence of the random variables $\widehat{\Delta\Theta}(\eta)$ and $v(\eta) = \eta - JA\widehat{\Delta\Theta}(\eta)$ (the assertion will be proved for an arbitrary $k \leq n$).

The $n \times k$-dimensional covariance matrix $cov[v(\eta), \widehat{\Delta\Theta}(\eta)]$ is zero, as

$$cov[v(\eta), \widehat{\Delta\Theta}(\eta)] =$$

$$= cov\{[I - JA(A'J'H^{-1}JA)^{-1} A'J'H^{-1}]\eta, (A'J'H^{-1}JA)^{-1} A'J'H^{-1}\eta\} =$$

$$= [I - JA(A'J'H^{-1}JA)^{-1} A'J'H^{-1}] var(\eta) H^{-1} JA (A'J'H^{-1}JA)^{-1} =$$

$$= \sigma^2 JA(A'J'H^{-1}JA)^{-1} - \sigma^2 JA(A'J'H^{-1}JA)^{-1} = 0$$

and, since $v(\eta) \sim N_n(0, \sigma^2[H - JA(A'J'H^{-1}JA)^{-1} A'J'])$ and $\hat{\Theta}(\eta) \sim N_k(\Theta, \sigma^2 (A'J'H^{-1}JA)^{-1})$, these random variables are stochastically independent.

This implies that the random variable

$$\{[\Theta - \hat{\Theta}(\eta)]' A'J'H^{-1} JA[\Theta - \hat{\Theta}(\eta)]/k\} \Big/ [v'(\eta) H^{-1} v(\eta)/(n-k)]$$

has the Fisher - Snedecor probability distribution with k and $n - k$ degrees of freedom (see p. 87).

This fact is important for constructing the confidence ellipsoid, which is in our case tri-axial. Since the random variable

$$\{[\Theta - \hat{\Theta}(\eta)]' A'J'H^{-1} JA[\Theta - \hat{\Theta}(\eta)]/k\} \Big/ [v'(\eta) H^{-1} v(\eta)/(n-3)]$$

has the Fisher — Snedecor probability distribution with 3 and $n - 3$ degrees of freedom

$$P\left\{\omega : \frac{[\Theta - \hat{\Theta}(y)]' A'J'H^{-1} JA[\Theta - \hat{\Theta}(y)]/3}{v'(y) H^{-1} v(y)/(n-3)} \leq F_{3,n-3}(1-\alpha)\right\} = 1 - \alpha,$$

where $y = \eta(\omega)$ and $F_{3,n-3}(1-\alpha)$ is the $(1-\alpha)$-quantile of the Fisher - Snedecor probability distribution with 3 and $n - 3$ degrees of freedom. Because of

$$P\left\{\omega : \frac{[\Theta - \hat{\Theta}(y)]' A'J'H^{-1} JA[\Theta - \hat{\Theta}(y)]/3}{v'(y) H^{-1} v(y)/(n-3)} \leq F_{3,n-3}(1-\alpha) \leq\right.$$

$$\leq \frac{3}{n-3} v'(y) H^{-1} v(y) F_{3,n-3}(1-\alpha) \Big\} = 1-\alpha,$$

we have obtained the ellipsoid

$$[\Theta - \hat{\Theta}(y)]' A' J' H^{-1} J A [\Theta - \hat{\Theta}(y)] \leq \frac{3}{n-3} v'(y) H^{-1} v(y) F_{3,n-3}(1-\alpha),$$

whose centre lies at the point $\Theta = (\Theta_1, \Theta_2, \Theta_3)'$, inside which the estimator $\hat{\Theta}(\eta) = \Theta_0 + (A'J'H^{-1}JA)^{-1} A'J'H^{-1}\eta$ of the parameter Θ is realized with the probability $1-\alpha$. This immediately implies that the sought confidence ellipsoid of the vector parameter Θ, estimated on the basis of an n-dimensional observation vector $\eta \sim N_n(JA\Delta\Theta, \sigma^2 H)$ given the condition of a known unit dispersion σ^2 at the confidence level $1-\alpha$, is

$$\mathcal{E}\Big\{u : [u - \hat{\Theta}(\eta)]' A' J' H^{-1} J A [u - \hat{\Theta}(\eta)] \leq$$

$$\leq \frac{3}{n-3} v'(\eta) H^{-1} v(\eta) F_{3,n-3}(1-\alpha) \Big\} \subset \mathcal{R}^3.$$

Its center is at the random point $\hat{\Theta}(\eta) = (\hat{\Theta}_1(\eta), \hat{\Theta}_2(\eta), \hat{\Theta}_3(\eta))'$ and the dimensions of its axes are random values as well.

The relations for confidence ellipsoids derived for $k = 3$ can evidently be generalized for any $k < n$.

Solutions of many practical problems do not require that we know the confidence ellipsoid of the whole k-dimensional vector parameter Θ estimated on the basis of a realization of the n-dimensional normal observation vector; it is often enough to determine the confidence ellipsoid of its suitable linear transformation $T\Theta$ (for the sake of completeness, it has to be stated that in measurement models in which the elements of the vector parameter Θ are bound by some conditions, the parametric space is formed by a certain linear manifold of the space $\mathcal{R}^k$ and the rank of the covariance matrix of the LS-estimator of the unknown vector parameter equals $k-q$, where k is the dimension of the unknown vector parameter Θ and q is the number of conditions binding its elements).

In what follows below, the linear transformations $T\Theta$ (T is a $m \times k$-dimensional matrix of coefficients of the linear transformation) are such that they ensure the matrix $var[T\hat{\Theta}(\eta)] = Tvar[\hat{\Theta}(\eta)]T'$ will be regular even if the matrix $var[\hat{\Theta}(\eta)]$ is not regular will be considered. This means that the following two assertions are true within all the 6 fundamental linear models of measurement.

Let the observation vector η have a normal probability distribution, also let its variance matrix be $var(\eta) = \sigma^2 H$ and the unit dispersion be a priori known; then the LS-estimator of the vector parameter Θ has the normal probability distribution $N_k(\Theta, \sigma^2 H_{\hat{\Theta}})$, where the $k \times k$-dimensional matrix $H_{\hat{\Theta}}$ in individual simple linear models of measurement has the form

Model 1 $H_{\hat{\Theta}} = (i'H^{-1}i)^{-1}$,

Model 2 $H_{\hat{\Theta}} = (A'J'H^{-1}JA)^{-1}$,

Model 3 $H_{\hat{\Theta}} =$

$$= (J'H^{-1}J)^{-1} - (J'H^{-1}J)^{-1}B'[B(J'H^{-1}J)^{-1}B']^{-1}B(J'H^{-1}J)^{-1},$$

Model 4 $H_{\hat{\Theta}} =$

$$= \begin{pmatrix} (J'H^{-1}J)^{-1} - (J'H^{-1}J)^{-1}B'Q_{11}B(J'H^{-1}J)^{-1}, & -(J'H^{-1}J)^{-1}B'Q_{12} \\ Q_{21}B(J'H^{-1}J)^{-1}, & -Q_{22} \end{pmatrix},$$

where $\begin{pmatrix} Q_{11}, & Q_{12} \\ Q_{21}, & Q_{22} \end{pmatrix} = \begin{pmatrix} B(J'H^{-1}J)^{-1}B', & C \\ C', & 0 \end{pmatrix}^{-1}$,

Model 5 $H_{\hat{\Theta}} =$

$$= (A'J'H^{-1}JA)^{-1} - (A'J'H^{-1}JA)^{-1}B'[B(A'J'H^{-1}JA)^{-1}B']^{-1}B(A'J'H^{-1}JA)^{-1},$$

Model 6 $H_{\hat{\Theta}} =$

$$\begin{pmatrix} (A'J'H^{-1}JA)^{-1} - (A'J'H^{-1}JA)^{-1}B'Q_{11}B(A'J'H^{-1}JA)^{-1}, & -(A'J'H^{-1}JA)^{-1}B'Q_{12} \\ -Q_{21}B(A'J'H^{-1}JA)^{-1}, & -Q_{22} \end{pmatrix},$$

where $\begin{pmatrix} Q_{11}, & Q_{12} \\ Q_{21}, & Q_{22} \end{pmatrix} = \begin{pmatrix} B(A'J'H^{-1}JA)^{-1}B', & C \\ C', & 0 \end{pmatrix}^{-1}$,

(see the relations (2.4.2) for Model 1, (2.4.4) for Model 2, (2.4.9) for Model 3, (2.4.15) for Model 4, (2.4.17) for Model 5 and (2.4.19) for Model 6 in their modified form for $var(\eta) = \Sigma = \sigma^2 H$). Let T be a $m \times k$-dimensional transformation matrix possessing the property $R(TH_{\hat{\Theta}}T') = m$ regardless of the regularity or irregularity of the matrix $H_{\hat{\Theta}}$. The vector $T\Theta$ is with probability $1-\alpha$ an element of the random ellipsoid

$$\mathcal{E} = \{u : [u - T\hat{\Theta}(\eta)]'(TH_{\hat{\Theta}}T')^{-1}[u - T\hat{\Theta}(\eta)] \leq \sigma^2 \chi^2_m(1-\alpha)\}, \tag{2.9.4}$$

where $\chi^2_m(1-\alpha)$ is the $(1-\alpha)$-quantile of the central chi-square probability distribution with m degrees of freedom.

If $m = 1$, instead of the transformation matrix T, a k-dimensional transformation vector t is considered; if $t'H_{\hat{\Theta}}t \neq 0$, the random variable $t'\hat{\Theta}(\eta)$ possesses the one-dimensional normal probability distribution $N(t'\Theta, \sigma^2 t' H_{\hat{\Theta}} t)$ and the quantity $t'\Theta$ is, with probability $1-\alpha$, included in the random interval

$$[t'\hat{\Theta}(\eta) - \sigma F_{0,1}^{-1}(1-\alpha/2)(t'H_{\hat{\Theta}}t)^{1/2}, t'\hat{\Theta}(\eta) + \sigma F_{0,1}^{-1}(1-\alpha/2)(t'H_{\hat{\Theta}}t)^{1/2}], \tag{2.9.5}$$

where $F_{0,1}^{-1}(1-\alpha/2)$ is the $(1-\alpha/2)$-quantile of the normal probability distribution $N(0,1)$.

Let $\eta \sim N_n(JA\Delta\Theta, \sigma^2 H)$ and the unit dispersion, being a priori unknown, be estimated using the relation $\hat{\sigma}^2(\eta) = v'(\eta)H^{-1}v(\eta)/(n+q-k)$. The vector $T\Theta$, where T is a $m \times k$-dimensional transformation matrix with the property $R(TH_{\hat{\Theta}}T') = m$, is with probability $1-\alpha$ an element of the random confidence ellipsoid

$$\mathcal{E} = \Big\{u : [u - T\hat{\Theta}(\eta)]'(TH_{\hat{\Theta}}T')^{-1}[u - T\hat{\Theta}(\eta)] \leq$$
$$\leq \frac{m}{n+q-k} v'(\eta)H^{-1}v(\eta)F_{m,n+q-k}(1-\alpha)\Big\} \subset \mathcal{R}^m, \tag{2.9.6}$$

where $F_{m,n+q-k}(1-\alpha)$ is the $(1-\alpha)$-quantile of the Fisher—Snedecor probability distribution with m and $n+q-k$ degrees of freedom.

If a transformation $t'\Theta$, where t is a k-dimensional transformation vector possessing the property $t'H_{\hat{\Theta}}t \neq 0$, is taken into account and the unit dispersion σ^2 is a priori unknown, the quantity $t'\Theta$ is with probability $1-\alpha$ an element of the random confidence interval

$$[t'\hat{\Theta}(\eta) - \{[1/(n+q-k)]v'(\eta)H^{-1}v(\eta)(t'H_{\hat{\Theta}}t)\}^{1/2} t_{S_{n+q-k}}(1-\alpha/2),$$
$$t'\hat{\Theta}(\eta) + \{[1/(n+q-k)]v'(\eta)H^{-1}v(\eta)(t'H_{\hat{\Theta}}t)\}^{1/2} t_{S_{n+q-k}}(1-\alpha/2)], \tag{2.9.7}$$

where $t_{S_{n+q-k}}(1-\alpha/2)$ is the $(1-\alpha/2)$-quantile of the Student's t random variable with $n+q-k$ degrees of freedom.

The following facts were taken into account for deriving the form of the confidence ellipsoid (2.9.4): the equivalent assumptions $\xi \sim N_n(Jf(\Theta_0) + JA\Delta\Theta, \sigma^2 H)$ and $\eta \sim N_n(JA\Delta\Theta, \sigma^2 H)$ imply $\hat{\Theta}(\eta) \sim N_k(\Theta, \sigma^2 H_{\hat{\Theta}})$, $T\hat{\Theta}(\eta) \sim N_m(T\Theta, \sigma^2 TH_{\hat{\Theta}}T')$ and $[T\Theta - T\hat{\Theta}(\eta)] \sim N_m(0, \sigma^2 TH_{\hat{\Theta}}T')$, where the matrix $TH_{\hat{\Theta}}T'$ is regular. Then the random variable $[T\Theta - \hat{\Theta}(\eta)]'(TH_{\hat{\Theta}}T')^{-1}[T\Theta - T\hat{\Theta}(\eta)]$, based on the Pearson lemma, has the central σ^2 chi-square probability distribution with m degrees of freedom. An application of the definition of the $(1-\alpha)$-quantile of the central σ^2 chi-square probability distribution with m degrees of freedom yields

$$P\{\omega : [T\Theta - T\hat{\Theta}(y)]'(TH_{\hat{\Theta}}T')^{-1}[T\Theta - T\hat{\Theta}(y)] \leq \sigma^2\chi^2_m(1-\alpha)\} = 1-\alpha,$$

which immediately implies that the set

$$\{u : [u - \hat{\Theta}(\eta)]'(TH_{\hat{\Theta}}T')^{-1}[u - T\hat{\Theta}(\eta)] \leq \sigma^2\chi^2_m(1-\alpha)\}$$

of the points of the m-dimensional space $\mathcal{R}^m$ is a confidence interval on the confidence level $1-\alpha$ of the vector $T\Theta$ estimated from a realization of a normally distributed observation vector η.

If the unit dispersion is not known a priori, its estimator $\hat{\sigma}^2(\eta) = v'(\eta)H^{-1}v(\eta)/(n+q-k)$ has to be applied. The assumption of the normality of the observation vector η implies $[T\Theta - T\hat{\Theta}(\eta)]'(TH_{\hat{\Theta}}T')^{-1}[T\Theta - T\hat{\Theta}(\eta)] \sim \sigma^2\chi^2(0,m)$ and $v'(\eta)H^{-1}v(\eta) \sim \sigma^2\chi^2(0, n+q-k)$, where simultaneously these random variables are stochastically independent, thus the random variable

$$\{[T\Theta - T\hat{\Theta}(\eta)]'(TH_{\hat{\Theta}}T')^{-1}[T\Theta - T\hat{\Theta}(\eta)]/m\}/[v'(\eta)H^{-1}v(\eta)/(n+q-k)]$$

has the Fisher — Snedecor probability distribution with m and $n+q-k$ degrees of freedom. By applying the definition of the $(1-\alpha)$-quantile of this probability distribution, we can write

$$P\{\omega : \{[T\Theta - T\hat{\Theta}(y)]'(TH_{\hat{\Theta}}T')^{-1}[T\Theta - T\hat{\Theta}(y)]/m\}\Big/$$

$$\Big/[v'(y)H^{-1}v(y)/(n+q-k)] \leq F_{m,n+q-k}(1-\alpha)\} =$$

$$= P\Big\{\omega : [T\Theta - T\hat{\Theta}(y)]'(TH_{\hat{\Theta}}T')^{-1}[T\Theta - T\hat{\Theta}(y)] \leq$$

$$\leq \frac{m}{n+q-k}v'(y)H^{-1}v(y)F_{m,n+q-k}(1-\alpha)\Big\} = 1-\alpha,$$

by which the assertion expressed by the relation (2.9.6) is proved.

If a transformation vector t is considered and the observation vector η is normal, $\hat{\Theta}(\eta) \sim N_k(\Theta, \sigma^2 H_{\hat{\Theta}})$, $t'\hat{\Theta}(\eta) \sim N(t'\Theta, \sigma^2 t'H_{\hat{\Theta}}t)$, where $t'H_{\hat{\Theta}}t >$

0, $t'\Theta - t'\hat{\Theta}(\eta) \sim N(0, \sigma^2 t' H_{\hat{\Theta}} t)$ and $[t'\Theta - t'\hat{\Theta}(\eta)]/\sqrt{t' H_{\hat{\Theta}} t} \sim N(0, \sigma^2)$; then

$$P\left\{\omega : \left|\frac{t'\Theta - t'\hat{\Theta}(y)}{\sqrt{t' H_{\hat{\Theta}} t}}\right| \leq \sigma F_{0,1}^{-1}(1-\alpha/2)\right\} =$$

$$= P\{\omega : t'\Theta - \sigma\sqrt{t' H_{\hat{\Theta}} t} F_{0,1}^{-1}(1-\alpha/2) \leq$$

$$\leq t'\hat{\Theta}(y) \leq t'\Theta + \sigma\sqrt{t' H_{\hat{\Theta}} t} F_{0,1}^{-1}(1-\alpha/2)\} = 1 - \alpha,$$

where $\sigma F_{0,1}^{-1}(1-\alpha/2)$ is the $(1-\alpha/2)$-quantile of the normal probability distribution $N(0, \sigma^2)$. Thus the relation (2.9.5) is proved.

If, under the same conditions, the unit dispersion is not known and is estimated by the expression $\hat{\sigma}^2(\eta) = v'(\eta) H^{-1} v(\eta)/(n+q-k)$, where $\eta \sim N_n(JA\Delta\Theta, \sigma^2 H)$, then $v'(\eta) H^{-1} v(\eta) \sim \sigma^2 \chi^2(n+q-k)$ and the random variable $[t'\Theta - t'\hat{\Theta}(\eta)]/\sqrt{t' H_{\hat{\Theta}} t v'(\eta) H^{-1} v(\eta)/(n+q-k)}$ possesses the Student's t probability distribution with $n+q-k$ degrees of freedom, thus

$$P\left\{\omega : \left|\frac{t'\Theta - t'\hat{\Theta}(y)}{\sqrt{t' H_{\hat{\Theta}} t v'(\eta) H^{-1} v(y)/(n+q-k)}}\right| \leq t_{S_{n+q-k}}(1-\alpha/2)\right\} =$$

$$= P\left\{\omega : t'\Theta - \sqrt{\frac{t' H_{\hat{\Theta}} t v'(y) H^{-1} v(y)}{n+q-k}}\, t_{S_{n+q-k}}(1-\alpha/2) \leq t'\hat{\Theta}(y) \leq\right.$$

$$\left.\leq t'\Theta + \sqrt{\frac{t' H_{\hat{\Theta}} t v'(y) H^{-1} v(y)}{n+q-k}}\, t_{S_{n+q-k}}(1-\alpha/2)\right\} \leq 1 - \alpha,$$

where $t_{S_{n+q-k}}(1-\alpha/2)$ is the $(1-\alpha/2)$-quantile of the Student's t probability distribution with $n+q-k$ degrees of freedom. This relation implies the validity of the assertion (2.9.7).

Example Consider the simplest case where we measure angles in a triangle. Each angle was measured once, independently of the measurements of the other angles. The problem is to determine the confidence interval on the confidence level $1-\alpha$ of each of the angles separately. For the sake of simplicity the unit dispersion is assumed to be known.

Model 3 in its simple form is applied here:

$$\Theta = (\Theta_1, \Theta_2, \Theta_3)', \ \eta = (\eta_1, \eta_2, \eta_3)', \ E[(\eta_1, \eta_2, \eta_3)'] = (\Theta_1, \Theta_2, \Theta_3)',$$

$$var(\eta) = \sigma^2 I,\ \eta \sim N_3(\Theta, \sigma^2 I),\ J = \begin{pmatrix} 1, & 0, & 0 \\ 0, & 1, & 0 \\ 0, & 0, & 1 \end{pmatrix},\ B = (1,1,1).$$

The transformation vector enabling us to determine the confidence interval of the angle Θ_1 is $t_1' = (1,0,0)$ (analogously $t_2' = (0,1,0)$ and $t_3' = (0,0,1)$).

In accordance with the relation (2.9.5), the expression $t'H_{\hat{\Theta}}t$ has to be determined and simultaneously the fulfilling of the condition $t'H_{\hat{\Theta}}t \neq 0$ to be verified: In our case

$$H_{\hat{\Theta}} = I - B'(BB')^{-1}B = (1/3)\begin{pmatrix} 2, & -1, & -1 \\ -1, & 2, & -1 \\ -1, & -1, & 2 \end{pmatrix}$$

and evidently $det(H_{\hat{\Theta}}) = 0$, however $t_1'H_{\hat{\Theta}}t_1 = 2/3 \neq 0$ (analogously $t_2'H_{\hat{\Theta}}t_2 = 2/3$ and $t_3'H_{\hat{\Theta}}t_3 = 2/3$), which means that we are entitled to apply the relation (2.9.5) for determining the confidence intervals of individual angles under the condition they are estimated using the relation (2.4.8).

At the conclusion we give the confidence interval of a unit dispersion and the unit standard deviation of a normally distributed observation vector η, under the condition that it is estimated using the expression $\hat{\sigma}^2(\eta) = v'(\eta)H^{-1}v(\eta)/(n+q-k)$. If $\eta \sim N_n(JA\Delta\Theta, \sigma^2 H)$, then $\hat{\sigma}^2(\eta) \sim [\sigma^2/(n+q-k)]\chi^2(0, n+q-k)$ and

$$P\left\{\omega : \frac{\sigma^2}{n+q-k}\chi^2_{n+q-k}(\alpha/2) \le \hat{\sigma}^2(y) \le \frac{\sigma^2}{n+q-k}\chi^2_{n+q-k}(1-\alpha/2)\right\} =$$

$$= P\left\{\omega : \frac{(n+q-k)\hat{\sigma}^2(y)}{\chi^2_{n+q-k}(1-\alpha/2)} \le \sigma^2 \le \frac{(n+q-k)\hat{\sigma}^2(y)}{\chi^2_{n+q-k}(\alpha/2)}\right\} =$$

$$= P\left\{\omega : \hat{\sigma}(y)\sqrt{\frac{n+q-k}{\chi^2_{n+q-k}(1-\alpha/2)}} \le \sigma \le \hat{\sigma}(y)\sqrt{\frac{n+q-k}{\chi^2_{n+q-k}(\alpha/2)}}\right\} = 1-\alpha,$$

which implies that the confidence interval of the unit dispersion estimated by $\hat{\sigma}^2(\eta) = v'(\eta)H^{-1}v(\eta)/(n+q-k)$ on the basis of a normally distributed observation vector has on the confidence level $1-\alpha$ the form

$$\left[\frac{\hat{\sigma}^2(\eta)(n+q-k)}{\chi^2_{n+q-k}(1-\alpha/2)}, \frac{\hat{\sigma}^2(\eta)(n+q-k)}{\chi^2_{n+q-k}(\alpha/2)}\right]$$

and the confidence interval of the unit standard deviation (under the same conditions) is

$$\left[\hat{\sigma}(\eta)\sqrt{\frac{(n+q-k)}{\chi^2_{n+q-k}(1-\alpha/2)}},\ \hat{\sigma}(\eta)\sqrt{\frac{(n+q-k)}{\chi^2_{n+q-k}(\alpha/2)}}\right]$$

($\chi^2_{n+q-k}(1-\alpha/2)$ is the $(1-\alpha/2)$-quantile of the central chi-square probability distribution with $n+q-k$ degrees of freedom and analogously $\chi^2_{n+q-k}(\alpha/2)$ is the $\alpha/2$-quantile of the the same probability distribution.

2.9.3 Estimated covariance matrix of a normal observation vector in a replicated model, and its influence on the first order parameter estimators

A replicated model of measurement, i.e., a model enabling a multiple independent realization of the observation vector, occurs frequently in solving many practical problems. This is the reason for dealing with this problem even if it exceeds the stated scope of the publication. The aim of what follows is to give the reader, in the form of rules, a review of results applicable in the framework of the regular measurement models.

In a replicated model we have at our disposal N independent realizations of the observation vector, which means that the input for data processing is a realization of an $N \times n$ dimensional hypervector $\xi = (\xi_1', \xi_2', \dots, \xi_N')'$, whose n-dimensional stochastically independent elements are identically distributed. We shall deal with normal random vectors whose covariance matrix is a priori unknown.

The aim of data processing in this case is to utilize all information obtained in this hypervector in order to determine optimum values for the second order parameters and, by means of them, to determine optimum values for the first order parameters.

The data processing procedure will be given for the replicated Model 2 of measurement (an indirect measurement of a vector parameter). In this case $\xi_i = Jf(\Theta_0) + JA\Delta\Theta + \varepsilon_i, \eta_i = \xi_i - Jf(\Theta_0), E[\eta_i] = JA\Delta\Theta, var(\xi_i) = var(\eta_i) = var(\varepsilon_i) = \Sigma, \eta_i \sim N_n(JA\Delta\Theta, \Sigma), i = 1, 2, ..., N$. As the $n \times n$ dimensional covariance matrix Σ is a priori unknown, it is unbiasedly estimated by the expression

$$\hat{\Sigma}(\eta) = [1/(N-1)]\sum_{i=1}^{N}(\eta_i - \bar{\eta})(\eta_i - \bar{\eta})',$$

where $\bar{\eta} = (1/N)\sum_{i=1}^{N} \eta_i \sim N_n(JA\Delta\Theta, \Sigma/N)$ and the random matrix $(N-1)\hat{\Sigma}(\eta)$ possesses the n dimensional Wishart probability distribution. It can be proved that the random variables $\bar{\eta}$ and $\hat{\Sigma}(\eta)$ are stochastically independent and simultaneously $R(\Sigma) = R(\hat{\Sigma})$ with probability one (the relation $N > n$ is assumed to be fulfilled).

The random variable

$$\hat{\hat{\Theta}}(\eta) = \Theta_0 + (A'J'\hat{\Sigma}^{-1}(\eta)AJ)^{-1}A'J'\hat{\Sigma}^{-1}(\eta)\bar{\eta} \tag{2.9.8}$$

is an unbiased estimator of the unknown vector parameter; its variance is

$$var(\hat{\hat{\Theta}}(\eta)) = \frac{N-2}{N[N-2-(n-k)]}(A'J'\Sigma^{-1}JA)^{-1}.$$

Let us recall that the unbiased estimator $\hat{\hat{\Theta}}(\eta)$ of the unknown parameter depends on an estimator of the covariance matrix, the variance matrix of this estimator depends on the unknown covariance matrix of the observation vector.

The fact that the observation vector is assumed to be normally distributed enables us to determine a confidence ellipsoid or a confidence interval (with a given confidence level) of a suitable linear transformation of the vector parameter Θ estimated by (2.9.8).

Let $\hat{v}(\eta) = \bar{\eta} - JA\Delta\hat{\hat{\Theta}}(\eta)$, where $\Delta\hat{\hat{\Theta}}(\eta)$ is given by (2.9.8). The random variable

$$\frac{[\Theta - \hat{\hat{\Theta}}(\eta)]'A'J'\hat{\Sigma}^{-1}(\eta)JA[\Theta - \hat{\hat{\Theta}}(\eta)]}{1 + \frac{N}{N-1}\hat{v}'(\eta)\hat{\Sigma}^{-1}(\eta)\hat{v}(\eta)} \frac{k(N-1)}{N(N-n)}$$

possesses the Fisher — Snedecor probability distribution with k and $N-k$ degrees of freedom, which implies the following form of the confidence ellipsoid (with the confidence level $1-\alpha$) of the parameter Θ, estimated using the relation (2.9.8)

$$\mathcal{E} = \{u : [u - \hat{\hat{\Theta}}(\eta)]'A'J'\hat{\Sigma}^{-1}(\eta)JA[u - \hat{\hat{\Theta}}(\eta)] \leq$$
$$\leq \frac{k(N-1)}{N(N-n)}\left(1 + \frac{N}{N-1}\hat{v}'(\eta)\hat{\Sigma}^{-1}(\eta)\hat{v}(\eta)\right) F_{k,N-n}(1-\alpha)\} \subset R^k,$$

where $F_{k,N-n}(1-\alpha)$ is the $(1-\alpha)$-quantile of the Fisher — Snedecor probability distribution with k and $N-n$ degrees of freedom. This ellipsoid is independent of the unknown covariance matrix Σ.

Let T be a $m \times k$ dimensional transformation matrix such that the matrix $T(A'J'\hat{\Sigma}^{-1}(\eta)JA)^{-1}T'$ is regular, thus $R[T(A'J'\hat{\Sigma}^{-1}(\eta)JA)^{-1}T'] = m > 1$. The random variable

$$\frac{[T\Theta - T\hat{\hat{\Theta}}(\eta)]'[T(A'J'\hat{\Sigma}^{-1}JA)^{-1}T'][T\Theta - T\hat{\hat{\Theta}}(\eta)]}{1 + \frac{N}{N-1}\hat{v}'(\eta)\hat{\Sigma}^{-1}(\eta)\hat{v}(\eta)} \frac{m(N-1)}{N[N-m-(n-k)]}$$

possesses the Fisher — Snedecor probability distribution with m and $N - m - n + k$ degrees of freedom. The confidence ellipsoid of the linear combination $T\Theta$ of the parameter Θ which is estimated using the relation (2.9.8) has the form

$$\mathcal{E} = \{u : [u - T\hat{\hat{\Theta}}(\eta)]'[T(A'J'\hat{\Sigma}^{-1}(\eta)JA)^{-1}T']^{-1}[u - T\hat{\hat{\Theta}}(\eta)] \leq$$

$$\leq \frac{m(N-1)}{N[N-m-(n-k)]}\left(1 + \frac{N}{N-1}\hat{v}'(\eta)\hat{\Sigma}^{-1}(\eta)\hat{v}(\eta)\right)\cdot$$

$$\cdot F_{m,N-m-n+k}(1-\alpha)\} \subset \mathcal{R}^k,$$

where $F_{m,N-m-n+k}(1-\alpha)$ is the $(1-\alpha)$-quantile of the Fisher — Snedecor probability distribution with m and $N - m - n + k$ degrees of freedom.

Let t be a k dimensional transformation vector such that $t'(A'J'\hat{\Sigma}^{-1}(\eta)J\cdot A)^{-1}t > 0$. Then the random variable

$$\frac{t'\Theta - t'\hat{\hat{\Theta}}(\eta)}{\sqrt{t'(A'J'\hat{\Sigma}^{-1}(\eta)JA)^{-1}t\left(1 + \frac{N}{N-1}\hat{v}'(\eta)\hat{\Sigma}^{-1}(\eta)\hat{v}(\eta)\right)}}\cdot$$

$$\cdot\sqrt{\frac{(N-1)}{N[N-1-(n-k)]}}$$

possesses the Student's t probability distribution with $N-1-n+k$ degrees of freedom. If the parameter Θ has been estimated by the relation (2.9.8), the confidence interval of the value $t'\Theta$ on the confidence level $1-\alpha$ is of the form

$$[t'\hat{\hat{\Theta}}(\eta) - \kappa t_{S_{N-1-n+k}}(1-\alpha/2), t'\hat{\hat{\Theta}}(\eta) + \kappa t_{S_{N-1-n+k}}(1-\alpha/2)],$$

where

$$\kappa = \sqrt{\frac{N-1}{N[N-1-(n-k)]}t'(A'J'\hat{\Sigma}^{-1}(\eta)JA)^{-1}t\left(1 + \frac{N}{N-1}\hat{v}'(\eta)\hat{\Sigma}^{-1}(\eta)\hat{v}(\eta)\right)};$$

$$(2.9.9)$$

$t_{S_{N-1-n+k}}(1-\alpha/2)$ is the $(1-\alpha/2)$-quantile of the Student's t probability distribution with $N-1-n+k$ degrees of freedom.

In the end a survey of all relations mentioned for separate regular linear models of measurement is given.

In all cases the n dimensional observation vector η is assumed to be normally distributed, its covariance matrix, which is a priori unknown, is estimated by the relation

$$\hat{\Sigma}(\eta) = [1/(N-1)]\sum_{i=1}^{N}(\eta_i - \bar{\eta})(\eta_i - \bar{\eta})', \ \bar{\eta} = (1/N)\sum_{i=1}^{N}\eta_i;$$

N denotes the number of replications of the observation vector η (the condition $N > n$ has to be fulfilled here); k is the dimension of the unknown parameter Θ, q is the number of conditions relating values of their components.

Model 1

$$\hat{\hat{\Theta}}(\eta) = \Theta_0 + (i'\hat{\Sigma}^{-1}(\eta)i)^{-1}i'\hat{\Sigma}^{-1}(\eta)\bar{\eta}, \ \hat{v}(\eta) = \bar{\eta} - i\Delta\hat{\hat{\Theta}}(\eta),$$

$$E[\hat{\hat{\Theta}}(\eta)] = \Theta,$$

$$D[\hat{\hat{\Theta}}(\eta)] = \tfrac{N-2}{N[N-2-(n-1)]}(i'\Sigma^{-1}i)^{-1}$$

and the confidence interval with the confidence level $1-\alpha$ is

$$[\hat{\hat{\Theta}}(\eta) - \kappa t_{S_{N-n-2}}(1-\alpha/2), \ \hat{\hat{\Theta}}(\eta) + \kappa t_{S_{N-n-2}}(1-\alpha/2)],$$

where $t_{S_{N-n-2}}(1-\alpha/2)$ is the $(1-\alpha/2)$-quantile of the Student's t probability distribution with $N-n-2$ degrees of freedom and

$$\kappa = \sqrt{\tfrac{N-1}{N(N-n)}(i'\hat{\Sigma}^{-1}(\eta)i)^{-1}\left(1 + \tfrac{N}{N-1}\hat{v}'(\eta)\hat{\Sigma}^{-1}(\eta)\hat{v}(\eta)\right)}.$$

Model 2

$$\hat{\hat{\Theta}}(\eta) = \Theta_0 + (A'J'\hat{\Sigma}^{-1}(\eta)JA)^{-1}A'J'\hat{\Sigma}^{-1}(\eta)\bar{\eta}, \ \hat{v}(\eta) = \bar{\eta} - JA\Delta\hat{\hat{\Theta}}(\eta),$$

$$E[\hat{\hat{\Theta}}(\eta)] = \Theta,$$

$$var[\hat{\hat{\Theta}}(\eta)] = \tfrac{N-2}{N[N-2-(n-k)]}(A'J'\Sigma^{-1}JA)^{-1},$$

the confidence ellipsoid with the confidence level $1-\alpha$ of the linear transformation $T\Theta$ of the vector parameter Θ (the $m \times k$ dimensional transformation matrix T has to possess the property that the matrix $T(A'J'\hat{\Sigma}^{-1}(\eta)J\cdot \cdot A)^{-1}T'$ is regular) is of the form

$$\mathcal{E} = \{u : [u - T\hat{\hat{\Theta}}(\eta)]'[T(A'J'\hat{\Sigma}^{-1}(\eta)JA)^{-1}T']^{-1}[u - T\hat{\hat{\Theta}}(\eta)] \leq$$

$$\leq \frac{m(N-1)}{N[N-m-(n-k)]}\left(1 + \frac{N}{N-1}\hat{v}'(\eta)\hat{\Sigma}^{-1}(\eta)\hat{v}(\eta)\right)\cdot$$

$$\cdot F_{m,N-m-n+k}(1-\alpha)\} \subset \mathcal{R}^m,$$

where $F_{m,N-m-n+k}(1-\alpha)$ is the $(1-\alpha)$-quantile of the Fisher- Snedecor random variable with m and $N-m-n+k$ degrees of freedom. Within Model 2 the matrix $A'J'\hat{\Sigma}^{-1}(\eta)JA$ is regular, we can put $T = I$ and we obtain a k dimensional confidence interval of the parameter Θ; the confidence interval of the value $t'\Theta$, where t is a k dimensional transformation vector such that $t'(A'J'\hat{\Sigma}^{-1}(\eta)JA)^{-1}t > 0$, is of the form

$$[t'\hat{\hat{\Theta}}(\eta) - \kappa t_{S_{N-1-n+k}}(1-\alpha/2),\ t'\hat{\hat{\Theta}}(\eta) + \kappa t_{S_{N-1-n+k}}(1-\alpha/2)],$$

where $t_{S_{N-1-n+k}}(1-\alpha/2)$ is the $(1-\alpha/2)$-quantile of the Fisher — Snedecor probability distribution with $N-1-n+k$ degrees of freedom and κ is defined by (2.9.9).

Model 3

$$\hat{\hat{\Theta}}(\eta) = \Theta_0 + (J'\hat{\Sigma}^{-1}(\eta)J)^{-1}J'\hat{\Sigma}^{-1}(\eta)\bar{\eta} - (J'\hat{\Sigma}^{-1}(\eta)J)^{-1}B'\cdot$$
$$\cdot[B(J'\hat{\Sigma}^{-1}(\eta)J)^{-1}B']^{-1}[b + B(J'\hat{\Sigma}^{-1}(\eta)J)^{-1}J'\hat{\Sigma}^{-1}(\eta)\bar{\eta}],$$
$$\hat{v}(\eta) = \bar{\eta} - J\Delta\hat{\hat{\Theta}}(\eta),$$
$$E[\hat{\hat{\Theta}}(\eta)] = \Theta,$$
$$var[\hat{\hat{\Theta}}(\eta)] = \frac{N-2}{N[N-2-(n+q-k)]}\{(J\Sigma^{-1}J)^{-1} - (J'\Sigma^{-1}J)^{-1}B'\cdot$$
$$\cdot[B(J'\Sigma^{-1}J)^{-1}B']^{-1}B(J'\Sigma^{-1}J)^{-1}\}.$$

Further, let $\hat{V}(\eta)$ denote a random matrix of the form

$$\hat{V}(\eta) = (J'\hat{\Sigma}^{-1}(\eta)J)^{-1} - (J'\hat{\Sigma}^{-1}(\eta)J)^{-1}B'[B(J'\hat{\Sigma}^{-1}(\eta)J)^{-1}B']^{-1}\cdot$$

$$\cdot B(J'\hat{\Sigma}^{-1}(\eta)J)^{-1}.$$

If the unknown parameter Θ has been estimated by $\hat{\hat{\Theta}}(\eta)$ and the transformation matrix T has the property that the matrix $T\hat{V}(\eta)T'$ is regular, the confidence ellipsoid of the vector $T\Theta$ has the form

$$\begin{gathered}\mathcal{E} = \{u : \ [u - T\hat{\hat{\Theta}}(\eta)]'(T\hat{V}(\eta)T')^{-1}[u - T\hat{\hat{\Theta}}(\eta)] \leq \\ \leq \frac{m(N-1)}{N[N-m-(n+q-k)]}\left(1 + \frac{N}{N-1}\hat{v}'(\eta)\hat{\Sigma}^{-1}(\eta)\hat{v}(\eta)\right) \cdot \\ \cdot F_{m,N-m-n-q+k}(1-\alpha)\} \subset \mathcal{R}^m\end{gathered} \tag{2.9.10}$$

and the confidence interval of the value $t'\Theta$ under the conditions that $\hat{\hat{\Theta}}(\eta)$ is the estimator of the unknown parameter Θ and, simultaneously the k-dimensional transformation vector t has the property $t'\hat{V}(\eta)t > 0$, is of the form

$$[t'\hat{\hat{\Theta}}(\eta) - \kappa t_{S_{N-1-n-q+k}}(1-\alpha/2),\ t'\hat{\hat{\Theta}}(\eta) + \kappa t_{S_{N-1-n-q+k}}(1-\alpha/2)], \tag{2.9.11}$$

where

$$\kappa = \sqrt{\frac{N-1}{N[N-1-(n+q-k)]}t'\hat{V}(\eta)t\left(1 + \frac{N}{N-1}v'(\eta)\hat{\Sigma}^{-1}(\eta)v(\eta)\right)} \tag{2.9.11'}$$

and $t_{S_{N-1-n-q+k}}(1-\alpha/2)$ is the $(1-\alpha/2)$-quantile of the Student's t probability distribution whose degrees of freedom are given by the index.

Model 4

$$\begin{pmatrix}\hat{\hat{\Theta}}_1(\eta)\\ \hat{\hat{\Theta}}_2(\eta)\end{pmatrix} = \begin{pmatrix}\Theta_{10}\\ \Theta_{20}\end{pmatrix} + \begin{pmatrix}(J'\hat{\Sigma}^{-1}(\eta)J)^{-1}\\ 0\end{pmatrix} J'\hat{\Sigma}^{-1}(\eta)\bar{\eta} -$$

$$- \begin{pmatrix}(J'\hat{\Sigma}^{-1}(\eta)J)^{-1}B'\hat{Q}_{11}(\eta)\\ \cdot\ \hat{Q}_{21}(\eta)\end{pmatrix}[b + B(J'\hat{\Sigma}^{-1}(\eta)J)^{-1}J'\hat{\Sigma}^{-1}(\eta)\bar{\eta}],$$

$$\hat{v}(\eta) = \bar{\eta} - J\Delta\hat{\hat{\Theta}}_1(\eta),$$

$$E\begin{pmatrix} \hat{\hat{\Theta}}_1(\eta) \\ \hat{\hat{\Theta}}_2(\eta) \end{pmatrix} = \begin{pmatrix} \Theta_1 \\ \Theta_2 \end{pmatrix},$$

$$var\begin{pmatrix} \hat{\hat{\Theta}}_1(\eta) \\ \hat{\hat{\Theta}}_2(\eta) \end{pmatrix} = \frac{\mathrm{N-2}}{\mathrm{N[N-2-(n+q-k)]}}\cdot$$

$$\cdot\begin{pmatrix} (J'\Sigma^{-1}J)^{-1} - (J'\Sigma^{-1}J)^{-1}B'Q_{11}B(J'\Sigma^{-1}J)^{-1}, & -(J'\Sigma^{-1}J)^{-1}B'Q_{12} \\ -Q_{21}B(J'\Sigma^{-1}J)^{-1}, & -Q_{22} \end{pmatrix},$$

where

$$\begin{pmatrix} Q_{11}, & Q_{12} \\ Q_{21}, & Q_{22} \end{pmatrix} = \begin{pmatrix} B(J'\Sigma^{-1}J)^{-1}B', & C \\ C', & 0 \end{pmatrix}^{-1}$$

and

$$\begin{pmatrix} \hat{Q}_{11}(\eta), & \hat{Q}_{12}(\eta) \\ \hat{Q}_{21}(\eta), & \hat{Q}_{22}(\eta) \end{pmatrix} = \begin{pmatrix} B(J'\hat{\Sigma}^{-1}(\eta)J)^{-1}B', & C \\ C', & 0 \end{pmatrix}^{-1};$$

if in the relation (2.9.10) for the confidence interval of a m-dimensional vector $T\Theta$ and in the relations (2.9.11) or (2.9.11′) for the confidence interval of the value $t'\Theta$ the matrix $\hat{V}(\eta)$ is substituted by the matrix

$$\hat{V}(\eta) = \begin{pmatrix} \hat{V}_{11}(\eta), & \hat{V}_{12}(\eta) \\ \hat{V}_{21}(\eta), & \hat{V}_{22}(\eta) \end{pmatrix},$$

where

$$\hat{V}_{11}(\eta) = (J'\hat{\Sigma}^{-1}(\eta)J)^{-1} - (J'\hat{\Sigma}^{-1}(\eta)J)^{-1}B'\hat{Q}_{11}(\eta)B(J'\hat{\Sigma}^{-1}(\eta)J)^{-1},$$
$$\hat{V}_{12}(\eta) = -(J'\hat{\Sigma}^{-1}(\eta)J)^{-1}B'\hat{Q}_{12}(\eta)$$
$$\hat{V}_{21}(\eta) = -\hat{Q}_{21}(\eta)B(J'\hat{\Sigma}^{-1}(\eta)J)^{-1},$$
$$\hat{V}_{22}(\eta) = -\hat{Q}_{22}(\eta),$$

we obtain immediately the corresponding confidence regions in Model 4.

Model 5

$$\hat{\hat{\Theta}}(\eta) = \Theta_0 + (A'J'\hat{\Sigma}^{-1}(\eta)JA)^{-1}A'J'\hat{\Sigma}^{-1}(\eta)\bar{\eta} - (A'J'\hat{\Sigma}^{-1}(\eta)JA)^{-1}\cdot$$
$$\cdot B'[B(A'J'\hat{\Sigma}^{-1}(\eta)JA)^{-1}B']^{-1}[b + B(A'J'\hat{\Sigma}^{-1}(\eta)JA)^{-1}A'J'\hat{\Sigma}^{-1}(\eta)\bar{\eta}],$$

$$\hat{v}(\eta) = \bar{\eta} - JA\Delta\hat{\hat{\Theta}}(\eta),$$

$$E[\hat{\hat{\Theta}}(\eta)] = \Theta,$$
$$var[\hat{\hat{\Theta}}(\eta)] = \frac{N-2}{N[N-2-(n+q-k)]}\{(A'J'\Sigma^{-1}JA)^{-1} - (A'J'\Sigma^{-1}JA)^{-1}B'\cdot$$
$$\cdot[B(A'J'\Sigma^{-1}JA)^{-1}B']^{-1}B(A'J'\Sigma^{-1}JA)^{-1}\}.$$

If in the relations (2.9.10), (2.9.11) and (2.9.11′) we put

$$\hat{V}(\eta) = (A'J'\hat{\Sigma}^{-1}(\eta)JA)^{-1} - (A'J'\hat{\Sigma}^{-1}(\eta)JA)^{-1}$$
$$B'[B(A'J'\hat{\Sigma}^{-1}(\eta)JA)^{-1}B']^{-1}B(A'J'\hat{\Sigma}^{-1}(\eta)JA)^{-1},$$

we obtain the relations for the confidence ellipsoid of the vector $T\Theta$ and for the value $t'\Theta$ in Model 5 for Θ being estimated by $\hat{\hat{\Theta}}(\eta)$.

Model 6

$$\begin{pmatrix} \hat{\hat{\Theta}}_1(\eta) \\ \hat{\hat{\Theta}}_2(\eta) \end{pmatrix} = \begin{pmatrix} \Theta_{10} \\ \Theta_{20} \end{pmatrix} + \begin{pmatrix} (A'J'\hat{\Sigma}^{-1}(\eta)JA)^{-1} \\ 0 \end{pmatrix} A'J'\hat{\Sigma}^{-1}(\eta)\bar{\eta} -$$
$$- \begin{pmatrix} (A'J'\hat{\Sigma}^{-1}(\eta)JA)^{-1}B'\hat{Q}_{11}(\eta) \\ -\hat{Q}_{21}(\eta) \end{pmatrix} [\mathrm{b} + \mathrm{B}(\mathrm{A}'\mathrm{J}'\hat{\Sigma}^{-1}(\eta)\mathrm{JA})^{-1}\mathrm{A}'\mathrm{J}'\hat{\Sigma}^{-1}(\eta)\bar{\eta}],$$

$$\hat{v}(\eta) = \bar{\eta} - JA\Delta\hat{\hat{\Theta}}_1(\eta),$$

$$E\begin{pmatrix} \hat{\hat{\Theta}}_1(\eta) \\ \hat{\hat{\Theta}}_2(\eta) \end{pmatrix} = \begin{pmatrix} \Theta_1 \\ \Theta_2 \end{pmatrix},$$

$$var\begin{pmatrix} \hat{\hat{\Theta}}_1(\eta) \\ \hat{\hat{\Theta}}_2(\eta) \end{pmatrix} = \frac{\mathrm{N}-2}{\mathrm{N}[\mathrm{N}-2-(\mathrm{n}+\mathrm{q}-\mathrm{k})]}\cdot$$
$$\cdot\begin{pmatrix} \begin{matrix}(A'J'\Sigma^{-1}JA)^{-1} - \\ -(A'J'\Sigma^{-1}JA)^{-1}B'Q_{11}B(A'J'\Sigma^{-1}JA)^{-1},\end{matrix} & -(A'J'\Sigma^{-1}JA)^{-1}B'Q_{12} \\ -Q_{21}B(A'J'\Sigma^{-1}JA)^{-1}, & -Q_{22} \end{pmatrix},$$

where

$$\begin{pmatrix} \hat{Q}_{11}(\eta), & \hat{Q}_{12}(\eta) \\ \hat{Q}_{21}(\eta), & \hat{Q}_{22}(\eta) \end{pmatrix} = \begin{pmatrix} B(A'J'\hat{\Sigma}^{-1}(\eta)JA)^{-1}B', & C \\ C', & 0 \end{pmatrix}^{-1}.$$

Putting in the relations (2.9.19), (2.9.11) and (2.9.11′)

$$\hat{V}(\eta) = \begin{pmatrix} \hat{V}_{11}(\eta), & \hat{V}_{12}(\eta) \\ \hat{V}_{21}(\eta), & \hat{V}_{22}(\eta) \end{pmatrix},$$

where

$$\hat{V}_{11}(\eta) = (A'J'\hat{\Sigma}^{-1}(\eta)JA)^{-1} -$$

$$-(A'J'\hat{\Sigma}^{-1}(\eta)JA)^{-1}B'\hat{Q}_{11}(\eta)B(A'J'\hat{\Sigma}^{-1}(\eta)JA)^{-1},$$

$$\hat{V}_{12}(\eta) = -(A'J'\hat{\Sigma}^{-1}(\eta)JA)^{-1}B'\hat{Q}_{12}(\eta),$$

$$\hat{V}_{21}(\eta) = -\hat{Q}_{21}(\eta)B(A'J'\hat{\Sigma}^{-1}(\eta)AJ)^{-1},$$

$$\hat{V}_{22}(\eta) = -\hat{Q}_{22}(\eta),$$

we get the confidence ellipsoid of the vector $T\Theta$ and the confidence interval of the value $t'\Theta$ in Model 6 for Θ estimated by $\hat{\hat{\Theta}}(\eta)$.

2.10 The universal model

So far we have dealt only with the regular linear models of measurement, i.e., models with a positive definite covariance matrix of the observation vector, with design matrices J, A and JA, respectively, possessing the full rank in columns and with matrices B and C occurring in conditions binding the values of the unknown vector parameters possessing the full rank in rows. All the conditions inherent in the regularity of the model were mentioned explicitly for each of the models described.

There exist many practical problems to be solved in which the regularity conditions are not fulfilled, mainly as far as the positive definiteness of the covariance matrix of the observation vector is concerned. Frequently, the problem of an ill conditioned covariance matrix of the observation vector occurs where the value of its determinant is close to zero. This situation is respected by the universal model.

Consider a theoretical model of an indirect measurement of a vector parameter

$$X = f(\Theta_0) + A\Delta\Theta,$$

where X is a p-dimensional vector of theoretical values of a directly observable quantity of the object being investigated, $\Theta = \Theta_0 + \Delta\Theta$ is a k-dimensional vector of theoretical values of the unknown parameter, $f(\cdot)$ is a p-dimensional function binding the theoretical values of the directly observable vector X with the theoretical values of the vector parameter

Θ, Θ_0 is its approximate value, and A is a $p \times k$-dimensional design matrix defined by the relation $A = \partial f(\Theta_0)/\partial\Theta'$ (see p. 115).

Let J be an $n \times p$-dimensional matrix reflecting the way of measuring the vector X, ξ be an n-dimensional random vector modelling the measurement (its realization is the n-tuple of measured data), and $\eta = Jf(\Theta_0)$. Then the stochastic model

$$\eta = JA\Delta\Theta + \varepsilon, \; E[\eta] = JA\Delta\Theta, \; E[\varepsilon] = 0, \; var(\eta) = var(\varepsilon) = \Sigma$$

is called the universal model.

Comparing this model with Model 2, which holds a universal position among the six fundamental models of measurement, we have to state that the essential difference between the regular Model 2 and the universal model consists in Model 2 lacking the regularity conditions found the universal model. It will be shown that this essential difference between these two models provides an essentially much simpler method of transforming other models of measurement to the universal form. The fact that each of the linear models of measurement may be transformed in a simple way into the form of the universal model convincingly demonstrates its title.

The problem is to derive the form of the observation vector η, the design matrices A, J, JA, and the covariance matrix of the observation vector after transforming individual models into the universal form. These values are denoted by an index UT_i, $i = 1, 2, \ldots, 6$; the subindex (subscript) i means the number of the model being transformed. The simple index in parentheses denotes the dimension of the vector considered, the double index the dimensions of the matrix. The meaning of the dimensions of vectors and matrices in individual linear measurement models are standard: p is the dimension of the vector X, k is the dimension of the vector $\Delta\Theta$, n is the dimension of the observation vector, q indicates the number of conditions, and l is the dimension of the subvector $\Delta\Theta_1$ in Model 6 of an incomplete indirect measurement of a vector parameter with a system of conditions.

Model 1 (a direct measurement of a scalar parameter)

$$\eta_{(n)} = i_{(n)}\Delta\Theta_{(1)} + \varepsilon_{(n)},\; E[\eta] = i\Delta\Theta,\; E[\varepsilon] = 0,\; var(\eta) = \Sigma_{(n,n)},$$
$$\eta_{UT_1(n)} = \eta_{(n)},\; A_{UT_1(1,1)} = 1,\; J_{UT_1(n,1)} = i_{(n)},\; (JA)_{UT_1} = i_{(n)},$$
$$var(\eta_{UT_1(n)}) = \Sigma_{(n,n)}.$$

Model 2 (an indirect measurement of a vector parameter)

$$\eta_{(n)} = J_{(n,p)}A_{(p,k)}\Delta\Theta_{(k)} + \varepsilon_{(n)},\; E[\eta] = JA\Delta\Theta,$$
$$E[\varepsilon] = 0,\; var(\eta) = \Sigma_{(n,n)},$$
$$\eta_{UT_2(n)} = \eta_{(n)},\; A_{UT_2(p,k)} = A_{(p,k)},\; J_{UT_2(n,p)} = J_{(n,p)},$$
$$(JA)_{UT_2(n,k)} = (JA)_{(n,k)},$$
$$var(\eta_{UT_2(n)}) = \Sigma_{(n,n)}.$$

Model 3 (a direct measurement of a vector parameter with a system of conditions)

$$\eta_{(n)} = J_{(n,k)}\Delta\Theta_{(k)} + \varepsilon_{(n)},\; b_{(q)} + B_{(q,k)}\Delta\Theta_{(k)} = 0_{(q)},$$
$$E[\eta] = J\Delta\Theta,\; E[\varepsilon] = 0,\; var(\eta) = \Sigma_{(n,n)},$$
$$\eta_{UT_3(n+q)} = \begin{pmatrix} \eta_{(n)} \\ -b_{(q)} \end{pmatrix},\quad A_{UT_3(k+q,k)} = \begin{pmatrix} I_{(k,k)} \\ B_{(q,k)} \end{pmatrix},$$
$$J_{UT_3(n+q,k+q)} = \begin{pmatrix} J_{(n,k)}, & 0_{(n,q)} \\ 0_{(q,k)}, & I_{(q,q)} \end{pmatrix},\quad (JA)_{UT_3(n+q,k)} = \begin{pmatrix} J_{(n,k)} \\ B_{(q,k)} \end{pmatrix},$$
$$var(\eta_{UT_3(n+q)}) = \begin{pmatrix} \Sigma_{(n,n)}, & 0_{(n,q)} \\ 0_{(q,n)}, & 0_{(q,q)} \end{pmatrix}.$$

Model 4 (a direct incomplete measurement of a vector parameter with a system of conditions)

$$\eta_{(n)} = J_{(n,p)}\Delta\Theta_{1(p)} + \varepsilon_{(n)}, \ \Delta\Theta_{(k)} = \begin{pmatrix} \Delta\Theta_{1(p)} \\ \Delta\Theta_{2(k-p)} \end{pmatrix},$$

$$b_{(q)} + (B_{(q,p)}, C_{(q,k-p)}) \begin{pmatrix} \Delta\Theta_{1(p)} \\ \Delta\Theta_{2(k-p)} \end{pmatrix} = 0_{(q)},$$

$$E[\eta] = J\Delta\Theta_1, \ E[\varepsilon] = 0, \ var(\eta) = \Sigma_{(n,n)},$$

$$\eta_{UT_4(n+q)} = \begin{pmatrix} \eta_{(n)} \\ -b_{(q)} \end{pmatrix}, \ A_{UT_4(p+q,k)} = \begin{pmatrix} I_{(p,p)}, & 0_{(p,k-p)} \\ B_{(q,p)}, & C_{(q,k-p)} \end{pmatrix},$$

$$J_{UT_4(n+q,p+q)} = \begin{pmatrix} J_{(n,p)}, & 0_{(n,q)} \\ 0_{(q,p)}, & I_{(q,q)} \end{pmatrix},$$

$$(JA)_{UT_4(n+q,k)} = \begin{pmatrix} J_{(n,p)}, & 0_{(n,k-p)} \\ B_{(q,p)}, & C_{(q,k-p)} \end{pmatrix},$$

$$var(\eta_{UT_4(n+q)}) = \begin{pmatrix} \Sigma_{(n,n)}, & 0_{(n,q)} \\ 0_{(q,n)}, & 0_{(q,q)} \end{pmatrix}.$$

Model 5 (an indirect measurement of a vector parameter with a system of conditions)

$$\eta_{(n)} = J_{(n,p)}A_{(p,k)}\Delta\Theta_{(k)} + \varepsilon_{(n)}, \ b_{(q)} + B_{(q,k)}\Delta\Theta_{(k)} = 0_{(q)},$$

$$E[\eta] = JA\Delta\Theta, \ E[\varepsilon] = 0, \ var(\eta) = \Sigma_{(n,n)},$$

$$\eta_{UT_5(n+q)} = \begin{pmatrix} \eta_{(n)} \\ -b_{(q)} \end{pmatrix}, \ A_{UT_5(p+q,k)} = \begin{pmatrix} A_{(p,k)} \\ B_{(q,k)} \end{pmatrix},$$

$$J_{UT_5(n+q,p+q)} = \begin{pmatrix} J_{(n,p)}, & 0_{(n,q)} \\ 0_{(q,p)}, & I_{(q,q)} \end{pmatrix}, \ (JA)_{UT_5(n+q,k)} = \begin{pmatrix} (JA)_{(n,k)} \\ B_{(q,k)} \end{pmatrix},$$

$$var(\eta_{UT_5(n+q)}) = \begin{pmatrix} \Sigma_{(n,n)}, & 0_{(n,q)} \\ 0_{(q,n)}, & 0_{(q,q)} \end{pmatrix}.$$

(putting $A = I_{(p,p)}$ and $p = k$ we obtain Model 3).

Model 6 (an indirect incomplete measurement of a vector parameter with a system of conditions)

$$\eta_{(n)} = J_{(n,p)}A_{(p,l)}\Delta\Theta_{1(l)} + \varepsilon_{(n)},\ \Delta\Theta_{(k)} = \begin{pmatrix} \Delta\Theta_{1(l)} \\ \Delta\Theta_{2(k-l)} \end{pmatrix},$$

$$b_{(q)} + (B_{(q,l)}, C_{(q,k-l)}) \begin{pmatrix} \Delta\Theta_{1(l)} \\ \Delta\Theta_{2(k-l)} \end{pmatrix} = 0_{(q)},$$

$$E[\eta] = JA\Delta\Theta_1,\ E[\varepsilon] = 0,\ var(\eta) = \Sigma_{(n,n)},$$

$$\eta_{UT_6(n+q)} = \begin{pmatrix} \eta_{(n)} \\ -b_{(q)} \end{pmatrix},\quad A_{UT_6(p+q,k)} = \begin{pmatrix} A_{(p,l)}, & 0_{(p,k-l)} \\ B_{(q,l)}, & C_{(q,k-l)} \end{pmatrix},$$

$$J_{UT_6(n+q,p+q)} = \begin{pmatrix} J_{(n,p)}, & 0_{(n,q)} \\ 0_{(q,p)}, & I_{(q,q)} \end{pmatrix},$$

$$(JA)_{UT_6(n+q,k)} = \begin{pmatrix} J_{(n,l)}, & 0_{(n,k-l)} \\ B_{(q,l)}, & C_{(q,k-l)} \end{pmatrix},$$

$$var(\eta_{UT_6(n+q)}) = \begin{pmatrix} \Sigma_{(n,n)}, & 0_{(n,q)} \\ 0_{(q,n)}, & 0_{(q,q)} \end{pmatrix}.$$

(if $A = I_{(l,l)}$ and $l = p$, Model 4 occurs).

This survey of simple transformations implies the following conclusion, which may be practicable: In linear estimation theory it suffices to derive relations for the universal model; the relations for all other linear models result from their transformation into the universal model.

The regularity conditions in each of the linear models of measurement considered ensure the estimability of the unknown parameters; therefore, we have no reason for dealing with this problem. If the conditions of regularity need not be fulfilled, the problem of their unbiased estimability arises. It is necessary to state conditions under which a linear combination $p'\Delta\Theta$ ($\Delta\Theta \in \mathcal{R}^k$) is unbiasedly estimable. In other words, it is necessary to formulate the condition concerning the k-dimensional vector p ensuring the existence of an unbiased estimator of the linear function $p'\Delta\Theta$, $\Delta\Theta \in \mathcal{R}^k$.

Within the universal model $\eta = JA\Delta\Theta + \varepsilon$, $E[\eta] = JA\Delta\Theta$, $E[\varepsilon] = 0$, $var(\eta) = \Sigma$, a linear combination $p'\Delta\Theta = \sum_{i=1}^k p_i\Delta\Theta_i$ of the components $\Delta\Theta_i$ of the k-dimensional vector parameter $\Delta\Theta$ is unbiasedly estimable if $p = (JA)'l$, $l \in \mathcal{R}^n$.

The assertion is a consequence of the following consideration: Due to the definition, a linear estimator $\widehat{p'\Delta\Theta}(\eta) = l'\eta + \lambda$, $l \in \mathcal{R}^n$, $\lambda \in \mathcal{R}^1$ of the linear function $p'\Delta\Theta$ is unbiasedly estimable if $l'JA\Delta\Theta + \lambda = p'\Delta\Theta$ for any $\Delta\Theta \in \mathcal{R}^k$, which immediately implies that $\lambda = 0$ and $p' = l'JA$.

Within a regular Model 2 the matrix JA has full rank in the columns ($R(JA) = k$), the vector p can acquire sequentially the form of the unit vectors of the k-dimensional space $\mathcal{R}^k$, i.e., $p = (1,0,\ldots,0)' = e_1$, $p = (0,1,\ldots,0)' = e_2$, $\ldots$, $p = (0,0,\ldots,1)' = e_k$, which means that each element of the k-dimensional vector $\Delta\Theta$ is unbiasedly estimable (compare with pp. 133–134).

The next simple example can serve as an illustration of the fact tnat the above mentioned choice of the coefficient p cannot be realized if the regularity condition $R(JA) = k$ of the model is not satisfied. Let

$$JA = \begin{pmatrix} 2, & 0, & -4 \\ 0, & 2, & 4 \\ 0, & 2, & 4 \\ 1, & 0, & -2 \end{pmatrix}, \quad R(JA) = 2 < 3,$$

the matrix JA has not the full rank in columns, i.e., the matrix $(JA)'$ does not possess the full rank in rows. In this case, there do not exist linear combinations of the columns of the matrix $(JA)'$ generating the unit vectors of the linear space $\mathcal{R}^3$. An $a' = (a_1, a_2, a_3, a_4)' \in \mathcal{R}^4$ satisfying the relation $(JA)'a = (1,0,0)'$ would have to be a solution of the system of equations

$$\begin{aligned} 2a_1 + a_4 &= 1, \\ 2a_2 + 2a_3 &= 0, \\ -4a_1 + 4a_2 + 4a_3 - 2a_4 &= 0 \end{aligned}$$

and this system has no solution as a consequence of the fact that there do not exist numbers a_1 and a_4 fulfilling simultaneously the relations $2a_1+a_4 = 1$ and $2a_1 + a_4 = 0$.

Analogously, the reader can convince himself that there do not exist vectors $b' = (b_1, b_2, b_3, b_4)$ and $c' = (c_1, c_2, c_3, c_4)$, b, $c, \in \mathcal{R}^4$ such that $(JA)'b = (0,1,0)'$ and $(JA)'c = (0,0,1)'$. This means that each of the components of the vector $\Delta\Theta$ of the unknown parameters is not estimable, but there exist unbiased estimators, e.g., of the linear combinations $\Delta\Theta_1 - 2\Delta\Theta_3$ (here $l = (1/2,0,0,0)'$) or $\Delta\Theta_2 + 2\Delta_3$ ($l = (0,1/2,0,0)'$), or $3\Delta\Theta_1 + 4\Delta\Theta_2 + 2\Delta\Theta_3$ ($l = (1,1,1,1)'$), etc.

For comparison, consider the matrix

$$(JA)' = \begin{pmatrix} 2, & 0, & 0, & 0 \\ 0, & 2, & 2, & 0 \\ -4, & 4, & 4, & -2 \end{pmatrix}$$

possessing the full rank in rows; it is not uneasy to convince oneself that $(JA)'a = (1,0,0)'$ if $a = (1/2, \alpha, -\alpha, -1)'$, where α is an arbitrary real number, $(JA)'b = (0,1,0)'$ if $b = (0, \beta, 1-\beta, 2)'$, where β is an arbitrary real number and $(JA)'c = (0,0,1)'$ if $c = (0, \gamma, -\gamma, -1/2)'$, where γ is an arbitrary real number.

For obtaining the efficient and unbiased estimator of an unbiasedly estimable linear function $p'\Delta\Theta$ of the unknown parameter $\Delta\Theta$, $\Delta\Theta \in \mathcal{R}^k$, $p \in \{p : p = (JA)'l, l \in \mathcal{R}^n\}$, within the universal model $\eta = JA\Delta\Theta + \varepsilon$, $E[\eta] = JA\Delta\Theta$, $E[\varepsilon] = 0$, $var(\eta) = \Sigma$, we have to be familiar with the notion of the generalized inverse (g-inverse) of a matrix. The fact that the regularity conditions within the universal model are given up means that the matrix inverses necessary for fulfilling the requirement of the joint efficiency of an unbiased estimator do not exist (e.g., the inverse of the covariance matrix of the observation vector if it is positive semidefinite or the inverse of the matrix $A'J'\Sigma^{-1}JA$ even if the matrix Σ is positive definite but $R[(JA)_{(n,k)}] < k < n$, etc.).

If the $n \times n$ matrix A in the linear system of equations $Ax = y$, where $x \in \mathcal{R}^n$, $y \in \mathcal{R}^n$, is regular, its solution is $x = A^{-1}y$ and the matrix A^{-1} satisfying the condition $AA^{-1} = A^{-1}A = I$ (the unit matrix of the type $n \times n$) is determined uniquely.

If the matrix A within the the equation system $Ax = y$ is a singular matrix of the type $n \times n$ or a non-square matrix of the type $m \times n$ such that $R(A_{(m,n)}) \leq \min(m,n)$, its inverse is not defined and the solution of the equation system, under the condition of its consistency, is given by means of a generalized inverse.

A matrix $G_{(n,m)}$ is said to be a generalized inverse (g-inverse) of the matrix $A_{(m,n)}$, $R(A_{(m,n)}) \leq \min(m,n)$ if $x = Gy$ is a solution of the linear system $Ax = y$ for each y for which this system is consistent.

This definition of a generalized inverse of a matrix explaining its genesis is equivalent with the following definition, which is used more frequently:

A matrix $G_{(n,m)}$ is said to be a generalized inverse of a matrix $A_{(m,n)}$, $R(A_{(m,n)}) \leq \min(m,n)$ if $AGA = A$. The generalized inverse of the matrix A is denoted A^-; the notation $\mathcal{A}^-$is used for the class of all generalized inverses of the matrix A.

Let A be a square and regular matrix. Then A^{-1} is a generalized inverse of the matrix A, as it satisfies the relation $AA^{-1}A = A$. In this case the g-inverse is determined uniquely, which is not true in general.

A general validity has the following assertion:

Any matrix A always possesses a generalized inverse.

Example The problem is to determine a g-inverse of the matrix $A_{(1,3)} = (1,0,1)$. The elements g_1, g_2 and g_3 of the matrix $G = (g_1, g_2, g_3)'$ will be determined by a direct application of the definition of a g-inverse. Consequently, $G = (g_1, g_2, 1 - g_1)'$, where g_1 and g_2 are arbitrary real numbers.

This example demonstrates that a matrix may possess even an uncountable set of g-inverses.

We shall see in the next section that this fact furnishes a basis for deriving a system of correctness checks of and reliability estimates for the first and second order parameters within the linear models of measurement.

For the sake of completeness, $\mathcal{A}^- = \{A_0^- + U - A_0^- AUAA_0^-,\ U$ an arbitrary $m \times n$ matrix, A_0^- an arbitrary but fixed g-inverse of the matrix $A\}$ is the class of all g-inverses of the matrix A.

For a given $y \neq 0$ the class $\{x : x = A^- y,\ A^- \in \mathcal{A}^-\}$ of all solutions of the equation system $Ax = y$ and, consequently, the class of all g-inverses of the matrix A will be restricted, if only its N-minimum solutions $x_N = A^-_{m(N)} y$ fulfilling the condition $x_N' N x_N = \min$ are taken into account; here N is a given positive semidefinite matrix of the type $n \times n$. The matrices $A^-_{m(N)}$ giving solutions x_N of the linear system $Ax = y,\ y \neq 0$, with the property $x_N' N x_N = \min$ are called the minimum N-seminorm g-inverses of the matrix A [8]; the symbol $\mathcal{A}^-_{m(N)}$ is used for denoting the class of all minimum N seminorm g-inverses of the matrix A. Self-evidently $\mathcal{A}^-_{m(N)} \subset$

---

[8] Let x be an element of an n-dimensional linear space $\mathcal{R}^n$. A real number $\|x\|$ which has the properties:

1. $\|x\| \geq 0$ for each $x \in \mathcal{R}^n$;
2. $\|x\| = 0$ if and only if $x = 0$ (the norm of a zero element of $\mathcal{R}^n$ equals zero; if the norm of an element is zero, this element must be the zero element of the linear space $\mathcal{R}^n$);
3. $\|ax\| = |a|\|x\|$ for any $a \in \mathcal{R}^1$ and any $x \in \mathcal{R}^n$;
4. $\|x + y\| \leq \|x\| + \|y\|$ for arbitrary elements x and y from $\mathcal{R}^n$

is said to be the norm of the element x.

There exist various ways of constructing a norm in the linear space $\mathcal{R}^n$; in each case, it is necessary to convince ourselves that the constructed number is actually a norm, i.e., whether it possesses the mentioned properties 1 - 4 of a norm.

From our point of view, the norm defined by the relation $\|x\| = \sqrt{x'Nx}$, where N is a positive definite matrix of the type $n \times n$, is interesting. We can easily convince ourselves that the number $\sqrt{x'Nx}$ has, for any $x \in \mathcal{R}^n$ and for a given $n \times n$-dimensional positive definite matrix N, the properties 1 - 4 of a norm. If the matrix N is only positive semidefinite, the number $\sqrt{x'Nx}$ possesses the properties 1, 3 and 4 only (the property 2 is not satified) and we are speaking of a seminorm.

Consider a positive semidefinite matrix $N = \begin{pmatrix} 0, & 0, & 0 \\ 0, & 0, & 0 \\ 0, & 0, & 1 \end{pmatrix}$. The number $\sqrt{x'Nx}$

$\mathcal{A}^-$ and $\{x_N : x_N = A^-_{m(N)}y,\ A^-_{m(n)} \in \mathcal{A}^-_{m(N)}\} \subset \{x : x = A^- y,\ A^- \in \mathcal{A}^-\}$, where $y \neq 0$ is given.

A minimum N seminorm g-inverse of the matrix A fulfils both the expressions $AA^-_{m(N)}A = A$ and $(A^-_{m(N)}A)'N = NA^-_{m(N)}A$.

Example The problem is to determine an N-minimum solution of the equation

$$(1,0,1)\begin{pmatrix} x_1 \\ x_2 \\ x_3 \end{pmatrix} = 1$$

if

$$N = \begin{pmatrix} 0, & 0, & 0 \\ 0, & 0, & 0 \\ 0, & 0, & 1 \end{pmatrix}.$$

In accordance with the results of the previous example, the set of all solutions of this equation is $\{x : (x_1, x_2, x_3)' = (g_1, g_2, 1-g_1),\ g_1 \in \mathcal{R}^1,\ g_2 \in \mathcal{R}^1\}$. Within this class, solutions are sought which also satisfy the condition

$$(g_1, g_2, 1-g_1)\begin{pmatrix} 0, & 0, & 0 \\ 0, & 0, & 0 \\ 0, & 0, & 1 \end{pmatrix}\begin{pmatrix} g_1 \\ g_2 \\ 1-g_1 \end{pmatrix} = min.$$

The quadratic form $(g_1, g_2, 1-g_1)\begin{pmatrix} 0, & 0, & 0 \\ 0, & 0, & 0 \\ 0, & 0, & 1 \end{pmatrix}\begin{pmatrix} g_1 \\ g_2 \\ 1-g_1 \end{pmatrix} = (1-g_1)^2$ attains its minimum for $g_{N1} = 1$ and each $g_{N2} \in \mathcal{R}^1$, thus the class of N-minimum solutions of the consistent system $(1,0,1)x = 1$ is $\{x_N : (x_{N1}, x_{N2}, x_{N3})' = (1, g_2, 0)',\ g_2 \in \mathcal{R}^1\}$; its amount is uncountable, nevertheless the class of N-minimum solutions is included in the set of all solutions of the consistent system $(1,0,1)x = y$.

The solution of the consistent system of equations $(JA)'l = p,\ p \neq 0$, obtained using these results is $l = [(JA)']^- p$, and the Σ-minimum solution of this system reads $l_\Sigma = [(JA)']^-_{m(\Sigma)}p$. This entitles us to make the following two important assertions:

---

defines a seminorm, since $(x_1, x_2, 0)\begin{pmatrix} 0, & 0, & 0 \\ 0, & 0, & 0 \\ 0, & 0, & 1 \end{pmatrix}\left(x_1, x_2, 0 \right) = 0$ for all elements $(x_1, x_2, 0) \in \mathcal{R}^3$, not only the zero elements.

Within the universal model $\eta = JA\Delta\Theta + \varepsilon$, $E[\eta] = JA\Delta\Theta$, $E[\varepsilon] = 0$, $var(\eta) = \Sigma$ an unbiased estimator of an unbiasedly estimable function $p'\Delta\Theta$, $\Delta\Theta \in \mathcal{R}^k$, $p \in \{(JA)'l,\ l \in \mathcal{R}^n\}$ is

$$\widetilde{p'\Delta\Theta}(\eta) = p'\{[(JA)']^-\}'\eta$$

and is characterized by the dispersion

$$D[\widetilde{p'\Delta\Theta}(\eta)] = p'\{[(JA)']^-\}'\Sigma[(JA)']^-p.$$

Within the universal model $\eta = JA\Delta\Theta + \varepsilon$, $E[\eta] = JA\Delta\Theta$, $E[\varepsilon] = 0$, $var(\eta) = \Sigma$ an unbiased and efficient estimator of an unbiasedly estimable function $p'\Delta\Theta$, $\Delta\Theta \in \mathcal{R}^k$, $p \in \{(JA)'l,\ l \in \mathcal{R}^n\}$ is

$$\widehat{p'\Delta\Theta}(\eta) = p'\{[(JA)']^-_{m(\Sigma)}\}'\eta \tag{2.10.1}$$

and its dispersion (being the minimum among all unbiased estimators) is

$$D[\widehat{p'\Delta\Theta}(\eta)] = p'\{[(JA)']^-_{m(\Sigma)}\}'\Sigma[(JA)']^-_{m(\Sigma)}p. \tag{2.10.2}$$

Finally, respecting that the Σ-minimum solution of a consistent system of equations need not be unique, we must introduce an important assertion which states the impact of this fact on the linear unbiased and efficient estimators of an unbiasedly estimable function $p'\Delta\Theta$.

Within the universal model $\eta = JA\Delta\Theta + \varepsilon$, $E[\eta] = JA\Delta\Theta$, $E[\varepsilon] = 0$, $var(\eta) = \Sigma$ the unbiased and efficient estimators of an unbiasedly estimable function $p'\Delta\Theta$, $\Delta\Theta \in \mathcal{R}^k$, $p \in \{(JA)'l,\ l \in \mathcal{R}^n\}$ of the forms $[\widehat{p'\Delta\Theta}(\eta)]_1 = p'\{[(JA)']^-_{m(\Sigma)1}\}'\eta$ and $[\widehat{p'\Delta\Theta}(\eta)]_2 = p'\{[(JA)']^-_{m(\Sigma)2}\}'$, where $[(JA)']^-_{m(\Sigma)1}$ and $[(JA)']^-_{m(\Sigma)2}$ are different Σ-minimum solutions of the consistent system of equations $(JA)'l = p$, $p \neq 0$, are eqiuvalent with probability 1.

The idea of proving this assertion is based on the following simple consideration. Take into account a random variable $[\widehat{p'\Delta\Theta}(\eta)]_1 - [\widehat{p'\Delta\Theta}(\eta)]_2$. As a consequence of its unbiasedness $E[[\widehat{p'\Delta\Theta}(\eta)]_1 - [\widehat{p'\Delta\Theta}(\eta)]_2] = 0$.

The dispersion of the difference considered is

$$D[[\widehat{p'\Delta\Theta}(\eta)]_1 - [\widehat{p'\Delta\Theta}(\eta)]_2] =$$

$$D[[\widehat{p'\Delta\Theta}(\eta)]_1] + D[[\widehat{p'\Delta\Theta}(\eta)]_2] - 2cov([\widehat{p'\Delta\Theta}(\eta)]_1, [\widehat{p'\Delta\Theta}(\eta)]_2),$$

where

$$D[[\widehat{p'\Delta\Theta}(\eta)]_i] = p'\{[(JA)']^-_{m(\Sigma)i}\}'\Sigma[(JA)']^-_{m(\Sigma)i}p, \; i = 1, 2,$$

and

$$cov([\widehat{p'\Delta\Theta}(\eta)]_1, [\widehat{p'\Delta\Theta}(\eta)]_2) = p'\{[(JA)']^-_{m(\Sigma)1}\}'\Sigma[(JA)']^-_{m(\Sigma)2}p.$$

Since

$$p'\{[(JA)']^-_{m(\Sigma)1}\}'\Sigma[(JA)']^-_{m(\Sigma)2}p = p'\{[(JA)']^-_{m(\Sigma)1}\}'\Sigma[(JA)']^-_{m(\Sigma)1}p =$$
$$= p'\{[(JA)']^-_{m(\Sigma)2}\}'\Sigma[(JA)']^-_{m(\Sigma)2}p, \text{ [9]}$$

we can see that

$$D[[\widehat{p'\Delta\Theta}(\eta)]_1 - [\widehat{p'\Delta\Theta}(\eta)]_2] = 0.$$

The Chebyshev inequality[10] immediately implies the assertion.

In all the previous assertions the covariance matrix $var(\eta) = \Sigma$ of the observation vector η was assumed to be known.

Consider now a universal model $\eta = JA\Delta\Theta + \varepsilon$, $E[\eta] = JA\Delta\Theta$, $E[\varepsilon] = 0$, $var(\eta) = \sigma^2 H$, where σ^2 is an unknown unit dispersion and H is a known positive semidefinite matrix. The linear unbiased and efficient estimator of an unbiasedly estimable function $p'\Delta\Theta$, $\Delta\Theta \in \mathcal{R}^k$, $p \in \{(JA)'l, \; l \in \mathcal{R}^n\}$ is

$$\widehat{p'\Delta\Theta}(\eta) = p'\{[(JA)']^-_{m(H)}\}'\eta$$

and it is characterized by the dispersion

$$D[\widehat{p'\Delta\Theta}(\eta)] = \sigma^2 p'\{[(JA)']^-_{m(H)}\}'H[(JA)']^-_{m(H)}p;$$

as far as the unbiased and invariant estimator of the unknown unit dispersion is concerned,

---

[9]The problem is to prove that

$$p'\{[(JA)']^-_{m(\Sigma)1}\}'\Sigma[(JA)']^-_{m(\Sigma)2}p = p'\{[(JA)']^-_{m(\Sigma)1}\}'\Sigma[(JA)']^-_{m(\Sigma)1}p =$$
$$p'\{[(JA)']^-_{m(\Sigma)2}\}'\Sigma[(JA)']^-_{m(\Sigma)2}p$$

for $p \in \{(JA)'l, \; l \in \mathcal{R}^n\}$.

The proof follows directly from the assertion:

Let G_1 and G_2 be different minimum N seminorm g-inverses of the matrix A. As $(G_1A)'N = NG_1A$ (cf.p. 242), $A'G_1'NG_2A = NG_1AG_2A = NG_1A$ and for the same reason also $A'G_1'NG_1A = NG_1AG_1A = NG_1A$, i.e., $A'G_1'NG_2A = A'G_1'NG_1A$. Similarly, $A'G_1'NG_2A = A'G_1'A'G_2'N = A'G_2'N$ and $A'G_2'NG_2A = A'G_2'A'G_2'N = A'G_2'N$ as a consequence of $(G_2A)'N = NG_2A$. This proves that $A'G_1'NG_2A = A'G_1'NG_2A = A'G_2'NG_2A$.

[10]For the Chebyshev inequality see the footnote on p. 114; here, this inequality is applied for expressing the probability of a complementary event.

$$\hat{\sigma}^2 = v'H^-v/[R((H, JA)) - R(JA)], \tag{2.10.3}$$

where the n-dimensional residual vector v is $v = \eta - JA\{[(JA)']^-_{m(H)}\}'\eta$ and H^- is an arbitrary g-inverse of the matrix H. Its dispersion is

$$D[\hat{\sigma}^2] = Tr[(H^- \otimes H^-)\Psi_v]/[R((H, JA)) - R(JA)]^2 - \sigma^4,$$

where Ψ_v is a matrix of the fourth central moments of the residual vector (see p. 194) of the form

$$\Psi_v = (P_U \otimes P_U)\Psi_\eta(P'_U \otimes P'_U),$$

where $P_U = I - JA\{[(JA)']^-_{m(H)}\}'$ and Ψ_η is the matrix of the fourth central moments of the observation vector η.

If $\eta \sim N_n(JA\Delta\Theta, \sigma^2 H)$, then $v'H^-v \sim \sigma^2\chi^2[R((H, JA)) - R(JA)]$. In this case, the dispersion of the unbiased invariant and efficient estimator of the unit dispersion of the form $\hat{\sigma}^2 = v'H^-v/[R((H, JA)) - R(JA)]$ is

$$D[\hat{\sigma}^2] = 2\sigma^4/[R((H, JA)) - R(JA)].$$

The estimator of the unit dispersion expressed by the relation (2.10.3) contains a g-inversion of the matrix H. We know that there exists a class of g-inverses of the matrix H, if it is not regular. The reason entitling us to apply the relation (2.10.3) for enumerating the estimate of the unit dispersion consists in the following assertion.

Within the universal model $\eta = JA\Delta\Theta + \varepsilon$, $E[\eta] = JA\Delta\Theta$, $E[\varepsilon] = 0$, $var(\eta) = \Sigma$, the unbiased and invariant (in the case of a normal observation vector η even efficient) estimators of the unit dispersion σ^2 of the form

$$\hat{\sigma}_1^2(\eta) = v'(\eta)H_1^-v(\eta)/[R((H, JA)) - R(A)] \tag{2.10.4}$$

and

$$\hat{\sigma}_2^2(\eta) = v'(\eta)H_2^-v(\eta)/[R((H, JA)) - R(A)] \tag{2.10.5}$$

(the explicit argument η at $\hat{\sigma}^2$ and v emphasizes that the estimators are taken into account), where H_1^- and H_2^- are two different g-inverses of the matrix H, equal with probability 1.

This assertion is an analogy of the one concerning the estimators of the first order parameters (see p. 243) and has the same impact on the numerical outputs.

The way of proving this important assertion is similar to that applied in the case of the first order parameters (first it has to be shown that the difference of random variables given by (2.10.4) and (2.10.5) is characterized by the zero mean value and the zero dispersion and then the Chebyshev inequality is used); the procedure involves some special knowledge of matrix calculus, concerning the g-inverses of matrices, which goes beyond the framework of this textbook and is therefore omitted.

The shapes of the confidence regions of the parameter σ^2, under the condition of a normally distributed observation vector η, can be easily derived by applying the same procedures used in the case of regular models.

2.11 Correctness checks in computing the first and second order parameters

An essential part of computing estimates of the first and second order parameters within individual measurement models is a check for the numerical correctness and reliability of the results obtained. The necessity of applying the correctness check becomes especialy salient in the cases where the results of processing carefully measured data form the input for large-scale critical projects, such as diagnosis, e.g., in medicine, or in govermental or other legal decision making.

2.11.1 Correctness checks of the estimates of the first order parameters

The basis for deriving a series of relations for checking the correctness of the estimates of the first order parameters is given by the proved assertion stating that, within the universal model $\eta = JA\Delta\Theta + \varepsilon$, $E[\eta] = JA\Delta\Theta$, $E[\varepsilon] = 0$, $var(\eta) = \sigma^2 H$, the linear unbiased and efficient estimators of an unbiasedly estimable function $p'\Delta\Theta$, $\Delta\Theta \in \mathcal{R}^k$, $p \in \{(JA)'l,\ l \in \mathcal{R}^n\}$ of the forms

$$[\widehat{p'\Delta\Theta}(\eta)]_1 = p'\{[(JA)']^-_{m(H)1}\}'\eta$$

and

$$[\widehat{p'\Delta\Theta}(\eta)]_2 = p'\{[(JA)']^-_{m(H)2}\}'\eta,$$

where $[(JA)']^-_{m(H)1}$ and $[(JA)']^-_{m(H)2}$ are different minimum H seminorm g-inverses of the matrix $(JA)'$, equal with probability 1.

This implies that the realizations of the unbiased and efficient estimators (i.e., the corresponding estimates) are invariant to the choice of the H minimum seminorm g-inverse of the matrix $(JA)'$, i.e., that

$$\widehat{p'\Delta\Theta}(y) = p'\{[(JA)']^-_{m(H)1}\}'y = p'\{[(JA)']^-_{m(H)2}\}'y =$$

$$\ldots = p'\{[(JA)']^-_{m(H)n}\}'y = \ldots.$$

The following explicit forms of the matrices $[(JA)']^-_{m(H)i}$

$$[(JA)']^-_{m(H)1} = [H + JA(JA)']^- JA\{(JA)'[H + JA(JA)']^- JA\}^- \tag{2.11.1}$$

and

$$[(JA)']^-_{m(H)2} = [H + JA(JA)']^- JA\{JA(JA)'[H + JA(JA)']^- JA\}^- JA \tag{2.11.2}$$

are useful and frequently applied here.

Further check relations can be obtained by applying any element of the class of all the minimum H seminorm g-inverses of the matrix $(JA)'$.

A sequence of other relations for the correctness and reliability check follows from the next assertion (it is given in its general form and proved, e.g., in [19], p. 162).

Consider the universal model $\eta = JA\Delta\Theta + \varepsilon$, $E[\eta] = JA\Delta\Theta$, $E[\varepsilon] = 0$, $var(\eta) = \sigma^2 H$ and an $n \times n$-dimensional matrix M such that:

1. $M = (H + JAUA'J')^- + K$;
2. $R(A'J'MJA) = R(A'J')$,

where U is a $k \times k$-dimensional matrix and K is an $n \times n$-dimensional matrix such that

3. $\{x : x = (H, JA)l,\ l \in \mathcal{R}^{n+k}\} =$
$= \{y : y = (H + JAUA'J')k,\ k \in \mathcal{R}^k\} =$
$= \{z : z = (H + JAU'A'J')m,\ m \in \mathcal{R}^k\} \subset \mathcal{R}^n;$

4. $MK'A = 0_{(n,k)},\ A'KA = 0_{(k,k)}$.

Then, for each realization of the linear unbiased and efficient estimator of an unbiasedly estimable function $p'\Delta\Theta,\ \Delta\Theta \in \mathcal{R}^k,\ p \in \{(JA)'l,\ l \in \mathcal{R}^n\}$,

$$p'\widehat{\Delta\Theta}(y) = p'\widehat{\Delta\Theta}_M(y), \tag{2.11.3}$$

where $\widehat{\Delta\Theta}_M(y)$ is a solution of the equation system $A'J'MJA\widehat{\Delta\Theta}_M(y) = A'J'My$.

The estimator $\widehat{\Delta\Theta}_N(\eta)$ of the unknown parameter $\Delta\Theta$ within the universal model $\eta = JA\Delta\Theta + \varepsilon,\ E[\eta] = JA\Delta\Theta,\ E[\varepsilon] = 0,\ var(\eta) = \sigma^2 H$, where H is an arbitrary matrix of the type $n \times n$ is said to be the N-LS estimator if $\widehat{\Delta\Theta}_N(y)$ is a solution of the equation system $A'J'NJA\widehat{\Delta\Theta}_N(y) = A'J'Ny$ for each realization y of the observation vector η.[11]

This assertion needs a special commentary. The reader surely noted that in the definition of the N-LS estimator no conditions concerning the unbiased estimability are established. However, we are not interested in the properties of the N-LS estimator of the parameter $\Delta\Theta$ in its entirety, but only in the properties of the linear combination $p'\widehat{\Delta\Theta}_N(\eta)$ and especially in the case of the M-LS estimator, where the matrix M possesses the above mentioned properties.

A further commentary concerns the choice of the matrix M in the check relation (2.11.3). It is evident that the matrix $K = 0_{(n,n)}$ has the property 4. As far as the matrix U is concerned, it can be proved that each positive definite matrix has the properties 1 and 3. The reader can easily convince himself that the matrix U of the form $U = DD'$, where D is an arbitrary matrix of the type $k \times l,\ k \le l$, such that $R(D) = k$, is positive definite. This simple rule enables us to construct a series of matrices U. Simultaneously it is possible to make use of the simplest type of a positive definite matrix, i.e., of the matrix $r^2 I_{(n,n)}$, where r is an arbitrary real number.

Summarizing our considerations, we see that the matrices

[11] The Σ^{-1}-LS-estimator of the unknown parameter within the regular Model 2 is its LS-estimator. A realization of the LS-estimator (the linear unbiased and jointly efficient estimator) of a parameter $\Delta\Theta$, due to the relation (2.4.3), is of the form $\widehat{\Delta\Theta}(y) = (A'J'\Sigma^{-1}JA)^{-1}A'J'\Sigma^{-1}y$ and this expression is a solution of the system of equations $A'J'\Sigma^{-1}JA\widehat{\Delta\Theta}(y) = A'J\Sigma^{-1}y$. The regularity conditions ensure the unbiased estimability of the vector $\Delta\Theta$ in its entirety (i.e., each of its components is estimable separately), which within the universal model is not true as a consequence of having abandoned them.

$$M_1 = [H + JA(JA)']^-,$$
$$M_r = [H + r^2 JA(JA)']^-, \ r \in \mathcal{R}^1, \ r \neq 0,$$
$$M_D = [H + JAD(JAD)']^-,$$

D being an arbitrary $k \times l$ matrix, such that $R(D) = k \leq l$, can be applied sequentially for calculating the $M{-}LS$ estimate and simultaneously

$$\widehat{p'\Delta\Theta}(y) = p'\widehat{\Delta\Theta}_{M_1}(y) = p'\widehat{\Delta\Theta}_{M_r}(y) = p'\widehat{\Delta\Theta}_{M_D}(y)$$

for each realization y of the observation vector η.

Another way enabling us to obtain a linear unbiased and efficient estimator of an unbiasedly estimable function $p'\Delta\Theta$, $\Delta\Theta \in \mathcal{R}^k$, $p \in \{(JA)'l, l \in \mathcal{R}^n\}$ in the universal model $\eta = JA\Delta\Theta + \varepsilon$, $E[\eta] = JA\Delta\Theta$, $E[\varepsilon] = 0$, $var(\eta) = \sigma^2 H$, utilizes the properties of the Pandora's box matrix.

The Pandora's box matrix in the linear universal model considered is called the matrix

$$\begin{pmatrix} C_1, & C_2 \\ C_3, & -C_4 \end{pmatrix} = \begin{pmatrix} H, & JA \\ (JA)', & 0 \end{pmatrix}^-, \tag{2.11.4}$$

which occurs when the problem of determining the linear unbiased and efficient estimator of an unbiasedly estimable function $p'\Delta\Theta$, $\Delta\Theta \in \mathcal{R}^k$, $p \in \{(JA)'l, l \in \mathcal{R}^n\}$ in the universal model considered is solved.

The next assertion is very important from the point of view of obtaining further relations for checking the reliability of a numerical realization of a linear unbiased and efficient estimator of an unbiasedly estimable function $p'\Delta\Theta$ of the components of the vector $\Delta\Theta \in \mathcal{R}^k$ within the universal model.

The random variables $p'C_2'\eta$ and $p'C_3\eta$, where C_2 and C_3 are the block matrices of the Pandora's box matrix (2.11.4) equal with probability 1; they represent simultaneously the linear unbiased and efficient estimators of the unbiasedly estimable function $p'\Delta\Theta$, $\Delta\Theta \in \mathcal{R}^k$, $p \in \{(JA)'l, l \in \mathcal{R}^n\}$, within the universal model

$$\eta = JA\Delta\Theta + \varepsilon, \ E[\eta] = JA\Delta\Theta, \ E[\varepsilon] = 0, \ var(\eta) = \sigma^2 H.$$

This assertion entitles us to write the following check relations

$$\widehat{p'\Delta\Theta}(y) = p'C_2'y = p'C_3 y. \tag{2.11.5}$$

The next check concerns the numerical reliability of the residuals v. This step, checking the correctness of the first order parameters, forms simultaneously the input for checking the correctness of the second order parameters.

The vector of residuals $v = y - JA\widehat{\Delta\Theta}(y)$ within the universal model $\eta = JA\Delta\Theta + \varepsilon$, $E[\eta] = JA\Delta\Theta$, $E[\varepsilon] = 0$, $var(\eta) = \sigma^2 H$, satisfies, for each realization y of the observation vector η and for each minimum H seminorm g-inverse of the matrix $(JA)'$, the condition

$$JA\{[(JA)']^-_{m(H)}\}'v = 0_{(n)} \tag{2.11.6}$$

(the vector $JA\Delta\Theta$ within the universal model is unbiasedly estimable, since each of its components is unbiasedly estimable; it suffices to realize that $\{JA\}'_{i,\cdot} = (JA)'e_i$, where e_i is the ith unit vector of the linear space $\mathcal{R}^n$, $i = 1, 2, \ldots, n$).

The relation (2.11.6) is implied by the fact that the realizations of the linear unbiased and efficient estimators of an unbiasedly estimable linear function of the parameter $\Delta\Theta$ within the universal model are invariant with respect to the choice of the minimum H seminorm g-inverse of the matrix $(JA)'$ (see p. 247). Thus, if $v = JA\{[(JA)']^-_{m(H)j}\}'y$, $j = 1, 2, \ldots$, then $JA\{[(JA)']^-_{m(H)i}\}'v = JA\{[(JA)']^-_{m(H)i}\}'(y - JA\{[(JA)']^-_{m(H)j}\}'y) = JA\{[(JA)']^-_{m(H)i}\}'y - JA\{[(JA)']^-_{m(H)i}\}'JA\{[(JA)']^-_{m(H)j}\}'y = 0$, as a consequence of $JA\{[(JA)']^-_{m(H)i}\}'JA = JA$ and, since $JA\{[(JA)']^-_{m(H)i}\}'y = JA\{[(JA)']^-_{m(H)j}\}'y$, $i, j = 1, 2, \ldots$, the relation (2.11.6) is true.

The essence of the above-mentioned checks in the universal model consists in the fact that, using various numerical procedures through the mediation of different intermediate results, we must obtain the same numerical values of the linear unbiased and efficient estimates (evidently subject to the conditions that the calculation is correct and the applied procedures stable) of an unbiasedly estimable function $p'\Delta\Theta$. This means, e.g., that the minimum H seminorm g-inverses of the matrix $(JA)'$ obtained using different numerical procedures differ, but a product of each of them by the realization y of the observation vector η must yield the same number which represents the linear unbiased and efficient estimator of the function $p'\Delta\Theta$, $\Delta\Theta \in \mathcal{R}^k$, $p \in \{(JA)'l,\ l \in \mathcal{R}^n\}$.

Consider regular Model 2 $\eta = JA\Delta\Theta + \varepsilon$, $E[\eta] = JA\Delta\Theta$, $E[\varepsilon] = 0$, $var(\eta) = \sigma^2 H$, $R(JA) = k \le n$, $R(H) = n$. As it is a special case of the universal model, all the check relations holding within the universal model can be applied in Model 2. Moreover, the regularity conditions that are fulfilled within it enable us to penetrate into the mechanism for deriving the check procedures which were introduced only within the universal model, since their derivation requires non-trivial knowledge of the matrix calculus. This obstacle does not exist within regular Model 2, where the regularity conditions ensure the unbiased estimability of each of the components of the vector $\Delta\Theta$ and the existence of the matrix inverses in the relation

$$\widehat{\Delta\Theta}(\eta) = (A'J'H^{-1}JA)^{-1}A'J'H^{-1}\eta \tag{2.11.7}$$

expressing the linear unbiased and jointly efficient estimator of the k-dimensional unknown parameter $\Delta\Theta$. The inverse of each regular matrix is determined uniquely, thus the matrix $(A'J'H^{-1}JA)^{-1}A'J'H^{-1}$ must be unique (with respect to the universal model this means that the class of the minimum H norm [12] consists of the only element; in accordance with the symbols used on p. 133 and for the sake of simplicity we shall denote it by L, i.e., $L = H^{-1}JA(A'J'H^{-1}JA)^{-1}$). The correctness check for the calculated estimate of the parameter $\Delta\Theta$ within the regular model considered concerns the correctness of the calculation of the matrix L, except for the check relation $JAL'v = 0$. This represents the essential difference between the checks within the universal and the regular models.

Repeating, in an abbreviated form, the procedure for deriving the relation (2.11.7) (pp. 133–134) we see that the problem is to find an $n \times k$ dimensional matrix L such that it minimizes the quadratic form $a'L'HLa$ for each $a \in \mathcal{R}^k$ (the estimator of the parameter $\Delta\Theta$ has to be jointly efficient) and simultaneously fulfils the condition $a'L'JA = a'$ (the estimator has to be unbiased).

The sought L is a solution of the system of equations

$$\begin{aligned} Hm + JA\lambda &= 0, \\ (JA)'m &= a \end{aligned} \tag{2.11.8}$$

[12] The matrix H is assumed to be positive definite, thus the norm instead of the seminorm has to be taken into account (see also the footnote on p. 241).

(in accordance with p. 133 $La = m$). By solving this system we get sequentially

$$m = -H^{-1}JA\lambda,$$

$$-(JA)'H^{-1}JA\lambda = a \Rightarrow \lambda = -[(JA)'H^{-1}JA]^{-1}a,$$

thus

$$m = H^{-1}JA[(JA)'H^{-1}JA]^{-1}a$$

and as $m = La$ and the equality $La = H^{-1}JA[(JA)'H^{-1}JA]^{-1}a$ has to be true for any $a \in \mathcal{R}^k$ we have

$$L = H^{-1}JA[(JA)'H^{-1}JA]^{-1}. \tag{2.11.9}$$

The solution of the equation system (2.11.8) is identical to that of

$$[H + JA(JA)']m + JA(\lambda - a) = 0,$$

$$(JA)'m = a \tag{2.11.10}$$

(this system arose from the original one by adding the JA multiple from the left of the second equation to the first). The solution of the system (2.11.10) yields

$$m = -[H + JA(JA)']^{-1}JA(\lambda - a),$$

$$-(JA)'[H + JA(JA)']^{-1}JA(\lambda - a) = a \Rightarrow$$

$$\Rightarrow (\lambda - a) = -\{(JA)'[H + JA(JA)']^{-1}JA\}^{-1}a,$$

which, for the same reason as in the previous case, gives

$$L = [H + JA(JA)']^{-1}JA\{(JA)'[H + JA(JA)']^{-1}JA\}^{-1}. \tag{2.11.11}$$

First the initial condition $(JA)'L = I$ ensuring that the unbiasedness of the estimator of the vector parameter $\Delta\Theta$ has been rewritten in the form $(JA)'m = a$, $m = La$, which has to be fulfilled for any $a \in \mathcal{R}^k$; its further modification consists in rewriting it in the form $JA(JA)'m = JAa$ which, naturally, must be fulfilled for each $a \in \mathcal{R}^k$.

The problem to find an $n \times k$ dimensional matrix that minimizes for any $a \in \mathcal{R}^k$ the quadratic form $a'L'HLa$, satisfying simultaneously the relation $JA(JA)'m = JAa$, leads to the solution of equation system

$$Hm + JA(JA)'\lambda = 0,$$

$$JA(JA)'m = JAa$$

(the auxiliary Lagrange function is of the form $\Phi(m) = m'Hm + 2\lambda'[JA(JA)'m - JAa]$).

The solution of this system is equivalent to that of the system

$$[H + JA(JA)']m + JA[(JA)'\lambda - a] = 0,$$

$$JA(JA)'m = JAa. \tag{2.11.12}$$

There holds

$$m = -[H + JA(JA)']^{-1}JA[(JA)'\lambda - a],$$

$$-JA(JA)'[H + JA(JA)']^{-1}JA[(JA)'\lambda - a] = JAa \Rightarrow$$

$$\Rightarrow (JA)'\lambda - a = -\{JA(JA)'[H + JA(JA)']^{-1}JA\}^{-}JAa$$

and, since this equality has to be valid for any $a \in \mathcal{R}^k$,

$$L = [H + JA(JA)']^{-1}JA\{JA(JA)'[H + JA(JA)']^{-1}JA\}^{-}JA. \tag{2.11.13}$$

The essence of this problem (the same problem has been solved in its three different modifications) implies that the matrices L given by the relations (2.11.9), (2.11.11) and (2.11.13) are equivalent, whereby the first relation enabling us to check the correctness of the calculated estimate of the unknown parameter within Model 2 is derived.

Comparing now the relation (2.11.11) for L with the relation (2.11.1) for $[(JA)']^{-}_{m(H)1}$ and the relation (2.11.13) for L with the relation (2.11.2) for $[(JA)']^{-}_{m(H)2}$ we have to note their similarity. In the case of the simple regular model the method of deriving these relations has been shown.

The matrix $JA(JA)'[H+JA(JA)']^{-1}JA = F$ whose g-inverse occurs in the relation (2.11.13) for L is of the type $n \times k$ and its rank is $R(F) = k \leq n$. For this reason there exists its g-inverse $G_{(k,n)}$ such that $G_{(k,n)}F_{(n,k)} = I_{(k,k)}$. One of elements of the class of the g-inverses of the mentioned type is $G = F'(FF')^{-1}$ and, by means of it, we obtain the following form of the sought matrix L

$$L = [H + JA(JA)']^{-1}JA(JA)'[H + JA(JA)']^{-1}JA(JA)'\{JA(JA)'[H+ \\ +JA(JA)']^{-1}JA(JA)'[H + JA(JA)']^{-1}JA(JA)'\}^{-1}JA, \tag{2.11.13'}$$

which does not contain an explicit g-inverse.

It is not easy even within the regular models to prove the assertion concerning the special M-LS estimators (p. 248), therefore proof is omitted here; neverthless the assertion will be applied to constructing further check relations within regular Model 2; simultaneously, the influence of regularity on their properties will be pointed out.

Let us consider regular Model 2 and an M-LS estimator within it; we choose, as analogous to the case of the universal model, the zero matrix for the matrix K and either the matrix r^2I, $r \in \mathcal{R}^1$, or a positive definite matrix DD' (D is an arbitrary $k \times l$ matrix, $R(D) = k \leq l$) for the matrix U; thus the matrix M is regular and its inverse exists. It is necessary to emphasize that, in contradistinction to the universal model, the zero matrix of the type $n \times n$ may be chosen for the matrix U within the regular model considered (within the universal model the zero matrix does not satisfy condition 3; the fact that within the regular model the condition $R(H) = n$ has to be fulfilled implies that any $k \times k$ matrix U, including the zero matrix, satisfies the property 3).

If $K = 0_{(n,n)}$ and $U = 0_{(k,k)}$, then $M = H^{-1}$ and an H^{-1}-LS estimator of the parameter $\Delta\Theta$ identifies with its LS estimator.

If $K = 0_{(n,n)}$ and $U = I_{(k,k)}$, then

$$M_1 = (H + JAA'J')^{-1};$$

the M_1-LS estimate of the vector parameter $\Delta\Theta$, due to its definition, is a solution of the system of equations

$$A'J'(H + JAA'J')^{-1}JA\widehat{\Delta\Theta}_{M_1}(y) = A'J'(H + JAA'J')^{-1}y,$$

thus, for each realization y of the observation vector η, it is of the form

$$\widehat{\Delta\Theta}_{M_1}(y) = [A'J'(H + JAA'J')^{-1}JA]^{-1}A'J'(H + JAA'J')^{-1}y;$$

the parameter $\Delta\Theta$ within regular Model 2 is unbiasedly estimable, therefore, as a consequence of (2.11.3), $\widehat{\Delta\Theta}_{M_1}(y) = \widehat{\Delta\Theta}(y)$ for any realization y of the observation vector η, which implies that

$$L'_{M_1} = [A'J'(H + JAA'J')^{-1}JA]^{-1}A'J'(H + JAA'J')^{-1};$$

this relation is identical with (2.11.11).

Applying an analogous procedure we can obtain further check relations

$$L_{M_r} = (H + r^2JAA'J')^{-1}JA[A'J'(H + r^2JAA'J')^{-1}JA]^{-1}, \ r \in \mathcal{R}^1. \tag{2.11.14}$$

and

$$L_{M_D} = (H + JADD'A'J')^{-1}JA[A'J'(H + JADD'A'J')^{-1}JA]^{-1}, \tag{2.11.15}$$

D is a $k \times l$ matrix such that $R(D) = k \leq l$, and simultaneously

$$L_{M_1} = L_{M_r} = L_{M_D} = L.$$

As far as the application of the Pandora's box matrix is concerned for the equivalent expressions $C'_2 y = C_3 y = \widehat{\Delta\Theta}(y)$ which are used for checking the correctness of the numerical values of the estimates of the first order parameters within regular Model 2, the procedure for deriving them is especially simple and instructive.

The solution of the equation system (2.11.8), written in its matrix form

$$\begin{pmatrix} H, & JA \\ (JA)', & 0 \end{pmatrix} \begin{pmatrix} m \\ \lambda \end{pmatrix} = \begin{pmatrix} 0 \\ a \end{pmatrix},$$

can be expressed as

$$\begin{pmatrix} m \\ \lambda \end{pmatrix} = \begin{pmatrix} H, & JA \\ (JA)', & 0 \end{pmatrix}^{-1} \begin{pmatrix} 0 \\ a \end{pmatrix}.$$

The regularity conditions of the model ensure the existence of the inverse of the block matrix on the right-hand side of the matrix relation in question. If we compare this block matrix with the relation (2.11.4), we can state that it is a Pandora's box matrix, thus

$$\begin{pmatrix} m \\ \lambda \end{pmatrix} = \begin{pmatrix} C_1, & C_2 \\ C_3, & -C_4 \end{pmatrix} \begin{pmatrix} 0 \\ a \end{pmatrix},$$

which implies $m = C_2 a$; the symmetry of the input matrix causes the symmetry of its inverse, thus simultaneously $m = C'_3 a$. As $m = La$ and the relations $La = C_2a$ and $La = C_3'a$ have to be fulfilled for each $a \in \mathcal{R}^k$, we can write

$$L = C_2 \tag{2.11.16}$$

and

$$L = C_3'. \tag{2.11.17}$$

The validity of the check relations $\widehat{\Delta\Theta}(y) = C_2'y = C_3y$ can be directly verified by means of the relations expressing the inverse of a block matrix (see p. 150). Due to them,

$$\begin{pmatrix} H, & JA \\ (JA)', & 0 \end{pmatrix}^{-1} =$$

$$= \begin{pmatrix} H^{-1} - H^{-1}JA(A'J'H^{-1}JA)^{-1}A'J'H^{-1}, & H^{-1}JA(A'J'H^{-1}JA)^{-1} \\ (A'J'H^{-1}JA)^{-1}A'J'H^{-1}, & -(A'J'H^{-1}JA)^{-1} \end{pmatrix}. \tag{2.11.18}$$

The essence of the check consists in the fact that the numerical values of the individual blocks, expressed in their explicit form, are calculated as a inverse of the matrix

$$\begin{pmatrix} H, & JA \\ (JA)', & 0 \end{pmatrix}$$

by the use of other numerical procedures.

The correctness check of residuals within regular Model 2 $\eta = JA\Delta\Theta + \varepsilon,\ E[\eta] = JA\Delta\Theta,\ E[\varepsilon] = 0,\ var(\eta) = \sigma^2 H,\ R(H) = n,\ R(JA) = k \leq n$ can be carried out using the simple formula

$$JAL_i'v = 0,\ i = 1, 2, \ldots, 8, \tag{2.11.19}$$

where $v = y - JA\widehat{\Delta\Theta}(y)$, $\widehat{\Delta\Theta}(y)$ is a realization of the linear unbiased and jointly efficient extimator of the vector parameter $\Delta\Theta$ and

$$L_1 = H^{-1}JA[(JA)'H^{-1}JA]^{-1}$$

(see the relation (2.11.9)),

$$L_2 = [H + JA(JA)']^{-1}JA\{(JA)'[H + JA(JA)']^{-1}JA\}^{-1}$$

(see the relation (2.11.11)),

$$L_3 = [H + JA(JA)']^{-1}JA\{JA(JA)'[H + JA(JA)']^{-1}JA\}^{-}JA$$

(see the relation (2.11.13)),

$$L_4 = [H + JA(JA)']^{-1}JA(JA)'[H + JA(JA)']^{-1}JA(JA)'\{JA(JA)'\cdot$$

$$\cdot[H + JA(JA)']^{-1}JA(JA)'[H + JA(JA)']^{-1}JA(JA)'\}^{-1}JA$$

(see the relation (2.11.13′)),

$$L_5 = L_{M_r} = [H + r^2JA(JA)']^{-1}JA\{(JA)'[H + r^2JA(JA)']^{-1}JA\}^{-1},$$

$$r \in \mathcal{R}^1,\ r \neq 0$$

(see the relation (2.11.14)),

$$L_6 = L_{M_D} = [H + JAD(JAD)']^{-1}JA\{(JA)'[H + JAD(JAD)']^{-1}JA\}^{-1},$$

D is any matrix of the type $k \times l$, $R(D) = k \leq l$ (see the relation (2.11.15)),

$$L_7 = C_2$$

(see the relation (2.11.16)) and

$$L_8 = C_3'$$

(see the relation (2.11.17)); C_2 and C_3 are submatrices of the Pandora's box matrix.

The reader can easily verify the validity of the relation (2.11.19) which was proved in its more general form on p. 250.

Summarizing the results obtained so far, we can state that within the **universal model** $\eta = JA\Delta\Theta + \varepsilon$, $E[\eta] = JA\Delta\Theta$, $E[\varepsilon] = 0$, $var(\eta) = \sigma^2 H$ the following relations can be used for checking the numerical correctness of the linear unbiased and efficient estimate of a linear unbiasedly estimable function $p'\Delta\Theta$ of the parameter $\Delta\Theta \in \mathcal{R}^k$, $p \in \{(JA)'l,\ l \in \mathcal{R}^n\}$

$$\widehat{p'\Delta\Theta}(y) = p'\{[(JA)']^{-}_{m(H)i}\}'y =$$

$$= p'([H + JA(JA)']^{-}JA\{(JA)'[H + JA(JA)']^{-}JA\}^{-})'y =$$

$$= p'([H + JA(JA)']^- JA\{JA(JA)'[H + JA(JA)']^- JA\}^- JA)'y =$$

$$= p'\widehat{\Delta\Theta}_{M_r}(y) =$$

$$= p'\widehat{\Delta\Theta}_{M_D}(y) =$$

$$= p'C_2'y =$$

$$= p'C_3y,$$

where y is a realization of the observation vector η, $[(JA)']^-_{m(H)i}$ an arbitrary minimum H seminorm g-inverse of the matrix $(JA)'$, $i = 1, 2, \ldots$, $\widehat{\Delta\Theta}_{M_r}$ is a solution of the system of equations $(JA)'M_rJA\widehat{\Delta\Theta}_{M_r}(y) = (JA)'M_ry$, $M_r = [H + r^2JA(JA)']^-$, $r \in \mathcal{R}^1$, $r \neq 0$, and analogously $\widehat{\Delta\Theta}_{M_D}(y)$ is a solution of the system of equations

$$(JA)'M_DJA\widehat{\Delta\Theta}_{M_D}(y) = (JA)'M_Dy, \; M_D = [H + JAD(JAD)']^-,$$

D is a matrix of the type $k \times l$ such that $R(D) = k \leq l$ and C_2 and C_3 are the blocks of the Pandora's box matrix.

Within **regular Model 2** $\eta = JA\Delta\Theta + \varepsilon$, $E[\eta] = JA\Delta\Theta$, $E[\varepsilon] = 0$, $var(\eta) = \sigma^2 H$, $R(H) = n$, $R(JA) = k \leq n$ the expressions checking the numerical correctness of the linear unbiased and jointly efficient estimate of the unknown parameter $\Delta\Theta$ (realize that the numerical correctness of the matrix L in the relation $\widehat{\Delta\Theta}(y) = L'y$ is checked in this case) are

$$L = H^{-1}JA(A'J'H^{-1}JA)^{-1} =$$

$$= [H + JA(JA)']^{-1}JA\{(JA)'[H + JA(JA)']^{-1}JA\}^{-1} =$$

$$= [H + JA(JA)']^{-1}JA\{JA(JA)'[H + JA(JA)']^{-1}JA\}^{-1}JA =$$

$$= [H + JA(JA)']^{-1}JA(JA)'[H + JA(JA)']^{-1}JA(JA)'\{JA(JA)'\cdot$$

$$\cdot[H + JA(JA)']^{-1}JA(JA)'[H + JA(JA)']^{-1}JA(JA)'\}^{-1}JA =$$

$$= [H + r^2JA(JA)']^{-1}JA\{(JA)'[H + r^2JA(JA)']^{-1}JA\}^{-1} =$$

$$= [H + JAD(JAD)']^{-1}JA\{(JA)'[H + JAD(JAD)']^{-1}JA\}^{-1} =$$

$$= C_2 =$$

$$= C_3',$$

where $r \in \mathcal{R}^1$, D is a matrix of the type $k \times l$ such that $R(D) = k \leq l$ and C_2 and C_3 are the submatrices of the Pandora's box matrix.

The check of the numerical correctness of the estimates of the first order parameters by means of the residual vector can be realized within the universal model using the relation (2.11.6) and within regular Model 2 using the relation (2.11.19).

2.11.2 Numerical correctness checks of the unit dispersion estimate

Consider the universal model $\eta = JA\Delta\Theta + \varepsilon$, $E[\eta] = JA\Delta\Theta$, $E[\varepsilon] = 0$, $var(\eta) = \sigma^2 H$; the series of check relations ensuring the numerical correctness of the unbiased and invariant estimator of the unit dispersion is based on the assertion that the estimators

$$\hat{\sigma}_1^2(\eta) = v'(\eta)H_1^- v(\eta)/[R((H, JA)) - R(JA)]$$

and

$$\hat{\sigma}_2^2(\eta) = v'(\eta)H_2^- v(\eta)/[R((H, JA)) - R(JA)],$$

where $v(\eta) = (I - JA\{[(JA)']^-_{m(H)}\}')\eta$ and H_1^- and H_2^- are different g-inverses of the matrix H, equal with probability 1 (see p. 245).

It means that the realizations of the unbiased and invariant estimators of the unit dispersion within the universal model are invariant to the choice of the g-inverse of the matrix H, therefore

$$\hat{\sigma}^2(y) = v'(y)H_1^- v(y)/[R((H, JA)) - R(JA)] =$$

$$= v'(y)H_2^- v(y)/[R((H, JA)) - R(JA)] =$$

$$= \ldots = v'(y)H_n^- v(y)/[R((H, JA)) - R(JA)] = \ldots$$

for each realization y of the observation vector η, where $H_i^- \in \mathcal{H}^-$, $i = 1, 2, \ldots$.

Further check relations follow from the assertion:

Within the universal model $\eta = JA\Delta\Theta + \varepsilon$, $E[\eta] = JA\Delta\Theta$, $E[\varepsilon] = 0$, $var(\eta) = \sigma^2 H$ the estimator

$$\hat{\sigma}_v^2(\eta) = v'(\eta)H_v^- v(\eta)/[R((H, JA)) - R(JA)], \tag{2.11.20}$$

where

$$H_v = (I - JA\{[(JA)']^-_{m(H)}\}')'H(I - JA\{[(JA)']^-_{m(H)}\}') =$$

$$= H - H[(JA)']^{-}_{m(H)}(JA)' = H - JA\{[(JA)']^{-}_{m(H)}\}'H$$

and

$$v(\eta) = (I - JA\{[(JA)']^{-}_{m(H)}\}')\eta$$

equals with probability 1 the estimator

$$\hat{\sigma}^2(\eta) = v'(\eta)H^{-}v(\eta)[R((H, JA)) - R(JA)],$$

which means that (2.11.20) represents the unbiased and invariant estimator of the unit dispersion; H_v^- in the relation (2.11.20) is an arbitrary g-inverse of the matrix H_v.

In consequence of this assertion, for any realization y of the observation vector η

$$\hat{\sigma}^2(y) = v'(y)H^{-}v(y)/[R((H, JA)) - R(JA)] =$$
$$= v'(y)H_v^{-}v(y)/[R((H, JA)) - R(JA)] = \hat{\sigma}_v^2(y), \tag{2.11.21}$$

where H^- is an arbitrary element of the class $\mathcal{H}^-$ of all the g-inverses of the matrix H and similarly H_v^- is an arbitrary element of the class $\mathcal{H}_v^-$ of all the g-inverses of the matrix H_v.

Another possibility to obtain independent expressions for the numerical value of the unbiased and invariant estimate of the unit dispersion within the universal model consists in applying the following assertion, which is an extended version of the assertion from p. 221 (we formulate it in its general form proved in [19], p. 174).

Consider the universal model $\eta = JA\Delta\Theta + \varepsilon$, $E[\eta] = JA\Delta\Theta$, $E[\varepsilon] = 0$, $var(\eta) = \sigma^2 H$. Let $M_{(n,n)}$, $U_{(k,k)}$ and $K_{(n,n)}$ be matrices satisfying the conditions

1. $M = (H + JAU'A'J')^{-} + K$,
2. $R(A'J'MJA) = R(A'J')$,
3. $\{x : x = (H, JA)l,\ l \in \mathcal{R}^{n+k}\} = \{y : y = (H + JAUA'J')k,\ k \in \mathcal{R}^k\} = \{z : z = (H + JAUA'J')m,\ m \in \mathcal{R}^k\} \subset \mathcal{R}^n$,
4. $MK'A = 0_{(n,k)}$, $A'KA = 0_{(k,k)}$,
5. $(\eta - JA\{[(JA)']^{-}_{m(H)}\}'\eta)'K(\eta - JA\{[(JA)']^{-}_{m(H)}\}'\eta) = 0$;

then the estimator

$$\hat{\sigma}^2_M(\eta) =$$
$$= [\eta - JA\widehat{\Delta\Theta}_M(\eta)]'M[\eta - JA\widehat{\Delta\Theta}_M(\eta)]/[R((H, JA)) - R(JA)], \tag{2.11.22}$$

where $\widehat{\Delta\Theta}_M(\eta)$ is the M-LS estimator of the vector parameter $\Delta\Theta$ (see p. 248), equals with probability 1 the estimator

$$\hat{\sigma}^2(\eta) = v'(\eta)H^- v(\eta)/[R((H, JA)) - R(JA)],$$

which means that (2.11.22) represents the unbiased and invariant estimator of the unit dispersion.

In consequence of this fact, for each realization y of the observation vector η

$$\hat{\sigma}^2(y) = v'(y)H^- v(y)/[R((H, JA)) - R(JA)] =$$
$$= [y - JA(A'J'MJA)^- A'J'My]'M[y - JA(A'J'MJA)^- A'J'My] =$$
$$= \hat{\sigma}^2_M(y), \tag{2.11.23}$$

where H^- is an arbitrary g-inverse of the matrix H and the matrix M satisfies the conditions $1 - 5$.

Choosing the matrices

$$M_r = [H + r^2 JA(JA)']^-, \; r \in \mathcal{R}^1, \; r \neq 0, \tag{2.11.24}$$

and

$$M_D = [H + JAD(JAD)']^-, \tag{2.11.25}$$

where D is an arbitrary matrix of the type $k \times l$ such that $R(D) = k \leq l$, for the matrix M (these matrices were used for checking the numerical correctness of the estimates of the first order parameters) in which the $n \times n$ zero matrix was chosen for the matrix K, we see that these matrices satisfy all the conditions required; thus the relation (2.11.23) with the matrices M_r and M_D of the forms (2.11.24) and (2.11.25), respectively, may be used for checking the numerical correctness of the unbiased and invariant estimate of the unit dispersion.

A very important and powerful system of check relations can be obtained using the properties of the Pandora's box matrix.

Within the universal model $\eta = JA\Delta\Theta + \varepsilon$, $E[\eta] = JA\Delta\Theta$, $E[\varepsilon] = 0$, $var(\eta) = \sigma^2 H$, the estimator of the form

$$\hat{\sigma}^2_{C_1}(\eta) = \eta' C_1 \eta / Tr(HC_1), \tag{2.11.26}$$

where C_1 is the upper left block of the Pandora's box matrix (2.11.4), equals with probability 1 the estimator

$$\hat{\sigma}^2(\eta) = v'(\eta) H^- v(\eta) / [R((H, JA)) - R(JA)],$$

which means that the random variable (2.11.26) represents the unbiased and invariant estimator of the unit dispersion within the universal model.

In consequence of this assertion, for each realization y of the observation vector η within the universal model

$$\hat{\sigma}^2(y) = v'(y) H^- v(y) / [R((H, JA)) - R(JA)] =$$

$$= y' C_1 y / Tr(HC_1) = \hat{\sigma}^2_{C_1}(y), \tag{2.11.27}$$

where H^- is any g-inverse of the matrix H and C_1 is the upper left block of the Pandora's box matrix.

The check relation (2.11.27) is of great importance; it enables us not only to check the numerical correctness of the expression $v'(y) H^- v(y)$ but, above all, it checks in an independent way the calculation of the ranks of the matrices (H, JA) and JA; they have to satisfy the equality $Tr(HC_1) = R((H, JA)) - R(JA)$.

All the mentioned assertions except the relation (2.11.23) will be proved now for their special case of a regular Model 2. The regularity conditions essentially simplify the procedures used; the idea of proving them is the same.

Consider regular Model 2 $\eta = JA\Delta\Theta + \varepsilon$, $E[\eta] = JA\Delta\Theta$, $E[\varepsilon] = 0$, $var(\eta) = \sigma^2 H$, $R(H) = n$, $R(JA) = k \leq n$. The unbiased and invariant estimator of the unit dispersion is

$$\hat{\sigma}^2(\eta) = v'(\eta) H^{-1} v(\eta) / (n - k)$$

(see (2.7.1); in order to distinguish between the estimators and the corresponding estimates, the arguments are supplied here), which is a special case of the expression (2.10.3). The matrix H is assumed to be regular, therefore its inverse is unique.

Now we shall prove that within regular Model 2 $\eta = JA\Delta\Theta + \varepsilon$, $E[\eta] = JA\Delta\Theta$, $E[\varepsilon] = 0$, $var(\eta) = \sigma^2 H$, $R(H) = n$, $R(JA) = k \leq n$,

$$\hat{\sigma}_v^2(\eta) = v'(\eta)H_v^- v(\eta)/(n-k); \tag{2.11.28}$$

here

$$\begin{aligned} H_v &= [I - JA(A'J'H^{-1}JA)^{-1}A'J'H^{-1}]'H \cdot \\ &\cdot [I - JA(A'J'H^{-1}JA)^{-1}A'J'H^{-1}] = \\ &= H - JA(A'J'H^{-1}JA)^{-1}A'J', \end{aligned}$$

$R(H_v) = n - k$ and H_v^- is an arbitrary g-inverse of the matrix H_v, is the unbiased and invariant estimator of the unit dispersion.

First, it is necessary to show that the expression $v'(y)H_v^- v(y)$ for each realization y of the observation vector η is invariant to the choice of the g-inverse of the matrix

$$H_v = H - JA(A'J'H^{-1}JA)^{-1}A'J'.$$

As the matrix H is positive definite,

$$v(y) = [I - JA(A'J'H^{-1}JA)^{-1}A'J'H^{-1}]y$$

and

$$H_v = [I - JA(A'J'H^{-1}JA)^{-1}A'J'H^{-1}]'H[I - H^{-1}JA(A'J'H^{-1}JA)^{-1}A'J'],$$

the residual vector $v(y)$ can be expressed in the form $H_v u_y$ (realize that $v(y)$ for a given y is a linear combination of the columns of the matrix $[I - JA(A'J'H^{-1}JA)^{-1}A'J']$ and, as a consequence of the facts that the matrix H is positive definite and

$$[I - H^{-1}JA(A'J'H^{-1}JA)^{-1}A'J'] = [I - JA(A'J'H^{-1}JA)^{-1}A'J'H^{-1}]',$$

$v(y)$ is simultaneously a linear combination of the columns of the matrix H_v). Using this format for the residual vector $v(y)$ we can write $v'(y)H_v^- v(y) = u_y' H_v H_v^- H_v u_y = u_y' H_v u_y$; the right-hand side of this relation does not contain any g-inverse of the matrix H_v, by which the invariance of $v'(y)H_v^- v(y)$ to the choice of the g-inverse of the matrix H_v is proved.

The equivalence

$$[H - JA(A'J'H^{-1}JA)^{-1}A'J']H^{-1}[H - JA(A'J'H^{-1}JA)^{-1}A'J'] =$$
$$= H - JA(A'J'H^{-1}JA)^{-1}A'J'$$

implies that H^{-1} represents one of the g-inverses of the matrix H_v. The relation $v'(y)H_v^-v(y)$ has been proved to be invariant to the choice of the g-inverse of the matrix H_v, which enables us to write

$$v'(y)H_v^-v(y)/(n-k) = v'(y)H^{-1}v(y)/(n-k) = \hat{\sigma}^2(y);$$

this relation is true for each realization y of the observation vector η, therefore the random variables

$$\hat{\sigma}_v^2(\eta) = v'(\eta)H_v^-v(\eta)/(n-k)$$

and

$$\hat{\sigma}^2(\eta) = v'(\eta)H^-v(\eta)/(n-k)$$

are equal with probability 1; thus the random variable $\hat{\sigma}_v^2(\eta)$ represents an unbiased and invariant estimator of the unit dispersion within regular Model 2.

Another explicit form of the matrix H_v^- is useful from the point of view of a concrete calculation of the unbiased and invariant estimate of the unit dispersion. It can be obtained by a simple modification of the expression $v'(y)'H^{-1}v(y)$:

$$v'(y)H^{-1}v(y) = y'[I - H^{-1}JA(A'J'H^{-1}JA)^{-1}A'J']H^{-1}\cdot$$
$$\cdot[I - JA(A'J'H^{-1}JA)^{-1}A'J'H^{-1}]y =$$
$$= y'[I - H^{-1}JA(A'J'H^{-1}JA)^{-1}A'J']H^{-1}y,$$

where

$$[I - H^{-1}JA(A'J'H^{-1}JA)^{-1}A'J'][I - H^{-1}JA(A'J'H^{-1}JA)^{-1}A'J'] =$$
$$= I - H^{-1}JA(A'J'H^{-1}JA)^{-1}A'J',$$

thus

$$y'[H^{-1} - H^{-1}JA(A'J'H^{-1}JA)^{-1}A'J'H^{-1}]y =$$
$$= y'[I - H^{-1}JA(A'J'H^{-1}JA)^{-1}A'J']\cdot$$
$$\cdot[H^{-1} - H^{-1}JA(A'J'H^{-1}JA)^{-1}A'J'H^{-1}]\cdot$$

$$\cdot[I - H^{-1}JA(A'J'H^{-1}JA)^{-1}A'J']y =$$
$$= v'(y)[H^{-1} - H^{-1}JA(A'J'H^{-1}JA)^{-1}A'J'H^{-1}]v(y);$$

the reader can easily convince himself that $H^{-1} - H^{-1}JA(A'J'H^{-1}JA)^{-1} \cdot A'J'H^{-1}$ is a g-inverse of the matrix H_v.

This result yields the next relation for checking the numerical stability of the unbiased and invariant estimate of the unit dispersion within the regular Model 2, namely

$$y'[H^{-1} - H^{-1}JA(A'J'H^{-1}JA)^{-1}A'J'H^{-1}]y =$$
$$= v'(y)[H^{-1} - H^{-1}JA(A'J'H^{-1}JA)^{-1}A'J'H^{-1}]v(y) \tag{2.11.29}$$

which has to be true for each realization y of the observation vector η. This relation is of great importance for checking the stability of the numerical procedure applied on a specific type of computer (the same number has to be obtained both by summing products of large numbers determined by realizations of the observation vector and by summing products of small numbers of the corresponding residuals).

For checking the numerical correctness of the estimators of the first order parameters, the matrix M which determines their M-LS estimator within regular Model 2 $\eta = JA\Delta\Theta + \varepsilon$, $E[\eta] = JA\Delta\Theta$, $E[\varepsilon] = 0$, $var(\eta) = \sigma^2 H$, $R(H) = n$, $R(JA) = k \leq n$, was chosen in the forms

$$M_r = (H + r^2 JAA'J')^{-1}, \; r \in \mathcal{R}^1, \; r \neq 0$$

and

$$M_D = [H + JAD(JAD)']^{-1},$$

D is an arbitrary matrix of the type $k \times l$, such that $R(D) = k \leq l$.

In both these cases the matrix K from the conditions on pp. 247–248 is a zero $n \times n$ matrix fulfilling the condition 5 (see p. 260), which, in our case, is of the form

$$[\eta - JA(A'J'H^{-1}JA)^{-1}A'J'H^{-1}\eta]'$$
$$K[\eta - JA(A'J'H^{-1}JA)^{-1}A'J'H^{-1}\eta] = 0;$$

therefore the random variable

$$\hat{\sigma}^2_M(\eta) = [\eta - JA\widehat{\Delta\Theta}_M(\eta)]'M[\eta - JA\widehat{\Delta\Theta}_M(\eta)]/(n-k), \; M = M_r, M_D,$$

where $\widehat{\Delta\Theta}_M(\eta)$ is the M-LS estimator of the parameter $\Delta\Theta$, represents the unbiased and invariant estimator of the unit dispersion within regular Model 2.

Next, it will be proved that within regular Model 2 $\eta = JA\Delta\Theta + \varepsilon$, $E[\eta] = JA\Delta\Theta$, $E[\varepsilon] = 0$, $var(\eta) = \sigma^2 H$, $R(H) = n$, $R(JA) = k \leq n$, the random variable

$$\hat{\sigma}^2_{C_1} = \eta' C_1 \eta / Tr(HC_1),$$

where C_1 is the upper left block of the Pandora's box matrix, is the unbiased and invariant estimator of the unit dispersion.

Due to (2.11.18),

$$C_1 = H^{-1} - H^{-1}JA(A'J'H^{-1}JA)^{-1}A'J'H^{-1}$$

and

$$\begin{aligned} Tr(HC_1) &= Tr[I_{(n,n)} - JA(A'J'H^{-1}JA)^{-1}A'J'H^{-1}] = \\ &= Tr(I_{(n,n)}) - Tr[JA(A'J'H^{-1}JA)^{-1}A'J'H^{-1}] = \\ &= Tr(I_{(n,n)}) - Tr[(A'J'H^{-1}JA)^{-1}A'J'H^{-1}JA] = \\ &= Tr(I_{(n,n)}) - Tr(I_{(k,k)}) = n - k \end{aligned}$$

(for calculating the trace of the matrix HC_1 the rules given on p. 194 were used). Thus,

$$\hat{\sigma}^2_{C_1}(\eta) = \eta'[H^{-1} - H^{-1}JA(A'J'H^{-1}JA)^{-1}A'J'H^{-1}]\eta/(n-k).$$

Applying now the relation (2.11.29) and realizing simultaneously that the matrix $H^{-1} - H^{-1}JA(A'J'H^{-1}JA)^{-1}A'J'H^{-1}$ is one of the g-inverses of the matrix $H_v = H - JA(A'J'H^{-1}JA)^{-1}A'J'$, we see that

$$\hat{\sigma}^2_{C_1}(\eta) = v'(\eta)H_v^- v(\eta)/(n-k),$$

which is an unbiased and invariant estimator of the unit dispersion within regular Model 2 (compare it with the relation (2.11.28)).

Summing up these results we can state the following two assertions:

Within the **universal model** $\eta = JA\Delta\Theta + \varepsilon$, $E[\eta] = JA\Delta\Theta$, $E[\varepsilon] = 0$, $var(\eta) = \sigma^2 H$, the unbiased and invariant estimate of the unit dispersion can be expressed in the following equivalent forms

$$\begin{aligned} \hat{\sigma}^2(y) &= v'(y)H_i^- v(y)/[R((H,JA)) - R(JA)] = \\ &= v'(y)H_{vi}^- v(y)/[R((H,JA)) - R(JA)] = \end{aligned}$$

$$= y'H_{vi}^{-}y/[R((H, JA)) - R(JA)] =$$
$$= [y - JA\widehat{\Delta\Theta}_{M_r}(y)]'M_r[y - JA\widehat{\Delta\Theta}_{M_r}(y)]/[R((H, JA)) - R(JA)] =$$
$$= [y - JA\widehat{\Delta\Theta}_{M_D}(y)]'M_D[y - JA\widehat{\Delta\Theta}_{M_D}(y)]/[R((H, JA)) - R(JA)] =$$
$$= y'C_1y/Tr(HC_1),$$

where $v(y) = (I - JA\{[(JA)']_{m(H)}^{-}\}')y$, H_i^{-}, $i = 1, 2, \ldots$, is an arbitrary g-inverse of the matrix H, H_{vi}^{-}, $i = 1, 2, \ldots$, an arbitrary g-inverse of the matrix $H_v = H - H[(JA)']_{m(H)}^{-}(JA)' = H - JA\{[(JA)']_{m(H)}^{-}\}'H$, $M_r = [H + r^2 JA(JA)']^{-}$, $r \in \mathcal{R}^1$, $r \neq 0$, $M_D = [H + JAD(JAD)']^{-}$, D an arbitrary matrix of the type $k \times l$ such that $R(D) = k < l$, $\widehat{\Delta\Theta}_M(y)$ the M-LS estimate of the parameter $\Delta\Theta$, $M = M_r$, M_D, C_1 is the left upper block of the Pandora's box matrix and $Tr(HC_1)$ is the trace of the matrix HC_1.

Within regular **Model 2** $\eta = JA\Delta\Theta + \varepsilon$, $E[\eta] = JA\Delta\Theta$, $E[\varepsilon] = 0$, $var(\eta) = \sigma^2 H$, $R(H) = n$, $R(JA) = k \leq l$, the unbiased and invariant estimate of the unit dispersion can be expressed in the following equivalent forms

$$\hat{\sigma}^2(y) = v'(y)H^{-1}v(y)/(n - k) =$$
$$= v'(y)[H^{-1} - H^{-1}JA(A'J'H^{-1}JA)^{-1}A'J'H^{-1}]v(y)/(n - k) =$$
$$= y'[H^{-1} - H^{-1}JA(A'J'H^{-1}JA)^{-1}A'J'H^{-1}]y/(n - k) =$$
$$= [y - JA\widehat{\Delta\Theta}_{M_r}(y)]'M_r[y - JA\widehat{\Delta\Theta}_{M_r}(y)]/(n - k) =$$
$$= [y - JA\widehat{\Delta\Theta}_{M_D}(y)]'M_D[y - JA\widehat{\Delta\Theta}_{M_D}(y)]/(n - k) =$$
$$= y'C_1y/Tr(HC_1),$$

where $v(y) = [I - JA(A'J'H^{-1}JA)^{-1}A'J'H^{-1}]y$, $M_r = (H + r^2 JAA'J')^{-1}$, $r \in \mathcal{R}^1$, $r \neq 0$, $M_D = [H + JAD(JAD)']^{-1}$, D an arbitrary matrix of the type $k \times l$ such that $R(D) = k < l$, $\widehat{\Delta\Theta}_M(y)$ is the M-LS estimate of the parameter $\Delta\Theta$, $M = M_r$, M_D, C_1 is the left upper block of the Pandora's box matrix, and $Tr(HC_1)$ is the trace of the matrix HC_1.

2.11.3 Numerical correctness checks of dispersion in the estimator of the function $p'\Delta\Theta$ and in the covariance matrix of the estimator of the vector function $(p_1, p_2, \ldots, p_n)'\Delta\Theta$

Finally, the relations are introduced for checking the dispersion of the estimator of an unbiasedly estimable function $p'\Delta\Theta$ of the first order vector parameter or of the vector of unbiasedly estimable functions $(p_1, p_2, \ldots, p_n)'$ $\Delta\Theta$ of the vector parameter $\Delta\Theta \in \mathcal{R}^k$, $p, p_1, \ldots, p_n \in \{(JA)'l, l \in \mathcal{R}^n\}$, within the universal model and regular Model 2.

The dispersion of an unbiased linear and efficient estimator of an unbiasedly estimable function $p'\Delta\Theta$, $\Delta\Theta \in \mathcal{R}^k$, $p \in \{(JA)'l, l \in \mathcal{R}^n\}$ within the universal model $\eta = JA\Delta\Theta + \varepsilon$, $E[\eta] = JA\Delta\Theta$, $E[\varepsilon] = 0$, $var(\eta) = \sigma^2 H$ is

$$D[\widehat{p'\Delta\Theta}(y)] = \sigma^2 p'\{[(JA)']^-_{m(H)}\}'H[(JA)']^-_{m(H)}p.$$

This relation may be expressed in the following five equivalent ways:

a)

$$D[\widehat{p'\Delta\Theta}(\eta)] = \sigma^2 p'(\{(JA)'[H + JA(JA)']^- JA\}^- - I)p;$$

it is the dispersion of the estimator $\widehat{p'\Delta\Theta}(\eta) = p'([H+JA(JA)']^- JA\{(JA)' \cdot [H + JA(JA)']^- JA\}^-)'\eta$ as

$$p'(\{(JA)'[H + JA(JA)']^- JA\}^-)'(JA)'\{[H + JA(JA)']^-\}'H \cdot$$

$$\cdot [H + JA(JA)']^- \times JA\{(JA)'[H + JA(JA)']^- JA\}^- =$$

$$= p'(\{(JA)'[H + JA(JA)']^- JA\}^- I)p$$

for any $p \in \{(JA)'l,\ l \in \mathcal{R}^n\}$;

b)

$$D[\widehat{p'\Delta\Theta}(\eta)] = \sigma^2 p'\{I - (JA)'[H + JA(JA)']^- JA\}\{JA(JA)'[H+$$

$$+JA(JA)']^- JA\}^- JAp;$$

this relation expresses the dispersion of the estimator $\widehat{p'\Delta\Theta}(\eta) = p'([H + JA(JA)']^- JA\{JA(JA)'[H + JA(JA)']^- JA\}^- JA)'\eta$, since for any $p \in \{(JA)'l,\ l \in \mathcal{R}^n\}$

$$p'(JA)'(\{JA(JA)'[H + JA(JA)']^- JA\}^-)'(JA)'\{[H + JA(JA)']^- H[H+$$

$$+JA(JA)']^- JA\{JA(JA)'[H + JA(JA)']^- JA\}^- JAp =$$

$$= p'\{I - (JA)'[H + JA(JA)']^- JA\}\{JA(JA)'[H + JA(JA)']^- JA\}^- JAp;$$

c)

$$D[\widehat{p'\Delta\Theta}(\eta)] = \sigma^2 p'\{[(JA)'M_r JA]^- - r^2 I\}p,$$

the dispersion of the estimator $\widehat{p'\Delta\Theta}(\eta) = \widehat{p'\Delta\Theta}_{M_r}(\eta) = p'[(JA)'M_rJA]^- JAM_r\eta$, $M_r = [H + r^2JA(JA)']^-$, $r \in \mathcal{R}^1$, $r \neq 0$, since for any $p \in \{(JA)'l,\ l \in \mathcal{R}^n\}$

$$p'[(JA)'M_rJA]^-(JA)'M_rHM_r'JA\{[(JA)'M_rJA]^-\}'p =$$
$$= p'\{[(JA)'M_rJA]^- - r^2I\}p;$$

d)

$$D[\widehat{p'\Delta\Theta}(\eta)] = \sigma^2p'\{[(JA)'M_DJA]^- - DD'\}p;$$

this relation represents the dispersion of the estimator $\widehat{p'\Delta\Theta}(\eta) = \widehat{p'\Delta\Theta}_{M_D}(\eta) = p'[(JA)'M_DJA]^-JAM_D\eta$, $M_D = [H+JAD(JAD)']^-$, where D is an arbitrary matrix of the type $k \times l$ such that $R(D) = k \leq l$, since for any $p \in \{(JA)'l,\ l \in \mathcal{R}^n\}$

$$p'[(JA)'M_DJA]^-(JA)'M_DHM_D'JA\{[(JA)'M_DJA]^-\}'p =$$
$$= p'\{[(JA)'M_DJA]^- - DD'\}p;$$

e)

$$D[\widehat{p'\Delta\Theta}(\eta)] = p'C_4p,$$

the dispersion of the estimator $\widehat{p'\Delta\Theta}(\eta) = p'C_3\eta = p'C_2'\eta$, where C_2, C_3 and C_4 are the blocks of the Pandora's box matrix.

Applying the analogous procedures, we can determine the covariance matrix of an unbiased linear and jointly efficient estimator of a vector of unbiasedly estimable functions $(p_1, p_2, \ldots, p_n)'\Delta\Theta = P\Delta\Theta$ of the vector parameter $\Delta\Theta \in \mathcal{R}^k$, $p_i \in \{(JA)'l,\ l \in \mathcal{R}^n\}$, $i = 1, 2, \ldots, n$; we obtain the following equivalent relations

$$var[\widehat{P\Delta\Theta}(\eta)] = \sigma^2P\{[(JA)']^-_{m(H)}\}'H[(JA)']^-_{m(H)}P' =$$
$$= \sigma^2P(\{(JA)'[H + JA(JA)']^-JA\}^- - I)P =$$
$$= \sigma^2P\{I-(JA)'[H+JA(JA)']^-JA\}\{JA(JA)'[H+JA(JA)']^-JA\}^-JAP' =$$
$$= \sigma^2P\{[(JA)'M_rJA]^- - r^2I\}P' =$$
$$= \sigma^2P\{[(JA)'M_DJA]^- - DD'\}P' =$$
$$= \sigma^2PC_4P',$$

where M_r, M_D, D and C_4 have the well known meaning.

It is necessary to realize that the equivalence of the given check relations analogous to that of the relations for checking the numerical correctness of the unbiased linear and efficient estimate of an unbiasedly estimable function $p'\Delta\Theta$, $\Delta\Theta \in \mathcal{R}^k$, $p \in \{(JA)'l,\ l \in \mathcal{R}^n\}$ within the universal model is valid for a given vector p that ensures the unbiased estimability.

The variance of an unbiased and jointly efficient estimator of the parameter $\Delta\Theta$ within regular Model 2 $\eta = JA\Delta\Theta+\varepsilon$, $E[\eta] = JA\Delta\Theta$, $E[\varepsilon] = 0$, $var(\eta) = \sigma^2 H$, $R(H) = n$, $R(JA) = k \leq n$, is

$$var[\widehat{\Delta\Theta}(\eta)] = \sigma^2[(JA)'H^{-1}JA]^{-1}.$$

The following expressions are equivalent to it

$$var[\widehat{\Delta\Theta}(\eta)] = \sigma^2(\{(JA)'[H + JA(JA)']^{-1}JA\}^{-1} - I) =$$

$$= \sigma^2\{I-(JA)'[H+JA(JA)']^{-1}JA\}\{JA(JA)'[H+JA(JA)']^{-1}JA\}^{-}JA =$$
$$= \sigma^2\{I - (JA)'[H + JA(JA)']^{-1}JA\}(JA)'[H + JA(JA)']^{-1}JA(JA)'\cdot$$
$$\cdot\{JA(JA)'[H + JA(JA)']^{-1}JA(JA)'[H + JA(JA)']^{-1}JA(JA)'\}^{-1}JA =$$
$$= \sigma^2\{[(JA)'M_rJA]^{-1} - r^2 I\} =$$
$$= \sigma^2\{[(JA)'M_DJA]^{-1} - DD'\} =$$
$$= \sigma^2 C_4;$$

here, $M_r = [H+r^2JA(JA)']^{-1}$, $r \in \mathcal{R}^1$, $r \neq 0$, $M_D = [H+JAD(JAD)']^{-1}$, D is an arbitrary matrix of the type $k \times l$ such that $R(D) = k \leq l$ and C_4 is the negative lower right block of the Pandora's box matrix.

The first equivalent relation represents the variance of the estimator

$$\widehat{\Delta\Theta}(\eta) = \{(JA)'[H + JA(JA)']^{-1}JA\}^{-1}(JA)'[H + JA(JA)']^{-1}\eta.$$

Its variance is

$$var[\widehat{\Delta\Theta}(\eta)] =$$
$$= \sigma^2\{(JA)'[H + JA(JA)']^{-1}JA\}^{-1}(JA)'[H + JA(JA)']^{-1}H\cdot$$
$$\cdot[H + JA(JA)']^{-1}JA\{(JA)'[H + JA(JA)']^{-1}JA\}^{-1}.$$

The matrix expression in this relation is a special case of the expression $[(JA)'M_rJA]^{-1}(JA)'M_rHM_rJA[(JA)'M_rJA]^{-1}$, $M_r = [H+r^2JA(JA)']^{-1}$, $r \in \mathcal{R}^1$, $r \neq 0$, this general form will be taken into account in what follows. Let $H + r^2JA(JA)' = S$, which means that $H = S - r^2JA(JA)'$. Then

$$[(JA)'M_rJA]^{-1}(JA)'M_rHM_rJA[(JA)'M_rJA]^{-1} =$$

$$= [(JA)'S^{-1}JA]^{-1}(JA)'S^{-1}[S - r^2JA(JA)']S^{-1}JA[(JA)'S^{-1}JA]^{-1} =$$

$$= [(JA)'S^{-1}JA]^{-1}(JA)'S^{-1}JA[(JA)'S^{-1}JA]^{-1}-$$
$$-r^2[(JA)'S^{-1}JA]^{-1}(JA)'S^{-1}JA(JA)'S^{-1}JA[(JA)'S^{-1}JA]^{-1} =$$
$$= [(JA)'M_rJA]^{-1} - r^2I;$$

thus actually

$$var[\widehat{\Delta\Theta}(\eta)] = \sigma^2(\{(JA)'[H + r^2JA(JA)']^{-1}JA\}^{-1} - I) =$$
$$= \sigma^2\{[(JA)'M_rJA]^{-1} - r^2I\}.$$

Consider now the estimator of the parameter $\Delta\Theta$ in the form

$$\widehat{\Delta\Theta}(\eta) = (JA)'(\{JA(JA)'[H+JA(JA)']^{-1}JA\}^-)'(JA)'[H+JA(JA)']^{-1}\eta.$$

Its variance is

$$var[\widehat{\Delta\Theta}(\eta)] = \sigma^2(JA)'(\{JA(JA)'[H + JA(JA)']^{-1}JA\}^-)'(JA)'\cdot$$
$$\cdot[H + JA(JA)']^{-1}H[H + JA(JA)']^{-1}JA\{JA(JA)'[H + JA(JA)']^{-1}\cdot$$
$$\cdot JA\}^-JA = \sigma^2V.$$

The matrix V will be modified using the substitution $S = H+JA(JA)'$:

$$V = (JA)'\{[JA(JA)'S^{-1}JA]^-\}'(JA)'S^{-1}\cdot$$
$$\cdot[S - JA(JA)']S^{-1}JA[JA(JA)'S^{-1}JA]^-JA =$$
$$= (JA)'\{[JA(JA)'S^{-1}JA]^-\}'(JA)'S^{-1}JA[JA(JA)'S^{-1}JA]^-JA-$$
$$-(JA)'\{[JA(JA)'S^{-1}JA]^-\}'(JA)'S^{-1}JA(JA)'S^{-1}JA\cdot$$
$$\cdot[JA(JA)'S^{-1}JA]^-JA = V^{(1)} + V^{(2)}.$$

The first summand of this relation can be modified using the following explicit form of the g-inverse of the matrix $JA(JA)'S^{-1}JA$:

$$[JA(JA)'S^{-1}JA]^- =$$
$$= (JA)'S^{-1}JA(JA)'[JA(JA)'S^{-1}JA(JA)'S^{-1}JA(JA)']^{-1}.$$

Using it we get

$$V^{(1)} =$$
$$= (JA)'\{[JA(JA)'S^{-1}JA]^-\}'(JA)'S^{-1}JA[JA(JA)'S^{-1}JA]^-JA =$$
$$= (JA)'[JA(JA)'S^{-1}JA(JA)'S^{-1}JA(JA)']^{-1}\cdot$$

$$\cdot JA(JA)'S^{-1}JA(JA)'S^{-1}JA(JA)'S^{-1}JA(JA)'\cdot$$
$$\cdot[JA(JA)'S^{-1}JA(JA)'S^{-1}JA(JA)']^{-1}JA =$$
$$= (JA)'S^{-1}JA(JA)'[JA(JA)'S^{-1}JA(JA)'S^{-1}JA(JA)']^{-1}JA =$$
$$= [JA(JA)'S^{-1}JA]^{-}JA.$$

An analogous modification of the second summand yields

$$V^{(2)} = -(JA)'S^{-1}JA[JA(JA)'S^{-1}JA]^{-}JA,$$

thus

$$V = V^{(1)} + V^{(2)} = [I - (JA)'S^{-1}JA][JA(JA)'S^{-1}JA]^{-}JA =$$

$$= \{I - (JA)'[H + JA(JA)']^{-1}JA\}\{JA(JA)'[H + JA(JA)']^{-1}JA\}^{-}JA,$$

whereby the validity of the third equivalent relation is proved; the fourth equivalent relation is its special case for

$$\{JA(JA)'[H + JA(JA)']^{-1}JA\}^{-} =$$

$$= (JA)'[H + JA(JA)']^{-1}JA(JA)'\{JA(JA)'[H + JA(JA)']^{-1}\cdot$$
$$\cdot JA(JA)'[H + JA(JA)']^{-1}JA(JA)'\}^{-1}.$$

The fifth equivalent relation was proved together with the first; the sixth relation $var[\widehat{\Delta\Theta}(\eta)] = \sigma^2\{[(JA)'M_DJA]^{-1} - DD'\}$ expresses the variance of the estimator of the parameter $\Delta\Theta$ of the form $\widehat{\Delta\Theta}(\eta) = [(JA)'M_DJA]^{-1}\cdot JAM_D\eta$, where $M_D = [H + JAD(JAD)']^{-1}$, D is an arbitrary matrix of the type $k \times l$ such that $R(D) = k \leq l$; here,

$$var[\widehat{\Delta\Theta}(\eta)] = \sigma^2[(JA)'M_DJA]^{-1}JAM_DHM_D(JA)'[(JA)'M_DJA]^{-1}.$$

Applying the usual substitution $H + JAD(JAD)' = S$, we can easily prove that

$$[(JA)'M_DJA]^{-1}JAM_DHM_D(JA)'[(JA)'M_DJA]^{-1} =$$
$$= [(JA)'M_DJA]^{-1} - DD'.$$

As far as the variance of the estimator of the unknown vector parameter within regular Model 2 is concerned, which is expressed by means of the block C_4 of the Pandora's box matrix, its validity can be proved using the expression (2.11.18) which, in the regular case, yields the explicit form of the individual blocks of the Pandora's box matrix.

The reader surely has noted the analogies between the corresponding expressions of the covariance matrix of the linear unbiased and jointly efficient estimator of unbiasedly estimable functions of the unknown vector parameter within the universal model and within the essentialy simpler regular Model 2. By proving in detail the equivalence of the expressions mentioned within regular Model 2, the idea of proving them within the universal model was fully demonstrated. However within it, in consequence of abandonning the regularity conditions, the detailed proving requires that we use advanced matrix calculus, a topic which exceeds the scope of this publication.

Chapter 3

ANALYSIS OF A SET OF MEASURED DATA: HYPOTHESIS TESTING

3.1 Preliminary definitions

3.1.1 Statistical hypotheses, statistical tests

A statistical hypothesis is any assumption or presumption (supposition of the null hypothesis) concerning the unknown probability distribution of some random variable. The class of probability distributions to be considered is usually characterized as either parametric or nonparametric, which depends on whether the hypothesis treats a value of a parameter or, when nonparametric, deals with such other matters as, e.g., the type of probability distribution, the homogeneity of the set of measured data, and so forth.

Furthermore, we restrict ourselves to the continuous probability distributions, which are uniquely characterized by the probability density; thus we shall deal with the class $\mathcal{F}$ of probability densities.

If a parametric case is taken into account, then $\mathcal{F} = \{f(\cdot, \Theta),\ \Theta \in \mathcal{T}\}$, where $f(\cdot, \Theta)$ is a known probability density and $\mathcal{T}$ the parametric space.

If a nonparametric case is taken into account, $\mathcal{F} = \{f_i(\cdot),\ i = 1, 2, \ldots, n\}$, where $f_i(\cdot)$, $i = 1, 2, \ldots, n$, are various types of probability density.

We shall see that the dividing line between the parametric and nonparametric hypotheses is not sharp; there exist cases where the unknown parameter must be estimated when testing a nonparametric hypothesis.

The expressed presumption is called the null hypothesis, the notation H_0 is used for it; in the parametric case the null hypothesis is, e.g., of the form $H_0 : \Theta = \Theta_0$, in the nonparametric case, e.g., $H_0 : f(\cdot) = f_0(\cdot)$, where Θ_0 and $f_0(\cdot)$ are a concrete value of the parameter and a concrete type of probability density, respectively.

The alternative hypothesis H_1 in a parametric space may be, e.g., H_1:

$\Theta = \Theta_1 (\neq \Theta_0)$, $H_1 : \Theta \in [\Theta_1, \Theta_2]$, $H_1 : \Theta < \Theta_0$ (Θ_0 is the value of the parameter presumed by the null hypothesis); the alternative hypothesis, self-evidently, depends on the nature of the problem solved and on our a priori information on it; in a nonparametric case, the alternative hypothesis may be, e.g., of the form $H_1 : f_1(\cdot) \neq f_0(\cdot)$, where $f_0(\cdot)$ represents the type of probability density presumed by the null hypothesis. The null hypothesis H_0 is tested against the alternative H_1.

A hypothesis (parametric or nonparametric) can be either simple or composite. The parametric hypothesis $H : \Theta = \Theta_0$, where Θ_0 is one concrete element of the parametric space $\mathcal{T}$, is simple. The parametric hypothesis of the form $H : \Theta \in \underline{\Theta} \subset \mathcal{T}$ ($\underline{\Theta}$ is a subset of $\mathcal{T}$ formed by those elements of the parameter Θ which are supposed to be acceptable by the hypothesis H, e.g., $\Theta \in [\Theta_1, \Theta_2]$ or $\Theta \leq \Theta_1$, where Θ_1 or Θ_2 are the concrete values of the parameter Θ) is a composite parametric hypothesis. If the nonparametric hypothesis treats a unique type of probability distribution from the set of all which can be taken into account, we speak of a simple nonparametric hypothesis, in the opposite case the composite nonparametric hypothesis is considered.

The justification for the expressed null hypothesis can be judged using results of a suitably chosen experiment by means of a suitably chosen test statistic τ. The test statistic τ is a random variable whose realization t is a known function of the measurement results $x_1, x_2, \ldots, x_n$ of the justifying experiment mentioned. The n-dimensional vector $(x_1, x_2, \ldots, x_n)'$ is a realization of the n-dimensional random variable $(\xi_1, \xi_2, \ldots, \xi_n)'$ whose components are stochastically independent and possess the same probability distribution (the random vector $(\xi_1, \xi_2, \ldots, \xi_n)'$ is a random sample with the size n and the test statistic is a sample random variable). The concrete test statistic can be used for testing a hypothesis if its probability distribution is known. The procedure by means of which, on the basis of a suitably chosen test statistic and by means of the experimental results, we judge the justification of the expressed hypothesis, is called the statistical test.

The testing procedure results in either a rejection of the null hypothesis if the outcome of the test statistic contradicts the null hypothesis or in its nonrejection (we can never speak about the correctness or validity of a null hypothesis; a hypothesis is correct as far as an event does not occur which stands in contradiction to it), when the outcome of the test statistic does not contradict the expressed null hypothesis.

The sample space $\mathcal{R}^k$ of the test statistic (in general the test statistic may be a k-dimensional random variable, however, most frequently one-dimensional test statistics occur; then the sample space equals $\mathcal{R}^1$) is decomposed into two disjoint sets, namely the acceptance region A and the critical region W ($A \cup W = \mathcal{R}^k$, $A \cap W = \emptyset$).

If the test statistic τ occurs inside the critical region, the null hypothesis is rejected; if it occurs inside the acceptance region, the null hypothesis is not rejected.

If the null hypothesis is correct and the test statistic occurs inside the critical region W, in consequence of which the correct null hypothesis is rejected, we make an error of the first kind.

If the null hypothesis is not correct, even if the test statistic occurs inside the acceptance region, in consequence of which the incorrect null hypothesis is not rejected, we make an error of the second kind.

The problem is to find a critical region such that the probability of an error of the first kind may not exceed the given limit α ($\alpha \in (0,1)$) called the significance level of the test (the risk of rejecting a correct hypothesis) and simultaneously the probability of the second kind error may be as small as possible.

3.1.2 The critical region for tests of the hypothesis H_0 against H_1 at a given significance level (the risk)

When deciding whether to reject or not to reject the null hypothesis, the chosen risk α representing the probability of rejecting a correct hypothesis is of fundamental importance.

Since the null and the alternative hypotheses, even in the case of nonparametric tests, can be reformulated using the applied test statistic to a statement on values of this statistic, the following considerations concern both the parametric and nonparametric tests.

Next, the following notation will be applied: Θ — the tested value, Θ_0 — the value of the tested quantity assumed by the null hypothesis, τ — the test statistic used, t — a realization of the test statistic.

At the given risk α, the decision on rejecting the null hypothesis H_0 depends also on the form of the alternative. In order to appreciate it we shall analyse the following three most frequent cases. We test

1. $H_0 : \Theta = \Theta_0$ against $H_1; \Theta \neq \Theta_0$;
2. $H_0 : \Theta = \Theta_0$ against $H_1; \Theta > \Theta_0$;
3. $H_0 : \Theta = \Theta_0$ against $H_1; \Theta < \Theta_0$.

The more the realized value t of the test statistic τ used in the first case differs from the value Θ_0 assumed by the null hypothesis (the larger the value of the distance $|t - \Theta_0|$ of the points t and Θ_0 on the real line is) the more we doubt the correctness of the null hypothesis. The reason for its rejection with respect to the given risk α arises when the probability of rejecting the null hypothesis (if it is correct) decreases under the value α. The known probability distribution of the test statistic τ enables us to determine the probability $P\{\omega : |\tau(\omega) - \Theta_0| \geq |t - \Theta_0|\}$. If this probability is larger than the chosen risk α, we have no reason to reject the null hypothesis considered. The null hypothesis H_0 against the alternative H_1 is rejected with the risk α if

$$P\{\omega : |\tau(\omega) - \Theta_0| \leq |t - \Theta_0|\} =$$

$$= P\{\omega : \tau(\omega) - \Theta_0 \leq -|t - \Theta_0|\} + P\{\omega : \tau(\omega) \geq +|t - \Theta_0|\} =$$

$$= P\{\omega : \tau(\omega) \leq \Theta_0 - |t - \Theta_0|\} + P\{\omega : \tau(\omega) \geq \Theta_0 + |t - \Theta_0|\} \leq \alpha. \tag{3.1.1}$$

In contradistinction to the first case, in the second case ($H_0 : \Theta = \Theta_0$ against $H_1 : \Theta > \Theta_0$) our doubt about the correctness of the formulated null hypothesis H_0 is the greater the larger the realized value t of the test statistic τ is (the larger realized value t of the test statistic τ is more advantegous for the alternative). A reason for rejecting the hypothesis $H_0 : \Theta = \Theta_0$ against $H_1 : \Theta > \Theta_0$ arises if

$$P\{\omega : \tau(\omega) - \Theta_0 \geq t - \Theta_0\} \leq \alpha. \tag{3.1.2}$$

In the third case ($H_0 : \Theta = \Theta_0$ against $H_1 : \Theta < \Theta_0$), analogous to the second case, a very small value t realized for the test statistic τ casts doubt on the correctness of the null hypothesis. If

$$P\{\omega : \tau(\omega) - \Theta_0 \leq t - \Theta_0\} \leq \alpha, \tag{3.1.3}$$

with the risk α the null hypothesis $H_0 : \Theta = \Theta_0$ tested against the alternative $H_1 : \Theta < \Theta_0$ is rejected.

Let us now look for numbers k_{1l} and k_{1r}, k_2 and k_3 for which the identically numbered inequalities (3.1.1), (3.1.2) and (3.1.3) for the given test statistic with an a priori known probability distribution and for the given significance level α become equalities.

First, for the sake of simplicity, consider the inequality (3.1.2), which corresponds to the case of testing the null hypothesis $H_0 : \Theta = \Theta_0$ against the alternative $H_1 : \Theta > \Theta_0$. The number k_2 satisfying the equality

$$P\{\omega : \tau(\omega) - \Theta_0 \geq k_2 - \Theta_0\} = P\{\omega : \tau(\omega) \geq k_2\} = \alpha$$

is such a number which the realized test statistic exceeds with the probability α. This number is called the critical value of the given probability distribution at the level α (it is the $(1 - \alpha)$-quantile of the probability distribution of the test statistic τ used), the notation $t_{crit}(\alpha)$ is used for it; thus $k_2 = t_{crit}(\alpha)$. If the realization of the test statistic obtained from the set of the experimental data is larger than or equals the number $t_{crit}(\alpha)$, the hypothesis $H_0 : \Theta = \Theta_0$ against the alternative $H_1 : \Theta > \Theta_0$ is rejected with the risk α.

The interval $[t_{crit}(\alpha), \infty)$ represents the critical region of the test of the hypothesis $H_0 : \Theta = \Theta_0$ against $H_1 : \Theta > \Theta_0$.

If the null hypothesis $H_0 : \Theta = \Theta_0$ is tested against the alternative $H_1 : \Theta < \Theta_0$ (the third case), the number k_3 is given by the relation

$$P\{\omega : \tau(\omega) - \Theta_0 \leq k_3 - \Theta_0\} = P\{\omega : \tau(\omega) \leq k_3\} = \alpha;$$

thus k_3 is that value of the test statistic which the test statistic does not exceed with the probability α. Since

$$P\{\omega : \tau(\omega) \geq k_3\} = 1 - P\{\omega : \tau(\omega) \leq k_3\} = 1 - \alpha,$$

we see, based on to the definition of the critical value of the probability distribution of the test statistic, that $k_3 = t_{crit}(1 - \alpha)$ (it is the α-quantile of the probability density of the test statistic τ). If the realization of the test statistic obtained from the measured data is smaller or equals $t_{crit}(1 - \alpha)$, the hypothesis $H_0 : \Theta = \Theta_0$ against the alternative $H_1 : \Theta < \Theta_1$ is rejected with the risk α.

The interval $(-\infty, t_{crit}(1 - \alpha)]$ is the critical region of the test of the null hypothesis $H_0 : \Theta = \Theta_0$ against the alternative $H_1 : \Theta < \Theta_0$.

Now we return to the first case of testing the null hypothesis $H_0 : \Theta = \Theta_0$ against the alternative $H_1 : \Theta \neq \Theta_0$. In this case the values k_{1l} and k_{1r} are sought such that

$$P\{\omega : \tau(\omega) \leq k_{1l}\} + P\{\omega : \tau(\omega) \geq k_{1r}\} = \alpha. \tag{3.1.4}$$

The values k_{1l} and k_{1r}, which are not determined by the relation (3.1.4) uniquely, are usually chosen in the form which simultaneously satisfies the equalities $P\{\omega : \tau(\omega) \leq k_{1l}\} = \alpha/2$ and $P\{\omega : \tau(\omega) \leq k_{1r}\} = 1 - \alpha/2$. Under this subsidiary condition, $k_{1l} = t_{crit}(1-\alpha/2)$ and $k_{1r} = t_{crit}(\alpha/2)$ (it is evident that the values $t_{crit}(1-\alpha+\beta)$ and $t_{crit}(\beta)$ for any $\beta < \alpha$ satisfy the equation (3.1.4), which ensures that the probability of the error of the first kind equals α). If the measured values yield a realization t of the test statistic smaller than or equal to $t_{crit}(1-\alpha/2)$ or larger than or equal to $t_{crit}(\alpha/2)$ (i.e., if the test statistic t occurs outside the interval $(t_{crit}(1-\alpha/2), t_{crit}(\alpha/2))$, the null hypothesis $H_0 : \Theta = \Theta_0$ against the alternative $H_1 : \Theta \neq \Theta_0$ is rejected on the significance level α.

The critical region of the test of the hypothesis $H_0 : \Theta = \Theta_0$ against the alternative $H_1 : \Theta \neq \Theta_0$ is the union $(-\infty, t_{crit}(1-\alpha/2)] \cup [t_{crit}(\alpha/2), \infty)$.

Example The problem is to test the hypothesis that the mean value of a normal probability distribution with known dispersion ($\sigma^2 = 1$) has a zero mean value, hence $H_0 : m = 0$.

This null hypothesis will be tested by the help of a realization $(x_1, x_2, \ldots, x_{10})'$ of the random sample $(\xi_1, \xi_2, \ldots, \xi_{10})'$. The test statistic is chosen in the form

$$\overline{\xi} = (1/10)\sum_{i=1}^{10} \xi_i;$$

this random variable, under the assumption that the null hypothesis is correct, possesses the normal probability distribution $N(0, 1/10)$ (in this case $\overline{\xi}$ is a suitable test statistic, as $\overline{\xi}$ represents an unbiased estimator of the parameter m, the dispersion of this estimator decreases with an increasing n).

Let $\overline{x} = 1.01$ and let us consider the significance levels $\alpha = 0.05$ and $\alpha = 0.01$.

1. Let the alternative $H_1 : m \neq 0$ be taken into account, i.e., the null hypothesis $H_0 : m = 0$ against the alternative $H_1 : m \neq 0$ is tested. In this case

$$P\{\omega : |\overline{\xi}(\omega) - 0| \geq |1.01 - 0|\} = P\{\omega : |\overline{\xi}(\omega)| \geq 1.01\} =$$

$$= P\{\omega : \overline{\xi}(\omega) \leq -1.01\} + P\{\omega : \overline{\xi}(\omega) \geq 1.01\} =$$

$$= P\{\omega : \sqrt{10}\,\overline{\xi}(\omega) \leq -1.01\sqrt{10}\} + P\{\omega : \sqrt{10}\,\overline{\xi}(\omega) \geq 1.01\sqrt{10}\} =$$

$$= P\{\omega : \sqrt{10}\,\overline{\xi}(\omega) \leq -3.19\} + P\{\omega : \sqrt{10}\,\overline{\xi}(\omega) \geq 3.19\} =$$

$$= 0.0007 + 1.0000 - 0.9993 = 0.0014 < 0.01 < 0.05,$$

which means that the measured data entitle us to reject the null hypothesis $H_0 : m = 0$ if the alternative is of the form $H_1 : m \neq 0$, on both the significance levels $\alpha_1 = 0.05$ and $\alpha_2 = 0.01$.

The critical values of the normal probability distribution $N(0,\ 1/10)$ on the significance levels 0.05 and 0.01 are

$$\overline{x}_{crit}(0.025) = F_{0,1}^{-1}(0.975)/\sqrt{10} = 1.960/\sqrt{10} = 0.620,$$

$$\overline{x}_{crit}(0.005) = F_{0,1}^{-1}(0.995)/\sqrt{10} = 2.576/\sqrt{10} = 0.815,$$

$$\overline{x}_{crit}(0.975) = -\overline{x}_{crit}(0.025) = -0.620,$$

$$\overline{x}_{crit}(0.995) = -\overline{x}_{crit}(0.005) = -0.815$$

($F_{0,1}^{-1}(\alpha)$ is the α-quantile of the normal probability distribution $N(0,1)$). The test statistic realized in our case by the value 1.01 lies outside the intervals $(\overline{x}_{crit}(0.975), \overline{x}_{crit}(0.025)) = (-0.620, 0.620)$ and $(\overline{x}_{crit}(0.995), \overline{x}_{crit}(0.005)) = (-0.815, 0.815)$, therefore the null hypothesis $H_0 : m = 0$ against the alternative $H_1 : m \neq 0$ is rejected at the significance levels 0.05 and 0.01, respectively.

2. Let the alternative be of the form $H_1 : m > 0$, hence the null hypothesis is tested against the alternative $H_1 : m > 0$. In this case

$$P\{\xi : \overline{\xi}(\omega) \geq 1.01\} = P\{\omega : \overline{\xi}(\omega)\sqrt{10} \geq 1.01\sqrt{10}\} =$$

$$= 0.0007 < 0.01 < 0.05.$$

The critical region for the test of the hypothesis $H_0 : m = 0$ against $H_1 : m > 0$ at the significance level 0.05 is $[\overline{x}_{crit}(0.05), \infty) = [0.520, \infty)$ and at the significance level 0.01 is of the form $[\overline{x}_{crit}(0.01), \infty) = [0.736, \infty)$. The value of the test statistic realized in both cases lies inside the corresponding critical region, thus with the risk of both 0.05 and 0.01 the hypothesis $H_0 : m = 0$ against $H_1 : m > 0$ is rejected.

3. Let us test the null hypothesis $H_0 : m = 0$ against the alternative $H_1 : m < 0$. Here

$$P\{\xi : \overline{\xi}(\omega) \leq 1.01\} = P\{\omega : \overline{\xi}(\omega)\sqrt{10} \leq 1.01\sqrt{10}\} =$$

$$= 0.9993 > 0.05 > 0.01.$$

The critical region for the test of the hypothesis $H_0 : m = 0$ against $H_1 : m < 0$ at the significance level 0.05 is $(-\infty, \overline{x}_{crit}(0.95)] = (-\infty, -0.520]$ and at the significance level 0.01 is $(-\infty, \overline{x}_{crit}(0.99)] = (-\infty, -0.736]$, which means that there does not exist any reason for refusing the hypothesis $H_0 : m = 0$ against the alternative $H_1 : m < 0$ at the significance levels considered.

3.1.3 The power function of a test

The power function of the test of the null hypothesis $H_0 : \Theta \in \underline{\Theta}$ against the alternative $H_1 : \Theta \in \mathcal{T} - \underline{\Theta}$ is called the function

$$\beta(\Theta; W) = \int_W f(t, \Theta) dt,$$

defined for each value of the parameter Θ from the parametric space $\mathcal{T}$; $f(\cdot, \Theta)$, $t \in \mathcal{R}^1$, is the probability density of the test statistic τ used (we restrict ourselves to the case which occurs most frequently, namely the case of a one-dimensional test statistic), W is the critical region of the hypothesis H_0 against the alternative H_1.

The values of the power function of the test of the null hypothesis $H_0 : \Theta \in \underline{\Theta}$ against the alternative $H_1 : \Theta :\in \mathcal{T} - \underline{\Theta}$ at the points $\Theta \in \underline{\Theta}$ for a given critical region W yield the probability of rejecting a true null hypothesis - the probability of an error of the first kind.

The values of the power function of the test of the null hypothesis $H_0 : \Theta \in \underline{\Theta}$ against the alternative $H_1 : \Theta :\in \mathcal{T} - \underline{\Theta}$ at the points $\Theta \in \mathcal{T} - \underline{\Theta}$ for a given critical region W yield the probability of rejecting an uncorrect null hypothesis; the difference $1 - \beta(\Theta; W)$, $\Theta \in \mathcal{T} - \underline{\Theta}$, represents the probability of non-rejecting an uncorrect null hypothesis — the probability of the error of the second kind.

The power function $\beta(\cdot; W)$ or the function $1 - \beta(\cdot; W)$, called the operational characteristic of the test of the null hypothesis H_0 against the alternative H_1, enables us to solve in a complex way the basic problem of finding a critical region W such that the probability of an error of the first kind may not exceed the chosen level of significance and, simultaneously, the probability of an error of the second kind may be minimized.

The requirement that the probability of an error of the first kind may not exceed an a priori given risk α implies that the critical region W must fulfil the relation

$$\sup\{\beta(\Theta; W) : \Theta \in \underline{\Theta}\} \leq \alpha; \tag{3.1.5}$$

in the special case of a simple null hypothesis $H_0 : \Theta = \Theta_0$ the condition (3.1.5) is of the form

$$\beta(\Theta_0; W) = \alpha.$$

However, not every critical region for which the probability of an error of the first kind does not exceed the given risk α ensures the minimum value of the the probability of an error of the second kind. Hence, in the next section we shall deal with the problem of minimizing the probability of an error of the second kind by a suitable choice of the critical region from the set of all critical regions which satisfy the relation (3.1.5).

But first we shall introduce a simple example illustrating the meaning of these new concepts.

Example Determine the power function of the test of the null hypothesis $H_0 : m = 0$ against the alternatives $H_1 : m \neq 0$, $H_1 : m > 0$ and $H_1 : m < 0$ from the previous section.

1. The power function of the null hypothesis $H_0 : m = 0$ against $H_1 : m \neq 0$ under the risk $\alpha = 0.05$ (Fig 3.1.1)

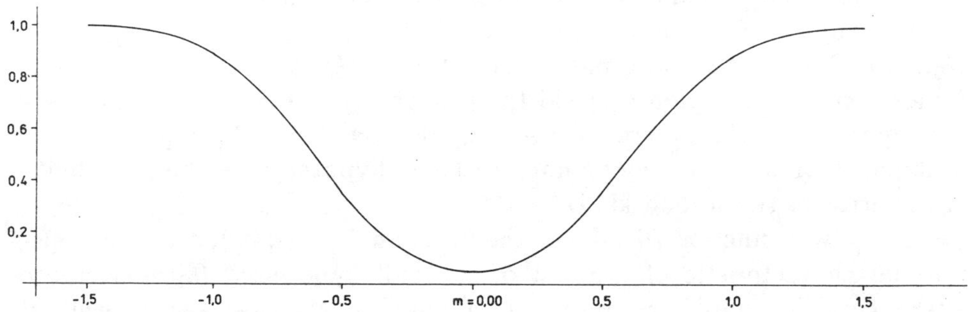

Fig. 3.1.1 The power function of the test of the null hypothesis that the mean value of a normal probability distributions whose dispersion equals 1 possesses a zero mean value, i.e., $H_0 : m = 0$ against the alternative $H_1 : m \neq 0$ if the test statistic is of the form $\bar{\xi} = (1/10)\sum_{i=1}^{10} \xi_i$

The critical region determined in the previous example is $W = (-\infty, \overline{x}_{crit}(0.975)] \cup [\overline{x}_{crit}(0.025), \infty) = (-\infty, -0.620] \cup [0.620, \infty)$, hence, since

$\bar{\xi} \sim N(0, 1/10)$, we may write

$$\beta(m; W) = \frac{\sqrt{10}}{\sqrt{2\pi}} \left(\int_{-\infty}^{-0.620} \exp[-10(\bar{x} - m)^2/2] d\bar{x} + \right.$$

$$\left. + \int_{0.620}^{\infty} \exp[-10(\bar{x} - m)^2/2] d\bar{x} \right) =$$

$$= 1 + \frac{1}{\sqrt{2\pi}} \left(\int_{-\infty}^{-1.960-\sqrt{10}m} \exp(-y^2/2) dy - \int_{-\infty}^{1.960-\sqrt{10}m} \exp(-y^2/2) dy \right).$$

In order to use properly the tabulated values of the normal probability distribution $N(0, 1)$ for calculating the values of the function $\beta(\cdot; W)$, it is necessary to modify them using the substitution $\sqrt{10}(\bar{x} - m) = y$.

The values of the power function of the test of the null hypothesis $H_0; m = 0$ against the alternative $H_1 : m \neq 0$ at $\alpha = 0.05$ are in Table 3.1.

Table 3.1

m	$-1.960 - \sqrt{10}m$	1st integral	$1.960 - \sqrt{10}m$	2nd integral	$\beta(m, W)$
1.50	−6.703	0	−2.783	0.003	0.997
1.00	−5.122	0	−1.202	0.115	0.885
0.50	−3.541	0.0002	+0.379	0.648	0.352
0.20	−2.592	0.005	+1.328	0.908	0.097
0.10	−2.276	0.012	+1.644	0.950	0.062
0.05	−2.118	0.015	+1.802	0.964	0.051
0.00	−1.960	0.025	+1.960	0.975	0.050
−0.05	−1.802	0.036	+2.118	0.985	0.051
−0.10	−1.644	0.050	+2.276	0.988	0.062
−0.20	−1.328	0.092	+2.592	0.995	0.097
−0.50	−0.379	0.352	+3.541	0.9998	0.352
−1.00	+1.202	0.885	+5.122	1	0.885
−1.50	+2.783	0.997	+6.703	1	0.997

2. The power function of the null hypothesis $H_0 : m = 0$ against $H_1 : m > 0$ under the risk $\alpha = 0.05$ (Fig 3.1.2)

The critical region is of the form

$$W = [\bar{x}_{crit}(0.05), \infty) = [0.520, \infty),$$

hence

$$\beta(m; W) = \frac{\sqrt{10}}{\sqrt{2\pi}} \int_{0.520}^{\infty} \exp[-10(\bar{x} - m)^2/2] d\bar{x} =$$

$$= 1 - \frac{1}{\sqrt{2\pi}} \int_{-\infty}^{1.645-\sqrt{10}m} \exp(-y^2/2)dy.$$

The values of the power function of the test of the null hypothesis $H_0 : m = 0$ against the alternative $H_1 : m > 0$ are contained in Table 3.2.

Table 3.2

m	$1.645 - \sqrt{10}m$	Integral	$\beta(m, W)$
1.50	−3.098	0.001	0.999
1.00	−1.517	0.06	0.94
0.50	+0.064	0.53	0.47
0.20	+1.013	0.84	0.16
0.10	+1.335	0.91	0.09
0.05	+1.487	0.93	0.07
0	+1.645	0.95	0.05
−0.05	+1.803	0.96	0.04
−0.10	+1.961	0.98	0.02
−0.20	+2.277	0.99	0.01
−0.50	+3.226	0.999	0.001
−1.00	+4.807	1	0
−1.50	+6.383	1	0

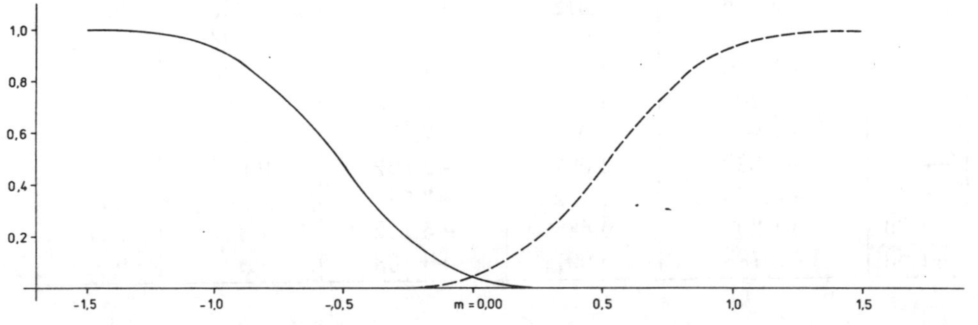

Fig. 3.1.2 The full line demonstrates the power function of the test of the null hypothesis that the mean value of a normal probability distribution whose dispersion equals 1 possesses a zero mean value, i.e., $H_0 : m = 0$ against the alternative $H_1 : m < 0$ if the test statistic is of the form $\bar{\xi} = \sum_{i=0}^{10} \xi_i$; the dashed line demonstrates (under the same conditions) the powerfunction of the test of the null hypothesis $H_0 : m = 0$ against the alternative $H_1 : m > 0$.

3. The power function of the test of the null hypothesis H_0 : $m = 0$ against the alternative $H_1 : m < 0$ at the significance level $\alpha = 0.05$ (Fig. 3.1.2)

The critical region is

$$W = (-\infty, \overline{x}_{crit}(0.95)] = (-\infty, -0.520],$$

thus

$$\beta(m; W) = \frac{\sqrt{10}}{\sqrt{2\pi}} \int_{-\infty}^{-0.520} \exp[-10(\overline{x} - m)^2/2] d\overline{x} =$$

$$= \frac{1}{\sqrt{2\pi}} \int_{-\infty}^{-1.645-\sqrt{10}m} \exp(-y^2/2) dy.$$

The values of the power function of the test of the null hypothesis $H_0 : m = 0$ against the alternative $H_1 : m < 0$ are contained in Table 3.3.

Table 3.3

m	$-1.645 - \sqrt{10}m$	$\beta(m, W)$
1.50	−6.383	0
1.00	−4.807	0
0.50	−3.226	0.001
0.20	−2.277	0.01
0.10	−1.961	0.02
0.05	−1.803	0.04
0	−1.645	0.05
−0.05	−1.487	0.07
−0.10	−1.335	0.09
−0.20	−1.013	0.16
−0.50	−0.064	0.47
−1.00	+1.517	0.94
−1.50	+3.098	0.999

3.1.4 The Neyman — Pearson lemma for a simple null and a simple alternative hypothesis

It is sufficient, for didactic reasons, to introduce this fundamental theorem and its proof for the case of a simple null and a simple alternative hypothesis, i.e., for the case where the parametric space $\mathcal{T}$ contains only two elements Θ_0 and Θ_1; here $\underline{\Theta} = \{\Theta_0\}$ and $\mathcal{T} - \underline{\Theta} = \{\Theta_1\}$. Self-evidently, the probability distributions $f(\cdot, \Theta_0)$ and $f(\cdot, \Theta_1)$ of the test statistic we applied are supposed to be known a priori. Under these assumptions the following statement can be proved.

Neyman — Pearson lemma Let α be the chosen significance level. Let the critical region W_0 of the null hypothesis being tested be of the form

$$W_0 = \{t : f(t, \Theta_1) \geq cf(t, \Theta_0)\}, \tag{3.1.6}$$

where the positive number c satisfies the requirement that the probability of an error of the first kind for the critical region may equal the chosen risk α, i.e.,

$$\beta(\Theta_0; W_0) = \int_{W_0 = \{t : f(t,\Theta_1) \geq cf(t,\Theta_0)\}} f(t, \Theta_0) dt = \alpha. \tag{3.1.7}$$

Then every other critical region W of the null hypothesis H_0 against the alternative H_1 such that

$$\beta(\Theta_0; W) = \int_W f(t, \Theta_0) dt = \alpha, \tag{3.1.8}$$

satisfies the relation

$$\beta(\Theta_1; W_0) = \int_{W_0} f(t, \Theta_1) dt \geq \int_W f(t, \Theta_1) dt = \beta(\Theta_1, W), \tag{3.1.9}$$

i.e., $1 - \beta(\Theta_1; W_0) \leq 1 - \beta(\Theta_1; W)$.

The assertion can be proved in the following way: The probability density is an integrable non-negative function, therefore

$$\int_{W_0} f(t, \Theta_1) dt - \int_W f(t, \Theta_1) dt =$$

$$= \int_{W_0 - W} f(t, \Theta_1) dt + \int_{W_0 \cap W} f(t, \Theta_1) dt -$$

$$-\int_{W-W_0} f(t,\Theta_1)dt - \int_{W_0\cap W} f(t,\Theta_1)dt,$$

which implies

$$\int_{W_0} f(t,\Theta_1)dt - \int_{W} f(t,\Theta_1)dt = \int_{W_0-W} f(t,\Theta_1)dt - \int_{W-W_0} f(t,\Theta_1)dt.$$

For each $t \in W_0$, due to (3.1.6), $f(t,\Theta_1) \geq cf(t,\Theta_0)$, which implies that this inequality must be fulfilled also for each $t \in W_0 - W$, where W is another arbitrary critical region possessing the property (3.1.8), thus

$$\int_{W_0-W} f(t,\Theta_0)dt \geq c\int_{W_0-W} f(t,\Theta_0)dt.$$

The inequality $f(t,\Theta_1) \geq cf(t,\Theta_0)$ satisfied for each $t \in W_0$ implies the inequality $f(t,\Theta_1) < cf(t,\Theta_0)$ which is fulfilled for each $t \in W_0^c$ ($W_0^c = \mathcal{R}^1 - W_0$ is the complement of the set W_0) must be fulfilled also for each $t \in W - W_0$ and this enables us to write

$$\int_{W-W_0} f(t,\Theta_1)dt < c\int_{W-W_0} f(t,\Theta_0)dt.$$

Both integral inequalities yield

$$\int_{W_0-W} f(t,\Theta_1)dt - \int_{W-W_0} f(t,\Theta_1)dt >$$

$$> c\left(\int_{W_0-W} f(t,\Theta_0)dt - \int_{W-W_0} f(t,\Theta_0)dt\right) =$$

$$= c\left(\int_{W_0} f(t,\Theta_0)dt - \int_{W} f(t,\Theta_0)dt\right) = c(\alpha - \alpha) = 0$$

(the relations (3.1.7) and (3.1.8) were applied here); thereby the proof of the Neyman — Pearson lemma is completed.

The choice of the critical region of the null hypothesis $H_0 : \Theta = \Theta_0$ against the alternative $H_1 : \Theta = \Theta_1$ in the form W_0 given by the relation (3.1.6) with the condition expressed by (3.1.7) respects the requirement of minimizing the probability of an error of the second kind (the probability of an error of the first kind equals, self-evidently, the a priori chosen significance level of the test). In this sense this critical region W_0 is considered to be optimum.

The test with the critical region W_0 shows the minimum probability of an error of the second kind among all the tests of the hypothesis H_0 :

$\Theta = \Theta_0$ against the alternative $H_1 : \Theta = \Theta_1$ at a given significance level α. The critical region W_0 determines the most powerful level-α test of the null hypothesis H_0 against H_1.

Example Test the null hypothesis H_0 that the sample of size n comes from the probability distribution $N(m_0, \sigma^2)$, where the dispersion $\sigma^2 > 0$ is supposed to be known a priori; the alternative is of the form $H_1 : m = m_1$, where $m_1 > m_0$. The problem is to determine the critical region of the most powerful level-α test of the null hypothesis $H_0 : m = m_0$ against $H_1 : m = m_1$ $(m_1 > m_0)$.

The problem will be solved by means of the Neyman — Pearson lemma. In our case $\mathcal{T} = \{m_0, m_1\}$, $\underline{\Theta} = \{m_0\}$, $\mathcal{T} - \underline{\Theta} = \{m_1\}$; the test statistic is the random variable $\bar{\xi} = (1/n) \sum_{i=1}^{n} \xi_i$ possessing the normal probability distribution $N(m_0, \sigma^2/n)$, which implies

$$f(\overline{x}, m_0) = \frac{\sqrt{n}}{\sigma\sqrt{2\pi}} \exp\left(-\frac{n(\overline{x} - m_0)^2}{2\sigma^2}\right)$$

and

$$f(\overline{x}, m_1) = \frac{\sqrt{n}}{\sigma\sqrt{2\pi}} \exp\left(-\frac{n(\overline{x} - m_1)^2}{2\sigma^2}\right).$$

In accordance with (3.1.6)

$$W_0 =$$

$$= \left\{\overline{x} : \frac{\sqrt{n}}{\sigma\sqrt{2\pi}} \exp\left(-\frac{n(\overline{x} - m_1)^2}{2\sigma^2}\right) \geq c \frac{\sqrt{n}}{\sigma\sqrt{2\pi}} \exp\left(-\frac{n(\overline{x} - m_0)^2}{2\sigma^2}\right)\right\} =$$

$$= \left\{\overline{x} : \exp\left(-\frac{n(\overline{x} - m_1)^2}{2\sigma^2}\right) \geq c \exp\left(-\frac{n\overline{x} - m_0)^2}{2\sigma^2}\right)\right\}.$$

First the determining inequality

$$\exp\left(-\frac{n(\overline{x} - m_1)^2}{2\sigma^2}\right) \geq c \exp\left(-\frac{n(\overline{x} - m_0)^2}{2\sigma^2}\right)$$

has to be modified.

By taking its algorithm, we get

$$-\frac{n(\overline{x} - m_1)^2}{2\sigma^2} + \frac{n(\overline{x} - m_0)^2}{2\sigma^2} \geq \ln c,$$

thus

$$2\overline{x}(m_1 - m_0) \geq \frac{2\sigma^2}{n} \ln c + m_1^2 - m_0^2. \tag{3.1.10}$$

This relation in consequence of the assumption $m_1 > m_0$ implies (the notation c_1 is used for c in this case)

$$\overline{x} \geq \frac{\sigma^2 \ln c_1}{n(m_1 - m_0)} + \frac{m_1^2 - m_0^2}{2(m_1 - m_0)},$$

hence

$$\overline{x} \geq \frac{\sigma^2 \ln c_1}{n(m_1 - m_0)} + \frac{m_1 + m_0}{2}.$$

The constant c_1 will be determined with the help of the constant $k_1 = \frac{\sigma^2 \ln c_1}{n(m_1 - m_0)} + \frac{m_1^2 - m_0^2}{2(m_1 - m_0)}$, where n, σ^2, m_0 and m_1 are known numbers. For determining the constant k_1 the requirement that the probability of an error of the first kind must equal the given significance level α has to be applied. This requirement is expressed by the relation (3.1.7) that in our case is of the form

$$\int_{W_0=\{\overline{x}:\overline{x}\geq k_1\}} f(\overline{x}; m_0) d\overline{x} = \frac{\sqrt{n}}{\sigma\sqrt{2\pi}} \int_{k_1}^{\infty} \exp\left(-\frac{n(\overline{x} - m_0)^2}{2\sigma^2}\right) d\overline{x} =$$

$$= \int_{\sqrt{n}(k_1 - m_0)/\sigma}^{\infty} \exp(-y^2/2) dy = \alpha$$

(the substitution $\sqrt{n}(\overline{x} - m_0)/\sigma = y$ was used here). Thus $\sqrt{n}(k_1 - m_0)/\sigma = F_{0,1}^{-1}(1-\alpha)$ and $k_1 = m_0 + \sigma F_{0,1}^{-1}(1-\alpha)/\sqrt{n}$, where $F_{0,1}^{-1}(1-\alpha)$ is the $(1-\alpha)$-quantile of the normal probability distribution $N(0,1)$. The critical region of the most powerful level-α test of the hypothesis $H_0 : m = m_0$ against $H_1 : m = m_1$ $(m_1 > m_0)$ is

$$W_0 = \{\overline{x} : \overline{x} \geq m_0 + \sigma F_{0,1}^{-1}(1-\alpha)/\sqrt{n}\} = [m_0 + \sigma F_{0,1}^{-1}(1-\alpha)/\sqrt{n}, \infty).$$

Let the alternative be $H_1 : m = m_1$, where $m_1 < m_0$. In order to determine the critical region of the most powerful test $H_0 : m = m_0$ against the alternative $H_1 : m = m_1$ $(m_1 < m_0)$, we proceed in the same way as we did previously up to the relation (3.1.10). This, in consequence of the assumption $m_1 < m_0$, implies (in this case the notation c_2 is used for c)

$$\overline{x} \leq \frac{\sigma^2 \ln c_2}{n(m_1 - m_0)} + \frac{m_1 + m_0}{2} = k_2,$$

therefore,

$$W_0 = \{\overline{x} : \overline{x} \leq k_2\} = (-\infty, k_2]$$

(as in the preceding case, the constant k_2 instead of c_2 is being determined). Applying the requirement that the probability of an error of the first kind must be equal to the given significance level α which expresses the relation (3.1.7), we obtain

$$\int_{W_0=\{\overline{x}:\overline{x}\leq k_2\}} f(\overline{x}, m_0) d\overline{x} = \frac{\sqrt{n}}{\sigma\sqrt{2\pi}} \int_{-\infty}^{k_2} \exp\left(-\frac{n(\overline{x}-m_0)^2}{2\sigma^2}\right) d\overline{x} =$$

$$= \frac{1}{\sqrt{2\pi}} \int_{-\infty}^{\sqrt{n}(k_2-m_0)/\sigma} \exp(-y^2/2) dy =$$

$$= 1 - \frac{1}{2\pi} \int_{\sqrt{n}(k_2-m_0)/\sigma}^{\infty} \exp(-y^2/2) dy = \alpha.$$

This means that $\sqrt{n}(k_2 - m_0)/\sigma = F_{0,1}^{-1}(\alpha) = -F_{0,1}^{-1}(1-\alpha)$, so that $k_2 = m_0 - \sigma F_{0,1}^{-1}(1-\alpha)/\sqrt{n}$.

The critical region of the most powerful level-α test of the hypothesis $H_0 : m = m_0$ against $H_1 : m = m_1$ $(m_1 < m_0)$ is of the form

$$W_0 = (-\infty, m_0 - \sigma F_{0,1}^{-1}(1-\alpha)/\sqrt{n}].$$

It is necessary to realize that the power function of the test $H_0 : m = m_0$ against $H_1 : m = m_1$ is defined at two points, namely m_1 and m_2, only.

3.1.5 Uniformly most powerful tests of a simple hypothesis H_0 against a composite hypothesis H_1

The most frequent case in hypothesis testing is the test of a simple null hypothesis against the composite alternative.

Let the parametric space $\mathcal{T}$ consist of a finite number of elements, thus $\mathcal{T} = \{\Theta_0, \Theta_1, \ldots, \Theta_k\}$, $k \geq 2$, the region $\underline{\Theta}$ is a one-element set, $\underline{\Theta} = \{\Theta_0\}$ and the region $\mathcal{T} - \underline{\Theta}$ contains the remaining k elements, $\mathcal{T} - \underline{\Theta} = \{\Theta_1, \Theta_2, \ldots, \Theta_k\}$. The simple null hypothesis $H_0 : \Theta = \Theta_0$ is tested against the composite alternative hypothesis $H_1 : \Theta \in \{\Theta_1, \Theta_2, \ldots, \Theta_k\}$.

The Neyman — Pearson lemma enables us to find the most powerful test of the simple hypothesis $H_0 : \Theta = \Theta_0$ against k simple partial alternatives $H_1^{(1)} : \Theta = \Theta_1, \ldots, H_1^{(k)} : \Theta = \Theta_k$. The corresponding critical regions $W_0^{(1)}, \ldots, W_0^{(k)}$ may differ, as it is self-evident.

If $W_0^{(1)} = W_0^{(2)} = \ldots = W_0^{(k)} = W_0$, the critical region W_0 represents the optimum critical region for testing the null hypothesis $H_0 : \Theta = \Theta_0$ against the alternative $H_1 : \Theta \in \{\Theta_1, \Theta_2, \ldots, \Theta_k\}$. The critical region W_0 determines the uniformly best test of the simple null hypothesis H_0 being considered against the composite alternative H_1.

The power function of the test of the null hypothesis H_0 against the composite alternative H_1 is in this case defined at $k+1$ points of the parametric space $\mathcal{T}$, $\beta(\Theta_0; W_0) = \alpha$, and for each value $\Theta \in \{\Theta_1, \Theta_2, \ldots, \Theta_k\} = \mathcal{T} - \underline{\Theta}$ we have $1-\beta(\Theta; W_0) \leq 1-\beta(\Theta; W)$, where W is any critical region $W \neq W_0$ such that $\beta(\Theta_0; W) = \alpha$.

However, there exists a series of cases in which the uniformly most powerful test of the simple null hypothesis H_0 against the composite alternative H_1 does not exist.

Example On the basis of a random sample $(\xi_1, \xi_2, \ldots, \xi_n)'$ the hypothesis is tested that it comes from the normal probability distribution $N(m_0, \sigma^2)$ (the dispersion σ^2 is supposed to be known a priori); the alternative H_1 is of the form $H_1 : m > m_0$. The problem is to determine the critical region of the uniformly most powerful test of the null hypothesis $H_0 : m = m_0$ against the alternative $H_1 : m > m_0$ (if it exists).

The parametric space $\mathcal{T}$ is formed by the interval $[m_0, \infty)$, $\underline{\Theta} = \{m_0\}$ and $\mathcal{T} - \underline{\Theta} = (m_0, \infty)$.

In the example illustrating the Neyman — Pearson lemma we determined the critical region of the most powerful level-α test of the hypothesis $H_0 : m = m_0$ against $H_1 : m = m_1$ $(m_1 > m_0)$ of the form

$$W_0 = [m_0 + \sigma F_{0,1}^{-1}(1-\alpha)/\sqrt{n}, \infty). \tag{3.1.11}$$

The critical region of the most powerful level-α test of the simple null hypothesis $H_0 : m = m_0$ against each simple partial hypothesis $H_1^{(i)} : m = m_i$, $m_i \in (m_0, \infty)$ has always had the same form as given by (3.1.11). This means that the critical region $W_0 = [m_0 + \sigma F_{0,1}^{-1}(1-\alpha)/\sqrt{n}, \infty)$ determines at the significance level α the uniformly most powerful test of the simple null hypothesis $H_0 : m = m_0$ against the composite hypothesis $H_1 : m > m_0$.

If, under the same conditions, the simple null hypothesis $H_0 : m = m_0$ against the composite alternative $H_1 : m < m_0$ is tested, the parametric space $\mathcal{T} = (-\infty, m_0]$, is taken into account; in this case $\underline{\Theta} = \{m_0\}$ and $\mathcal{T} - \underline{\Theta} = (-\infty, m_0)$. In the example we mentioned concerning the Neyman — Pearson lemma, we derived the critical region of the most powerful level-α

test of the null hypothesis $H_0 : m = m_0$ against the alternative $H_1 : m = m_1,\ m_1 < m_0$ of the form

$$W_0 = (-\infty, m_0 - \sigma F_{0,1}^{-1}(1-\alpha)/\sqrt{n}]. \tag{3.1.12}$$

The critical region of the most powerful test of the simple null hypothesis $H_0 : m = m_0$ against each simple partial alternative $H_1^{(j)} : m = m_j,\ m_j \in (-\infty, m_0)$ always has the same form as given by (3.1.12). Therefore, the critical region $W_0 = (-\infty, m_0 - \sigma F_{0,1}^{-1}(1-\alpha)/\sqrt{n}]$ determines at the significance level α the uniformly most powerful test of the simple null hypothesis $H_0 : m = m_0$ against the composite alternative $H_1 : m < m_0$.

At the end, the example is shown where the uniformly most powerful test of a simple null hypothesis against a composite alternative does not exist.

Analogous to the two preceding cases, let us consider the random sample $(\xi_1, \xi_2, \ldots, \xi_n)'$ and test by using it the hypothesis that it comes from the normal probability distribution $N(m_0, \sigma^2)$, where $\sigma^2 > 0$ is known a priori. Thus, we test the simple null hypothesis $H_0 : m = m_0$ against the composite alternative $H_1 : m \neq m_0$.

In this case, the parametric space $\mathcal{T}$ is presented by the interval $(-\infty, \infty)$, $\underline{\Theta} = \{m_0\}$ and $\mathcal{T} - \underline{\Theta} = (-\infty, \infty) - \{m_0\}$.

The critical region of the most powerful test of the simple hypothesis $H_0 : m = m_0$ against any simple partial alternative $H_1^{(i)} : m = m_i,\ m_i > m_0$, is $W_0^{(i)} = [m_0 + \sigma F_{0,1}^{-1}(1-\alpha)/\sqrt{n}, \infty)$; the critical region of the most powerful test of the simple null hypothesis $H_0 : m = m_0$ against any simple partial alternative $H_1^{(j)} : m = m_j,\ m_j < m_0$, is $W_0^{(j)} = (-\infty, m_0 - \sigma F_{0,1}^{-1}(1-\alpha)/\sqrt{n}]$. It is evident that $W_0^{(i)} \neq W_0^{(j)}$, which means that the uniformly most powerful test of the simple null hypothesis $H_0 : m = m_0$ against the composite alternative $H_1 : m \neq m_0$ does not exist.

3.2 Tests verifying the linear hypothesis on the first order parameters in linear models of estimation theory

The processing of measured data is often performed by applying one of the six fundamental linear models dealt with in Sections 2.2, 2.3 and

2.4. Many further situations of a dynamic nature, which are modelled using stochastic functions, also result in an application of these models. There are two main reasons for this; the first consists in the fact that the continuous function often may be measured only at discrete points on their definition region, the second is the fact that the measured data are usually processed with the help of digital computers.

In general, the hypothesis on the numerical values of the components of the vector parameter $\Theta = (\Theta_1, \Theta_2, \ldots, \Theta_k)'$ estimated within one of the fundamental linear models of measurements is of the form $h(\Theta) = d,\ d \in \mathcal{R}^m$; after linearizing the m-dimensional vector function $h(\cdot)$, $h_{lin}(\Theta) = h(\Theta_0) + [\partial h(\Theta_0)/\partial\Theta]\Delta\Theta)$ and denoting $\partial h(\Theta_0)/\partial\Theta = H$ (the $m \times k$-dimensional matrix such that $R(H) = m \leq k$) and $d - h(\Theta_0) = h$ (the m-dimensional vector) the hypothesis in question assumes the linear form $H\Delta\Theta = h$. This parametric hypothesis expresses not only the assumption on the numerical values of individual parameters but also on their linear combinations.

In order to verify this parametric hypothesis we will perform an experiment resulting in an n-dimensional vector $(x_1, x_2, \ldots, x_n)'$, which is a realization of the random vector $\xi = (\xi_1, \xi_2, \ldots, \xi_n)'$. The random vector ξ is supposed to have the normal probability distribution $N_n(Jf(\Theta_0) + JA\Delta\Theta, \sigma^2 I)$[1] (all the assertions are stated for Model 2 because of its universal position in linear estimation theory).

The test verifying the linear hypothesis on the parameters within linear models of measurement is based on the next assertion.

Let $\xi = (\xi_1, \xi_2, \ldots, \xi_n)' \sim N_n(Jf(\Theta_0) + JA\Delta\Theta, \sigma^2 I)$ and $\eta = (\eta_1, \eta_2, \ldots, \eta_n) \sim N_n(JA\Delta\Theta, \sigma^2 I)$, $\eta = \xi - Jf(\Theta_0)$ be observation vectors within Model 2 (see p. 133); $\Theta = \Theta_0 + \Delta\Theta$ is the unknown k-dimensional vector whose elements are treated by the hypothesis $H\Delta\Theta = h$. Then the random variable

$$\frac{(R_1^2 - R_0^2)/m}{R_0^2/(n-k)}, \tag{3.2.1}$$

where

$$R_0^2 = \min\{(\eta - JA\Delta\Theta)'(\eta - JA\Delta\Theta),\ \Delta\Theta \in \mathcal{R}^k\} = \eta'(I - JA(A'J'JA)^{-1}A'J')\eta \tag{3.2.2}$$

and

$$R_1^2 = \min\{(\eta - JA\Delta\Theta)'(\eta - JA\Delta\Theta),\ \Delta\Theta \in \mathcal{T}\} =$$

[1] All the assertions can be generalized for an arbitrary form of the covariance matrix $var(\xi) = var(\eta)$.

$$= \eta'\{I - JA(A'J'JA)^{-1}J'A' + JA(A'J'JA)^{-1}H'[H(A'J'JA)^{-1}H]^{-1}\cdot$$
$$\cdot H(A'J'JA)^{-1}A'J'\}\eta - 2h'[H(A'J'JA)^{-1}H']^{-1}\cdot$$
$$\cdot H(A'J'JA)^{-1}A'J'\eta + h'[H(A'J'JA)^{-1}H']^{-1}h, \tag{3.2.3}$$

where $\mathcal{T} = \{\Delta\Theta : H\Delta\Theta = h\} \subset \mathcal{R}^k$, $H\Delta\Theta = h$ is the relation between the components of the vector parameter $\Delta\Theta$ assumed by the null hypothesis, H is an $m \times k$-dimensional matrix such that $R(H) = m \leq k$ and h is an m-dimensional vector, possesses the noncentral Fisher - Snedecor probability distribution with m and $n-k$ degrees of freedom and with the noncentrality parameter

$$\delta = (H\Delta\Theta - h)'[H(A'J'JA)^{-1}H']^{-1}(H\Delta\Theta - h).$$

In order to understand the essence of this important test, we shall show the supporting steps for proving the assertion on which the test is based.

1. The random variable $R_0^2 = \min\{(\eta - JA\Delta\Theta)'(\eta - JA\Delta\Theta),\ \Delta\Theta \in \mathcal{R}^k\} = (\eta - JA\widehat{\Delta\Theta})'(\eta - JA\widehat{\Delta\Theta})$, where $\widehat{\Delta\Theta} = (A'J'JA)^{-1}A'J'\eta$ is the LS-estimator of the parameter $\Delta\Theta$, has, under the condition $\eta \sim N_n(JA\Delta\Theta, \sigma^2 I)$, the central σ^2 chi-square probability distribution with $n-k$ degrees of freedom.

2. The random variable

$$R_1^2 = \min\{(\eta - JA\Delta\Theta)'(\eta - JA\Delta\Theta),\ \Delta\Theta \in \{\Delta\Theta : H\Delta\Theta = h\}\} =$$
$$= (\eta - JA\widehat{\widehat{\Delta\Theta}})'(\eta - JA\widehat{\widehat{\Delta\Theta}}),$$

where

$$\widehat{\widehat{\Delta\Theta}} = \{(A'J'JA)^{-1} - (A'J'JA)^{-1}H'[H(A'J'JA)^{-1}H']^{-1}H(A'J'JA)^{-1}\}\cdot$$
$$\cdot A'J'\eta + (A'J'JA)^{-1}H'[H(A'J'JA)^{-1}H']^{-1}h$$

is the LS-estimator of the parameter $\Delta\Theta$ whose components are bound by the condition $H\Delta\Theta = h$, has, under the condition $\eta \sim N_n(JA\Delta\Theta, \sigma^2 I)$, the noncentral σ^2 chi-square probability distribution with $n-k+m$ degrees of freedom and with the noncentrality parameter

$$\delta = (H\Delta\Theta - h)'[H(A'J'JA)^{-1}H']^{-1}(H\Delta\Theta - h).$$

3. Let R_0^2 and R_1^2 be random variables defined in assertions 1 and 2, respectively. The random variable $R_1^2 - R_0^2$ possesses the noncentral σ^2 chi-square probability distribution with m degrees of freedom and with the noncentrality parameter $\delta = (H\Delta\Theta - h)'[H(A'J'JA)^{-1}H']^{-1}(H\Delta\Theta - h)$.

Let $R_1^2 - R_0^2$ be the random variable defined in the assertion 3 and R_0^2 the random variable from the assertion 1. The random variables $R_1^2 - R_0^2 \sim \sigma^2\chi^2(m, (H\Delta\Theta - h)'[H(A'J'JA)^{-1}H']^{-1}(H\Delta\Theta - h))$ and $R_0^2 \sim \sigma^2\chi^2(n - k)$ are stochastically independent, the random variable defined by (3.2.1) possesses the noncentral Fisher — Snedecor probability distribution with m and $n - k$ degrees of freedom and with the parameter of noncentrality

$$\delta = (H\Delta\Theta - h)'[H(A'J'JA)^{-1}H']^{-1}(H\Delta\Theta - h).$$

This fundamental assertion enables us to test the null hypothesis $H_0 : H\Delta\Theta = h$ against the alternative $H_1 : H\Delta\Theta \neq h$. If the test statistic $\frac{(R_1^2 - R_0^2)/m}{R_0^2/(n-k)}$ is realized by the value $\frac{(r_1^2 - r_0^2)/m}{r_0^2/(n-k)} > F_{m,(n-k)}(\alpha)$, where $F_{m,(n-k)}(\alpha)$ is the critical value of the central Fisher - Snedecor random variable with m and $n - k$ degrees of freedom at the significance level α, then the null hypothesis $H_0 : H\Delta\Theta = h$ is rejected with the risk α (if the assumption $H\Delta\Theta = h$ stated by the null hypothesis is fulfilled, the parameter of the noncentrality of the random variable R_1^2 and thereby also of the test random variable $[(R_1^2 - R_0^2)/m]/[R_0^2/(n - k)]$ equals zero).

Example a) The hypothesis on a value of the angle in a triangle. Let $x = (54.00g, 100.29g, 45.08g)'$ be a realization of the three-dimensional random vector $\xi \sim N_3(I\Theta, I)$ which is a result of a direct measurement of the angles Θ_1, Θ_2 and Θ_3 in a plane triangle.

The problem is to test the null hypothesis $H_0 : \Theta_2 = 100g$ against the alternative $H_1 : \Theta_2 \neq 100g$.

The realization of the test statistic can be determined in two ways, either by using the explicit relations for Model 2 (this requires us to transform Model 3 into Model 2) or by calculating the corresponding quadratic forms within Model 3.

1. **Transformation into Model 2** In accordance with the formulation of the problem, Model 3 is considered (a direct measurement with a condition). The condition binding the parameters Θ_1, Θ_2 and Θ_3 reads $(1,1,1)\Theta = 200g$, hence $B = (1,1,1)$, $b = -200g$; $\Theta = (\Theta_1, \Theta_2, \Theta_3)'$, $n = 3$, $q = 1$, $J = I$ (every angle was measured only once). An application of the transform relations from pp. 123–124 enables us to transform Model 3

into regular Model 2 with unknown parameters Θ_2 and Θ_3. Thus, in our case,

$$\Theta^{(1)} = \Theta_1,\ \Theta^{(2)} = (\Theta_2, \Theta_3)',\ B^{(1)} = 1,\ B^{(2)} = (1,1),$$

$$x_{T_3} = x + \begin{pmatrix} [B^{(1)}]^{-1}b \\ 0 \end{pmatrix} =$$

$$= \begin{pmatrix} 54.00 \\ 100.29 \\ 45.08 \end{pmatrix} + \begin{pmatrix} -200.00 \\ 0 \\ 0 \end{pmatrix} = \begin{pmatrix} -146.00 \\ 100.29 \\ 45.08 \end{pmatrix},$$

$$A_{T_3} = \begin{pmatrix} [B^{(1)}]^{-1}B^{(2)} \\ I \end{pmatrix} = \begin{pmatrix} -1, & -1 \\ 1, & 0 \\ 0, & 1 \end{pmatrix},\ A'_{T_3}A_{T_3} = \begin{pmatrix} 2, & 1 \\ 1, & 2 \end{pmatrix}$$

and

$$(A'_{T_3}A_{T_3})^{-1} = (1/3)\begin{pmatrix} 2, & -1 \\ -1, & 2 \end{pmatrix}.$$

Now we are prepared to calculate the realization of the test statistic $[(R_1^2 - R_0^2)/m]/[R_0^2/(n-k)]$ on the basis of the realization x of the random variable ξ

$$r_0^2 = x'_{T_3}[I - A_{T_3}(A'_{T_3}A_{T_3})^{-1}A'_{T_3}]x_{T_3} = \begin{pmatrix} -146.00 \\ 100.29 \\ 45.08 \end{pmatrix}' \left\{ \begin{pmatrix} 1, & 0, & 0 \\ 0, & 1, & 0 \\ 0, & 0, & 1 \end{pmatrix} - \right.$$

$$\left. - (1/3)\begin{pmatrix} -1, & -1 \\ 1, & 0 \\ 0, & 1 \end{pmatrix}\begin{pmatrix} 2, & -1 \\ -1, & 2 \end{pmatrix}\begin{pmatrix} -1, & 1, & 0 \\ -1, & 0, & 1 \end{pmatrix}\right\}\begin{pmatrix} -146.00 \\ 100.29 \\ 45.08 \end{pmatrix} =$$

$$= \begin{pmatrix} -146.00 \\ 100.29 \\ 45.08 \end{pmatrix}\begin{pmatrix} 1/3, & 1/3, & 1/3 \\ 1/3, & 1/3, & 1/3 \\ 1/3, & 1/3, & 1/3 \end{pmatrix}\begin{pmatrix} -146.00 \\ 100.29 \\ 45.08 \end{pmatrix} = 0.1323;$$

as $R(A_{T_3}) = k_{T_3} = 2$ the number of degrees of freedom of the random variable R_0^2 realized by the value $r_0^2 = 0.1323$ equals $n - k_{T_3} = 3 - 2 = 1$. Next r_1^2 has to be determined. The condition $\Theta_2 = 100.00g$ stated by the null hypothesis will be recorded as $(1,0)\Theta^{(2)} - 100 = 0$, which means that $H = (1,0)$, $h = 100$ and $m = 1$, therefore

$$H(A'_{T_3}A_{T_3})^{-1}H' = (1/3)(1,0)\begin{pmatrix} 2, & -1 \\ -1, & 2 \end{pmatrix}\begin{pmatrix} 1 \\ 0 \end{pmatrix} = 2/3,$$

$$[H(A'_{T3}A_{T3})^{-1}H']^{-1} = 3/2,$$

$$r_1^2 = x'_{T3}\{I - A_{T3}(A'_{T3}A_{T3})^{-1}A'_{T3} + A_{T3}(A'_{T3}A_{T3})^{-1}H'[H(A'_{T3}A_{T3})^{-1}H']^{-1}\cdot$$
$$\cdot H(A'_{T3}A_{T3})^{-1}A'_{T3}\}x_{T3} - 2h'[H(A'_{T3}A_{T3})^{-1}H']^{-1}H(A'_{T3}A_{T3})^{-1}A'_{T3}x_{T3} +$$
$$+h'[H(A'_{T3}A_{T3})^{-1}H']^{-1}h,$$

where

$$x'_{T3}[I - A_{T3}(A'_{T3}A_{T3})^{-1}A'_{T3}]x_{T3} = 0.1323,$$

$$x'_{T3}A_{T3}(A'_{T3}A_{T3})^{-1}H'[H(A'_{T3}A_{T3})^{-1}H']^{-1}H(A'_{T3}A_{T3})^{-1}A'_{T3}x_{T3} =$$

$$(1/6)\begin{pmatrix}-146.00\\100.29\\45.08\end{pmatrix}'\begin{pmatrix}-1, & -1\\1, & 0\\0, & 1\end{pmatrix}\begin{pmatrix}2, & -1\\-1, & 2\end{pmatrix}\begin{pmatrix}1\\0\end{pmatrix}\begin{pmatrix}1, & 0\end{pmatrix}\cdot$$

$$\cdot\begin{pmatrix}2, & -1\\-1, & 2\end{pmatrix}\begin{pmatrix}-1, & 1, & 0\\-1, & 0, & 1\end{pmatrix}\begin{pmatrix}-146.00\\100.29\\45.08\end{pmatrix} =$$

$$(1/6)\begin{pmatrix}-146.00\\100.29\\45.08\end{pmatrix}'\begin{pmatrix}1, & -2, & 1\\-2, & 4, & -2\\1, & -2, & 1\end{pmatrix}\begin{pmatrix}-146.00\\100.29\\45.08\end{pmatrix} = 15\ 150.3750,$$

$$2h'[H(A'_{T3}A_{T3})^{-1}H']^{-1}H(A'_{T3}A_{T3})^{-1}A'_{T3}x_{T3} =$$

$$= 100\begin{pmatrix}1, & 0\end{pmatrix}\begin{pmatrix}2, & -1\\-1, & 2\end{pmatrix}\begin{pmatrix}-1, & 1, & 0\\-1, & 0, & 1\end{pmatrix}\begin{pmatrix}-146.00\\100.29\\45.08\end{pmatrix} =$$

$$= 30\ 150.0000$$

and

$$h'[H(A'_{T3}A_{T3})^{-1}H']^{-1}h = (3/2)10\ 000.0000 = 15\ 000.0000,$$

so that

$$r_1^2 = 0.1323 + 15\ 150.3750 - 30\ 150.0000 + 15\ 000.0000 = 0.5073.$$

The realization of the test statistic $[(R_1^2 - R_0^2)/m]/[R_0^2/(n-k)]$ results in the value

$$\frac{(r_1^2 - r_0^2)/m}{r_0^2/(n - k_{T3})} = 2.834, \ m = 1, \ n - k_{T3} = 1.$$

2. Direct calculation using the quadratic forms The LS-estimator of the unknown vector parameter Θ within Model 3 is

$$\hat{\Theta} = -B'(BB')^{-1}b + [I - B'(BB')^{-1}B]x,$$

where in our case $B = (1,1,1)$, $b = -200$, $x = (54.00,\ 100.29,\ 45.08)'$, $n = 3$ and $q = 1$. As $BB' = 3$, we have $B'(BB')^{-1}b = -(1/3)(200,\ 200,\ 200)'$, hence

$$\begin{pmatrix} \hat{\Theta}_1 \\ \hat{\Theta}_2 \\ \hat{\Theta}_3 \end{pmatrix} = (1/3)\begin{pmatrix} 200 \\ 200 \\ 200 \end{pmatrix} + \left[\begin{pmatrix} 1, & 0, & 0 \\ 0, & 1, & 0 \\ 0, & 0, & 1 \end{pmatrix} - \right.$$

$$\left. - (1/3)\begin{pmatrix} 1, & 1, & 1 \\ 1, & 1, & 1 \\ 1, & 1, & 1 \end{pmatrix}\right]\begin{pmatrix} 54.00 \\ 100.29 \\ 45.08 \end{pmatrix} = \begin{pmatrix} 54.21 \\ 100.50 \\ 45.29 \end{pmatrix}$$

and $r_0^2 = (x - \hat{\Theta})'(x - \hat{\Theta}) = (0.21, 0.21, -0.21)(0.21, 0.21, -0.21)' = 0.1323$. The number of degrees of freedom of the random variable R_0^2 is $n - q - 1 = 3 - 1 - 1 = 1$.

The realization of the random variable R_1^2 is given by the relation $r_0^2 = (x - \hat{\hat{\Theta}})'(x - \hat{\hat{\Theta}})$, where $\hat{\hat{\Theta}}$ is the LS-estimator of the parameter $\Theta = (\Theta_1, \Theta_2, \Theta_3)'$ that, in addition to the condition $\Theta_1 + \Theta_2 + \Theta_3 = 200g$, fulfils the condition $\Theta_2 = 100g$ stated by the null hypothesis H_0. The LS-estimator of the parameter Θ within Model 3, when the condition is now of the form $G\Theta - g = 0$, where $G = \begin{pmatrix} 1, & 1, & 1 \\ 0, & 1, & 0 \end{pmatrix}$ and $g = (200, 100)'$, is

$$\hat{\hat{\Theta}} = -G'(GG')^{-1}g + [I - G'(GG')^{-1}G]x = (54.46, 100.00, 45.54)'$$

and

$$r_1^2 = (x - \hat{\hat{\Theta}})'(x - \hat{\hat{\Theta}}) =$$

$$= (0.46, -0.29, 0.46)(0.46, -0.29, 0.46)' = 0.5073.$$

The number of degrees of freedom of the random variable R_1^2 equals $n - q - 1 + m = 2$.

The test statistic $[(R_1^2 - R_0^2)/m]/[R_0^2/(n - q - 1)]$ is realized by the value 2.834 ($m = 1$ and $n - q - 1 = 1$).

Comparing this value with the critical values of the Fisher — Snedecor random variable with 1 and 1 degrees of freedom, which at the significance levels $\alpha_1 = 0.01$, $\alpha_2 = 0.025$ and $\alpha_3 = 0.05$ are $F_{1,1}(0.01) = 4\ 050$, $F_{1,1}(0.025) = 648$ and $F_{1,1}(0.05) = 161$, respectively, we see that there does not exist any reason for rejecting the null hypothesis $H_0 : \Theta_2 = 100g$ against the alternative $H_1 : \Theta_2 \neq 100g$.

b) The hypothesis on the identical slopes of two straight lines At the intercept on the t axis between the points 1 to 8 we measure the ordinate values of the first straight line and at its intercept between the points 9 to 15 we measure the ordinate values of the second straight line. The ordinate of both straight lines at the point $t = 8$ must be equal. The result of the measurement is the vector $x = (1.2, 1.2, 1.6, 1.9, 2.2, 2.2, 2.6, 2.5, 3.2, 3.6, 4.1, 4.5, 5.0, 5.5, 5.8)'$ which is a realization of the random vector $\xi = (\xi_1, \xi_2, \dots, \xi_{15})' \sim N_{15}(A\Theta, I)$, where $\Theta = (\Theta_1, \Theta_2, \Theta_3, \Theta_4)'$ (the parameters Θ_1 and Θ_2 are related to the first straight line, Θ_3 and Θ_4 are parameters of the second line), the matrix A is of the form

$$A = \begin{pmatrix} 1, & 1, & 1, & 1, & 1, & 1, & 1, & 1, & 0, & 0, & 0, & 0, & 0, & 0, & 0 \\ 1, & 2, & 3, & 4, & 5, & 6, & 7, & 8, & 0, & 0, & 0, & 0, & 0, & 0, & 0 \\ 0, & 0, & 0, & 0, & 0, & 0, & 0, & 0, & 1, & 1, & 1, & 1, & 1, & 1, & 1 \\ 0, & 0, & 0, & 0, & 0, & 0, & 0, & 0, & 9, & 10, & 11, & 12, & 13, & 14, & 15 \end{pmatrix}'$$

$J = I$. The condition has the form $(1, 8, -1, -8)\Theta = 0$, thus $B = (1, 8, -1, -8)$, $b = 0$, $q = 1$.

The problem is to test the hypothesis H_0 that both straight lines have the same slope, i.e., that in both intercepts of the t axis the same straight line occurs.

First, Model 5 (an indirect measurement with a condition) describing our situation must be transformed into the Model 2 (the parameter Θ_1 is determined by the condition $B\Theta = 0$, the unknown vector parameter to be determined is $(\Theta_2, \Theta_3, \Theta_4)'$)

$$B = (B^{(1)}, B^{(2)}),\ B^{(1)} = 1,\ B^{(2)} = (8, -1, -8),\ \Theta^{(1)} = \Theta_1,$$

$$\Theta^{(2)} = (\Theta_2, \Theta_3, \Theta_4)',\ A_{T_5} = A \begin{pmatrix} -[B^{(1)}]^{-1} B^{(2)} \\ I \end{pmatrix} =$$

$$= A \begin{pmatrix} -8, & 1, & 8 \\ 1, & 0, & 0, \\ 0, & 1, & 0 \\ 0, & 0, & 1 \end{pmatrix},$$

so that

$$A_{T_5} =$$

$$\begin{pmatrix} -7, & -6, & -5, & -4, & -3, & -2, & -1, & 0, & 0, & 0, & 0, & 0, & 0, & 0, & 0 \\ 1, & 1, & 1, & 1, & 1, & 1, & 1, & 1, & 1, & 1, & 1, & 1, & 1, & 1, & 1 \\ 8, & 8, & 8, & 8, & 8, & 8, & 8, & 8, & 9, & 10, & 11, & 12, & 13, & 14, & 15 \end{pmatrix}.$$

Transformation of the input data

$$x_{T_5} = x + A \begin{pmatrix} -[B^{(1)}]^{-1}b \\ 0 \end{pmatrix} = x$$

(realize that $b = 0$).

Calculation of the unknown parameters

$$\hat{\Theta}^{(2)} = (A'_{T_5} A_{T_5})^{-1} A'_{T_5} x = \begin{pmatrix} -5\,292, & -20\,384, & 1\,568 \\ -3\,976, & -10\,192, & 784 \\ -2\,660, & 0, & 0 \\ -1\,344, & 10\,192, & -784 \\ -28, & 20\,384, & -1\,568 \\ 1\,288, & 30\,576, & -2\,352 \\ 2\,604, & 40\,786, & -3\,136 \\ 3\,920, & 50\,960, & -3\,920 \\ 3\,136, & 36\,512, & -2\,604 \\ 2\,352, & 22\,064, & -1\,288 \\ 1\,568, & 7\,616, & 28 \\ 784, & -6\,832, & 1\,344 \\ 0, & -21\,280, & 2\,660 \\ -784, & -35\,728, & 3\,976 \\ -1\,568, & -50\,176, & 5\,292 \end{pmatrix}' \begin{pmatrix} 1.2 \\ 1.2 \\ 1.6 \\ 1.9 \\ 2.2 \\ 2.2 \\ 2.6 \\ 2.5 \\ 3.2 \\ 3.6 \\ 4.1 \\ 4.5 \\ 5.0 \\ 5.5 \\ 5.8 \end{pmatrix} \Rightarrow$$

$$\Rightarrow \hat{\Theta}_2 = 0.2211,\ \hat{\Theta}_3 = -0.9263,\ \hat{\Theta}_4 = 0.4539,$$

and

$$\hat{\Theta}_1 = -[B^{(1)}]^{-1}b - [B^{(1)}]^{-1}B^{(2)}\hat{\Theta}^{(2)} =$$

$$= (-8, 1, 8)(0.2211, -0.9263, 0.4539)' = 0.9368.$$

Calculation of the realization of the test statistic R_0^2 with its degrees of freedom

$$r_0^2 = (x - A\hat{\Theta})'(x - A\hat{\Theta}) = 0.140\,395,$$

and $R(A_{T_5}) = 3$, therefore the number of degrees of freedom of the random variable R_0^2 is $n - R(A_{T_5}) = 12$.

For determining the value of the estimate r_1^2 it is necessary to repeat the previous calculation with respect to the condition stated by the null hypothesis $H_0 : \Theta_2 = \Theta_4$. It implies $H = (1, 0, -1)$ and $h = 0$.

Calculation of the correcting term for satisfying the condition (see p. 144, the relation (2.4.16))

$$\delta\hat{\hat{\Theta}}^{(2)} = (A'_{T_5}A_{T_5})^{-1}H'[H(A'_{T_5}A_{T_5})^{-1}H']^{-1}h-$$

$$-(A'_{T_5}A_{T_5})^{-1}H'[H(A'_{T_5}A_{T_5})^{-1}H']^{-1}H(A'_{T_5}A_{T_5})^{-1}A_{T_5}x =$$

$$= (420.7448)^{-1}\begin{pmatrix} -144\ 060, & -1\ 690\ 304, & 144\ 060 \\ -99\ 960, & -1\ 172\ 864, & 99\ 960 \\ -55\ 860, & -655\ 424, & 55\ 860 \\ -11\ 760, & -137\ 984, & 11\ 760 \\ 32\ 340, & 379\ 456, & -32\ 340 \\ 76\ 440, & 896\ 896, & -76\ 440 \\ 120\ 540, & 1\ 414\ 336, & -120\ 540 \\ 164\ 640, & 1\ 931\ 776, & -164\ 640 \\ 120\ 540, & 1\ 414\ 336, & -120\ 540 \\ 76\ 440, & 896\ 896, & -76\ 440 \\ 32\ 340, & 379\ 456, & -32\ 340 \\ -11\ 760, & -137\ 984, & -11\ 760 \\ -55\ 860, & -655\ 424, & 55\ 860 \\ -99\ 960, & -1\ 172\ 864, & 99\ 960 \\ -144\ 060, & -1\ 690\ 304, & 144\ 060 \end{pmatrix}' \begin{pmatrix} 1.2 \\ 1.2 \\ 1.6 \\ 1.9 \\ 2.2 \\ 2.2 \\ 2.6 \\ 2.5 \\ 3.2 \\ 3.6 \\ 4.1 \\ 4.5 \\ 5.0 \\ 5.5 \\ 5.8 \end{pmatrix} =$$

$$= (0.1164, 1.3663, -0.1164)',$$

i.e., $\hat{\hat{\Theta}}_2 = 0.3375$, $\hat{\hat{\Theta}}_3 = 0.4400$, $\hat{\hat{\Theta}}_4 = 0.3375 \Rightarrow \hat{\hat{\Theta}}_1 = (-8, 1, 8)\hat{\hat{\Theta}} = 0.4400$.

Calculation of the realization of the test statistic R_1^2

$$r_1^2 = (x - A\hat{\hat{\Theta}})'(x - \hat{\hat{\Theta}}) = 1.102\ 250.$$

Calculation of the realization of the test statistic $[(R_1^2 - R_0^2)/m]/[R_0^2(n - k_{T_5})]$ and its number of degrees of freedom

$$\frac{(r_1^2 - r_0^2)/m}{r_0^2/(n - k_{T_5})} = \frac{0.961\ 855}{0.140\ 395/12} = 82.213,$$

since $m = 1$, $n = 15$ and $k_{T_5} = R(A_{T_5}) = 3$.

Comparing the value obtained with the critical values of the Fisher—Snedecor random variable with 1 and 12 degrees of freedom, which on the significance levels $\alpha_1 = 0.01$, $\alpha_2 = 0.025$, and $\alpha_3 = 0.05$ are sequentially $F_{1,12}(0.01) = 9.33$, $F_{1,12}(0.025) = 6.55$ and $F_{1,12}(0.05) = 4.75$, we see that already at the significance level 0.01 the hypothesis has to be rejected.

3.3 Tests on stochastic relations

In connection with stuThe random vardies of the stochastic relations between two random variables, the most frequent problem consists in judging whether there exists or does not exist a stochastic relation between them, if the existing relation can be characterized by a correlation coefficient of a certain value, or whether two different probability distributions may be characterized by the same value of the correlation coefficient.

The tests for testing hypotheses of this nature are based on the next three assertions that are a consequence of the fundamental theorem on the probability distribution of the sample correlation coefficient, see p. 103.

Let $((\xi_1,\eta_1)',(\xi_2,\eta_2)'\dots,(\xi_n,\eta_n)')'$, $n \geq 3$ be a random sample from the normal probability distribution $N_2\left((m_\xi,m_\eta)',\Sigma_0 = \begin{pmatrix} \sigma_\xi^2, & 0 \\ 0, & \sigma_\eta^2 \end{pmatrix}\right)$, $\sigma_\xi^2 > 0$, $\sigma_\eta^2 > 0$. Then the random variable

$$T = \frac{\hat{\rho}_{\xi,\eta}}{\sqrt{1-\hat{\rho}^2_{\xi,\eta}}}\sqrt{n-2}, \tag{3.3.1}$$

where $\hat{\rho}_{\xi,\eta}$ is the sample correlation coefficient defined by the relation (1.6.10), has the Student's t probability distribution with $n-2$ degrees of freedom ([1], p. 117).

This assertion enables us to test the null hypothesis that the correlation coefficient of a two-dimensional normal probability distribution equals zero, i.e., $H_0 : \rho_{\xi,\eta} = 0$ against the alternative $H_1 : \rho_{\xi,\eta} \neq 0$. If $|t| \geq t_{crit}(n-2,\alpha)$, where α is the given risk and t a realization of the test statistic T, the null hypothesis is rejected at the significance level α.

Let $((\xi_1,\eta_1)',(\xi_2,\eta_2)',\dots,(\xi_n,\eta_n)')'$, $n \geq 3$ be a random sample from the normal probability distribution $N_2\left((m_\xi,m_\eta)',\ \Sigma_0 = \begin{pmatrix} \sigma_\xi^2, & 0 \\ 0, & \sigma_\eta^2 \end{pmatrix}\right)$, $\sigma_\xi^2 > 0$, $\sigma_\eta^2 > 0$. Then the random variable

$$\zeta = \hat{\rho}^2_{\xi,\eta}, \tag{3.3.2}$$

where $\hat{\rho}^2_{\xi,\eta}$ is the sample correlation coefficient defined by the relation (1.6.10), has the beta probability distribution with the parameters $p = 1/2$ and $q = (n-2)/2$, which means that its probability density is of the form

$$f(z;n) = \begin{cases} \frac{z^{-1/2}(1-z)^{(n-4)/2}}{B(1/2,(n-2)/2)}, & z \in (0,1), \\ 0, & z \notin (0,1). \end{cases}$$

This assertion, similar to the previous one, enables us to test the null hypothesis, that the correlation coefficient $\rho_{\xi,\eta}$ of a two-dimensional normal probability distribution equals 0, against the alternative $H_1 : \rho_{\xi,\eta} \neq 0$.

Let $((\xi_1,\eta_1)', (\xi_2,\eta_2)', \ldots, (\xi_n,\eta_n)')'$, where n is a sufficiently large number, be a random sample from the two-dimensional normal probability distribution $N_2\left((m_\xi, m_\eta)',\ \Sigma = \begin{pmatrix} \sigma^2_\xi, & \rho_{\xi,\eta}\sigma_\xi\sigma_\eta \\ \rho_{\xi,\eta}\sigma_\xi\sigma_\eta, & \sigma^2_\eta \end{pmatrix}\right)$, $\sigma^2_\xi > 0$, $\sigma^2_\eta > 0$, $\rho_{\xi,\eta} \in (-1,1)$. Then the random variable

$$Z = \frac{1}{2}\ln\frac{1+\hat{\rho}_{\xi,\eta}}{1-\hat{\rho}_{\xi,\eta}}, \tag{3.3.3}$$

where $\hat{\rho}_{\xi,\eta}$ is the sample correlation coefficient defined by the relation (1.6.10), has approximately the normal probability distribution $N\left(\frac{1}{2}\ln\frac{1+\rho_{\xi,\eta}}{1-\rho_{\xi,\eta}} + \frac{\rho_{\xi,\eta}}{2(n-1)},\ \frac{1}{n-3}\right)$ (the approximation is reliable for $n \geq 10$).

If $\rho_{\xi,\eta} = 0$, the random variable Z defined by the relation (3.3.3) has approximately the normal probability distribution $N(0,\ 1/(n-3))$. In this case the test statistic

$$U = \sqrt{n-3}Z = \frac{\sqrt{n-3}}{2}\ln\frac{1+\hat{\rho}_{\xi,\eta}}{1-\hat{\rho}_{\xi,\eta}},$$

that under the mentioned assumptions possesses approximately the normal probability distribution $N(0,1)$, is more convenient (see [1], p. 118).

The assertion on the approximately normal distribution of Z from (3.3.3) forms the basis for testing the null hypothesis $H_0 : \rho_{\xi,\eta} = \rho_0$ against the alternative $H_1 : \rho_{\xi,\eta} \neq \rho_0$.

The test statistic Z defined by the relation (3.3.3) enables us to test the hypothesis that two normal probability distributions also possess the same correlation coefficient. Let $((\xi_{1,1}, \eta_{1,1})', (\xi_{2,1}, \eta_{2,1})', \ldots, (\xi_{n_1,1}, \eta_{n_1,1})')'$ be a random sample from the two-dimensional normal probability distribution $N_2(m_1,\ \Sigma_1)$ and $((\xi_{1,2}, \eta_{1,2})', (\xi_{2,2}, \eta_{2,2})', \ldots, (\xi_{n_2,2}, \eta_{n_2,2})')'$ be a random sample from the two-dimensional normal probability distribution $N_2(m_2,\ \Sigma_2)$, where $m_i = (m_{\xi_i}, m_{\eta_i})'$

$$\Sigma_i = \begin{pmatrix} \sigma^2_{\xi_i}, & \rho_i\sigma_{\xi_i}\sigma_{\eta_i} \\ \rho_i\sigma_{\xi_i}\sigma_{\eta_i}, & \sigma^2_{\eta_i} \end{pmatrix},\ \sigma^2_{\xi_i} > 0,\ \sigma^2_{\eta_i} > 0,$$

$$\rho_i \in (0,1),\ i = 1,2,\ \rho_1 = \rho_2.$$

Then the random variable

$$Z_1 - Z_2 = \frac{1}{2}\left(\ln\frac{1+\hat\rho_1}{1-\hat\rho_1} - \ln\frac{1+\hat\rho_2}{1-\hat\rho_2}\right),$$

where $\hat\rho_1$ and $\hat\rho_2$ are sample correlation coefficients corresponding to the first and the second sample, respectively, possesses approximately the normal probability distribution $N(0,\ [1/(n_1-3)] + [1/(n_2-3)])$ and the random variable

$$U = \frac{Z_1 - Z_2}{\left(\frac{1}{n_1-3} + \frac{1}{n_2-3}\right)^{1/2}}$$

has approximately the normal probability distribution $N(0,1)$ (see [1], p. 118).

Example The experiment resulted in 92 pairs of numbers that in the x-axis component occur on the interval [0.1, 1.6] with the step 0.1 and in the y-axis component on the interval [8.8, 10.2] with the step 0.2. For the frequencies of individual pairs see Table 3.4

Table 3.4

$x \backslash y$	8.8	9.0	9.2	9.4	9.6	9.8	10.0	10.2
1.0			1	1		1		
1.1	1	2	2	2	2	1		
1.2	2	4	4	5	4	1	1	1
1.3	1	3	6	7	3	2	3	2
1.4		1	3	3	2	3	3	1
1.5		1	1	1	2	3	2	
1.6				1	1	1	1	

The problem is, on the basis of the realization of the random sample from the two-dimensional normal probability distribution whose results are described by the table of frequencies, to test the null hypothesis $H_0 : \rho_{\xi,\eta} = 0$ against the alternative $H_1 : \rho_{\xi,\eta} \neq 0$.

Using the relation

$$r_{\xi,\eta} = \frac{\sum_{i=1}^{n} x_i y_i - n\overline{x}\overline{y}}{\left[\left(\sum_{i=1}^{n} x_i^2 - n\overline{x}^2\right)\left(\sum_{i=1}^{n} y_i^2 - n\overline{y}^2\right)\right]^{1/2}}$$

we calculate the realization of the sample correlation coefficient $\hat{\rho}_{\xi,\eta}$:

$$\sum_{i=1}^{92} x_i = 119.3, \ \overline{x} = 1.296\ 74, \ n\overline{x}^2 = 154.70,$$

$$\sum_{i=1}^{92} y_i = 871.4, \ \overline{y} = 9.471\ 74, \ n\overline{y}^2 = 8\ 253.67,$$

$$\sum_{i=1}^{92} x_i y_i = 1\ 131.36, \ \sum_{i=1}^{92} x_i^2 = 156.51,$$

$$\sum_{i=1}^{92} y_i^2 = 8\ 265.72, \ n\overline{x}\overline{y} = 1\ 129.98,$$

hence

$$r_{\xi,\eta} = \frac{1\ 131.36 - 1\ 129.98}{[(156.51 - 154.70)(8\ 265.72 - 8\ 253.67)]^{1/2}} = 0.295$$

and the test statistic T is realized in the value

$$t = \frac{r_{\xi,\eta}}{(1 - r_{\xi,\eta})^{1/2}}\sqrt{n-2} = 2.929.$$

The random variable T has the Student's t probability distribution with 90 degrees of freedom in our case. Its critical values at the significance levels $\alpha_1 = 0.01$ and $\alpha_2 = 0.05$ are $t_{crit}(90;\ 0.01) = 2.632$ and $t_{crit}(90;\ 0.05) = 1.987$, respectively. The value realized in the test statistic is larger than the critical values, therefore the null hypothesis $H_0 : \rho_{\xi,\eta} = 0$ against the alternative $H_1 : \rho_{\xi,\eta} \neq 0$ is rejected at both significance levels.

The test statistic U realized by the value

$$u = \frac{\sqrt{n-3}}{2} \ln \frac{1 + r_{\xi,\eta}}{1 - r_{\xi\eta}} = 2.868.$$

The random variable U possesses asymptotically the normal probability distribution $N(0,1)$. The value realized for the test statistic lies inside the critical region $(-\infty, -2.576] \cup [2.576, \infty)$ of the normal probability distribution $N(0,1)$ at the significance level $\alpha = 0.01$ (it obviously lies inside the critical region $(-\infty, -1.960] \cup [1.960, \infty)$ of this random variable, which correponds to the significance level $\alpha = 0.05$).

The test statistics T and U depend on the size of the random sample on the basis of which the sample correlation coefficient is estimated. The importance of its value when judging the significance of the value realized for the sample correlation coefficient is illustrated by the next example.

Example Let $r_{\xi,\eta} = 0.295$. The null hypothesis $H_0 : \rho_{\xi\eta} = 0$ is tested against the alternative $H_1 : \rho_{\xi,\eta} \neq 0$ for the sample sizes $n = 5$, $n = 10$, $n = 20$, $n = 50$ and $n = 100$. The test statistics are the random variables T and U. Their realizations, together with the critical values and the critical regions at the significance levels $\alpha = 0.01$ and $\alpha = 0.05$, are in Table 3.5

Table 3.5

size of the sample n	$t = \frac{r_{\xi,\eta}\sqrt{n-2}}{\sqrt{1-r^2_{\xi,\eta}}}$	$t_{crit}(n-2:0.01)$ $t_{crit}(n-2;0.05)$	$u = \frac{\sqrt{n-3}}{2} \ln \frac{1+r_{\xi,\eta}}{1-r_{\xi,\eta}}$
5	0.535	5.841 3.182	
10	0.873	3.355 2.306	0.804
20	1.315	2.878 2.101	1.254
50	2.139	2.682 2.011	2.084
100	3.056	2.627 1.984	2.994

If the sizes of the sample equal $n = 5$, $n = 10$ and $n = 20$, there is no reason to reject the null hypothesis $H_0 : \rho_{\xi,\eta} = 0$ against the alternative $H_1 : \rho_{\xi,\eta} \neq 0$ on the basis of the realized values of the test statistics T and U (the value of the test statistic U for $n = 5$ was not calculated, as it possesses the asymptotic normal probability distribution). This means that the correlation coefficient realized by the value 0.295 in these cases cannot be considered to be essentially different from zero. Only when $n = 50$ at the significance level $\alpha = 0.05$ the null hypothesis $H_0 : \rho_{\xi,\eta} = 0$ may be rejected and the correlation coefficient realized by the mentioned value 0.295 can be considered to differ from zero.

Next some extremum cases will be analysed, illustrating clearly that for a rigorous interpretation of the magnitude of the correlation coefficient the size of the random sample, on the basis of which its sample value was determined, has to be known.

Let $n = 10$ and the realization of the empirical correlation coefficient acquire the value $r_{\xi,\eta} = 0.6$. The null hypothesis $H_0 : \rho_{\xi,\eta} = 0$ against the alternative $H_1 : \rho_{\xi,\eta} \neq 0$ is tested. The test statistic T is realized in the value $t = 2.121$. The critical value of the Student's t random variable with $n - 2 = 8$ degrees of freedom at the significance level $\alpha = 0.05$ is $t_{crit}(8; 0.05) = 2.306$, which means that there exists no reason to reject the null hypothesis and the sample correlation coefficient realized from the sample with the size $n = 10$ by a relatively high value 0.6 cannot be considered to be significantly different from zero (self-evidently, with the risk $\alpha = 0.05$). The same conclusion can be obtained using the test statistic U. It is realized by the value $u = 1.834$ and this value lies outside the critical region at the significance level $\alpha_1 = 0,05$, and more so at the significance level $\alpha_2 = 0.01$.

On the other hand, let the realized value of the sample correlation coefficient be $r_{\xi,\eta} = 0.1$ and let the size of the random sample used to determine it be $n = 700$. If we now test the null hypothesis $H_0 : \rho_{\xi,\eta} = 0$ against the alternative $H_1 : \rho_{\xi,\eta} \neq 0$, the test statistic T is realized by the value $t = 2.655$ and the test statistic U by the value $u = 2.645$. As $t = 2.655 > 2.576 = t_{crit}(698,\ 0.05)$ and similarly $u = 2.645 \in (-\infty, -2.576) \cup (2.576, \infty)$, the null hypothesis $H_0 : \rho_{\xi,\eta} = 0$ against the alternative has to be rejected at the significance level $\alpha = 0.05$; we may state that the empirical correlation coefficient realized by the value 0.1 from the sample of the size $n = 700$ significantly differs from zero.

For the sake of completeness, let us determine the confidence region of the sample correlation coefficient estimated from the random sample $((\xi_1, \eta_1)', (\xi_2, \eta_2)', \ldots, (\xi_n, \eta_n)')'$ from the two-dimensional probability distribution $N_2\left((m_\xi, m_\eta)', \begin{pmatrix} \sigma_\xi^2, & \rho_{\xi,\eta}\sigma_\xi\sigma_\eta \\ \rho_{\xi,\eta}\sigma_\xi\sigma_\eta, & \sigma_\eta^2 \end{pmatrix}\right)$, $\sigma_\xi^2 > 0$, $\sigma_\eta^2 > 0$, $\rho_{\xi,\eta} \in (-1, 1)$ at the significance level $1 - \alpha$.

The value of the correlation coefficient is a realization of the random variable (1.6.10) expressing the sample correlation coefficient. Due to the definition of the confidence region (p. 213), the problem is to determine the random variables $\hat{\rho}_{1\alpha}$ and $\hat{\rho}_{2\alpha}$ such that

$$P\{\omega : \hat{\rho}_{1\alpha}(\omega) \leq \rho_{\xi,\eta} \leq \hat{\rho}_{2\alpha}(\omega)\} = 1 - \alpha, \ \alpha \in (0, 1).$$

The solution of the problem using the exact probability distribution of the sample correlation coefficient $\hat{\rho}_{\xi,\eta}$ whose probability density is given by (1.6.15), would be very time-consuming. An essentially simpler and for the formulated purpose sufficiently accurate way of determining the bounds of the confidence interval is based on the fact that under the given assumptions concerning the random sample $((\xi_1, \eta_1)', (\xi_2, \eta_2)', \ldots, (\xi_n, \eta_n)')'$ the random variable

$$Z = \frac{1}{2} \ln \frac{1 + \hat{\rho}_{\xi,\eta}}{1 - \hat{\rho}_{\xi,\eta}} \tag{3.3.3}$$

has approximately the normal probability distribution $N((1/2)\ln[(1+\rho_{\xi,\eta})/(1 - \rho_{\eta\xi})],\ 1/(n-3))$.

The function $z(\rho_{\xi,\eta}) = (1/2)\ln[(1+\rho_{\xi,\eta})/(1-\rho_{\xi,\eta})]$, $\rho_{\xi,\eta} \in (-1,1)$ is strictly increasing, therefore

$$P\{\omega : \hat{\rho}_{1\alpha}(\omega) \leq \rho_{\xi,\eta} \leq \hat{\rho}_{2\alpha}(\omega)\} =$$

$$= P\left\{\omega : \frac{1}{2}\ln\frac{1+\hat{\rho}_{1\alpha}(\omega)}{1-\hat{\rho}_{1\alpha}(\omega)} \leq \frac{1}{2}\ln\frac{1+\rho_{\xi,\eta}}{1-\rho_{\xi,\eta}} \leq \frac{1}{2}\ln\frac{1+\hat{\rho}_{2\alpha}(\omega)}{1-\hat{\rho}_{2\alpha}(\omega)}\right\} =$$

$$= P\{\omega : Z_{1\alpha}(\omega) \leq z(\rho_{\xi,\eta}) \leq Z_{2\alpha}(\omega)\}.$$

The random variable Z has approximately the normal probability distribution $N\Big((1/2)\ln[(1+\rho_{\xi,\eta})/(1-\rho_{\xi,\eta})] = z(\rho_{\xi,\eta}),\ 1/(n-3)\Big)$, which means that the bounds Z_1 and Z_2 of the confidence interval may be determined using the same procedure that was applied for determining the bounds limits of the confidence interval of the mean value estimated from an n-dimensional sample by means of a normal probability distribution under the assumption of an a priori known dispersion (see p. 215).

Thus, let $Z_1 = Z - \delta_\alpha$, $Z_2 = Z + \delta_\alpha$, where Z is given by (3.3.3), and $\delta_\alpha = (n-3)^{-1/2} F_{0,1}^{-1}(1-\alpha/2)$, where $F_{0,1}^{-1}(1-\alpha/2)$ is the $(1-\alpha/2)$-quantile of the normal probability distribution $N(0,1)$.

At the given significance level $1-\alpha$

$$P\left\{\omega : \frac{1}{2}\ln\frac{1+\hat{\rho}_{\xi,\eta}(\omega)}{1-\hat{\rho}_{\xi,\eta}(\omega)} - \frac{F_{0,1}^{-1}(1-\alpha/2)}{\sqrt{n-3}} \leq \frac{1}{2}\ln\frac{1+\rho_{\xi,\eta}}{1-\rho_{\xi,\eta}} \leq \right.$$

$$\left. \leq \frac{1}{2}\ln\frac{1+\hat{\rho}_{\xi,\eta}(\omega)}{1-\hat{\rho}_{\xi,\eta}(\omega)} + \frac{F_{0,1}^{-1}(1-\alpha/2)}{\sqrt{n-3}}\right\} = 1-\alpha. \tag{3.3.4}$$

The mapping $\rho_{\xi,\eta} = (e^{2z} - 1)/(e^{2z} + 1)$, $z \in (-\infty, \infty)$ (the inverse mapping of the mapping $z = (1/2)\ln[(1 + \rho_{\xi,\eta})/(1 - \rho_{\xi,\eta})]$)) is also a strictly increasing function, therefore it can be applied to the individual elements in the parentheses of the relation (3.3.4) without affecting their ordering.

Since

$$\frac{1}{2}\ln\frac{1+\hat{\rho}_{\xi,\eta}}{1-\hat{\rho}_{\xi,\eta}} - \frac{F_{0,1}^{-1}(1-\alpha/2)}{\sqrt{n-3}} = \frac{1}{2}\ln\frac{1+\hat{\rho}_{\xi,\eta}}{1-\hat{\rho}_{\xi,\eta}}\exp\left(-\frac{2F_{0,1}^{-1}(1-\alpha/2)}{\sqrt{n-3}}\right),$$

the lower bound $\hat{\rho}_{1\alpha}(\cdot)$, $\omega \in \Omega$, of the confidence interval of the correlation coefficient is

$$\hat{\rho}_{1\alpha} = \frac{\exp\left\{\ln\left[\frac{1+\hat{\rho}_{\xi,\eta}}{1-\hat{\rho}_{\xi,\eta}}\exp\left(-\frac{2F_{0,1}^{-1}(1-\alpha/2)}{\sqrt{n-3}}\right)\right]\right\} - 1}{\exp\left\{\ln\left[\frac{1+\hat{\rho}_{\xi,\eta}}{1-\hat{\rho}_{\xi,\eta}}\exp\left(-\frac{2F_{0,1}^{-1}(1-\alpha/2)}{\sqrt{n-3}}\right)\right]\right\} + 1} =$$

$$= \frac{\frac{1+\hat{\rho}_{\xi,\eta}}{1-\hat{\rho}_{\xi,\eta}}\exp\left(-\frac{2F_{0,1}^{-1}(1-\alpha/2)}{\sqrt{n-3}}\right) - 1}{\frac{1+\hat{\rho}_{\xi,\eta}}{1-\hat{\rho}_{\xi,\eta}}\exp\left(-\frac{2F_{0,1}^{-1}(1-\alpha/2)}{\sqrt{n-3}}\right) + 1} =$$

$$= \frac{(1+\hat{\rho}_{\xi,\eta})\exp\left(-\frac{2F_{0,1}^{-1}(1-\alpha/2)}{\sqrt{n-3}}\right) - 1 + \hat{\rho}_{\xi,\eta}}{(1+\hat{\rho}_{\xi,\eta})\exp\left(-\frac{2F_{0,1}^{-1}(1-\alpha/2)}{\sqrt{n-3}}\right) + 1 - \hat{\rho}_{\xi,\eta}}$$

and analogously

$$\hat{\rho}_{2\alpha} = \frac{(1+\hat{\rho}_{\xi,\eta})\exp\left(\frac{2F_{0,1}^{-1}(1-\alpha/2)}{\sqrt{n-3}}\right) - 1 + \hat{\rho}_{\xi,\eta}}{(1+\hat{\rho}_{\xi,\eta})\exp\left(\frac{2F_{0,1}^{-1}(1-\alpha/2)}{\sqrt{n-3}}\right) + 1 - \hat{\rho}_{\xi,\eta}}.$$

Example Let the random samples with the sizes $n = 10$, $n = 20$, $n = 50$ and $n = 100$ be taken into account. Let the empirical correlation coefficient in all these cases be $r_{\xi,\eta} = 0.295$. The problem is to determine the confidence regions of the correlation coefficient for the mentioned sizes n at the confidence regions $1 - \alpha_1 = 0.95$ and $1 - \alpha_2 = 0.99$.

Using (3.3.4) we obtain

$$z_{1\alpha} = \frac{1}{2}\ln\frac{1+r_{\xi\eta}}{1-r_{\xi\eta}} - \frac{F_{0,1}^{-1}(1-\alpha/2)}{\sqrt{n-3}},$$

$$z_{1\alpha} = \frac{1}{2}\ln\frac{1+r_{\xi\eta}}{1-r_{\xi\eta}} + \frac{F_{0,1}^{-1}(1-\alpha/2)}{\sqrt{n-3}}$$

and by means of them the realizations $r_{1\alpha}$ and $r_{2\alpha}$ of the random bounds $\hat{\rho}_{1\alpha}$ and $\hat{\rho}_{2\alpha}$ of the confidence region

$$r_{1\alpha} = \frac{e^{2z_{1\alpha}}-1}{e^{2z_{1\alpha}}+1}, \quad r_{2\alpha} = \frac{e^{2z_{2\alpha}}-1}{e^{2z_{2\alpha}}+1}.$$

The results of the calculation at the significance levels 0.95 and 0.99 for the given sizes of the random sample are contained in Table 3.6.

Table 3.6

Size of the sample	Realization of the confidence interval for	
n	$1-\alpha=0.95$	$1-\alpha=0.99$
10	[−0.411,+0.780]	[−0.585,+0.856]
20	[−0.170,+0.652]	[−0.310,+0.730]
50	[+0.018,+0.530]	[−0.072,+0.591]
100	[+0.105,+0.465]	[+0.042,+0.512]

We can observe a close connection between the realized confidence interval of the correlation coefficient and the result of testing the null hypothesis $H_0 : \rho_{\xi,\eta} = 0$ against the alternative $H_1 : \rho_{\xi,\eta} \neq 0$, given on p. 306. In both cases the same input data were used for judging the significance of the correlation coefficient.

The next assertion has the fundamental consequence for evaluating the statistical significance of the coefficient of the multiple correlation estimated from the random sample of size n from a certain $(p+1)$-dimensional normal probability distribution, i.e., for testing the null hypothesis $H_0 : \rho_{\eta,\xi} = 0$ against the alternative $H_1 : \rho_{\eta,\xi} \neq 0$.

Let $((\eta_1, \xi_1')', (\eta_2, \xi_2')' \ldots, (\eta_n, \xi_n')')$ be a random sample from a $(p+1)$-dimensional normal probability distribution
$N_{p+1}\left((m_\eta, m_\xi')', \begin{pmatrix} \sigma_\eta^2, & 0 \\ 0, & \Sigma_\xi \end{pmatrix}\right)$, $\sigma_\eta^2 > 0$ and Σ_ξ is a positive definite matrix (i.e., $\rho_{\eta,\xi} = 0$), $n > p+1$. Then the random variable

$$Z = \frac{n-p-1}{p}\frac{\hat{\rho}_{\eta,\xi}^2}{1-\hat{\rho}_{\eta,\xi}^2},$$

where $\hat{\rho}_{\eta,\xi}$ is the sample coefficient of the multiple correlation defined by the relation (1.6.12), possesses the Fisher - Snedecor probability distribution with p and $n-p-1$ degrees of freedom ([1], p. 125).

Example The problem is to test the null hypothesis $H_0 : \rho_{\eta,(\zeta,\xi)'} = 0$ against the alternative $H_1 : \rho_{\eta,(\zeta,\xi)'} \neq 0$; at our disposal is the random sample of size 7 from the three-dimensional normal probability distribution whose results are in Table 3.7.

Table 3.7

i	1	2	3	4	5	6	7	$\sum_{i=1}^{7}$
y	4	3	4	1	6	4	5	27
z	4	2	4	1	5	3	4	23
x	10	8	12	3	15	12	13	73

In this case $n = 7$, $p = 2$. The relation (1.6.12) defining the coefficient of the multiple correlation in our case when $p = 2$ reads

$$\hat{\rho}^2_{\eta,(\zeta,\xi)'} = \hat{\Sigma}_{\eta,(\zeta,\xi)'}\hat{\Sigma}^{-1}_{(\zeta,\xi)'}\hat{\Sigma}_{(\zeta,\xi)',\eta}/\hat{\sigma}^2_{\eta} =$$

$$= (\hat{\rho}_{\eta,\zeta}, \hat{\rho}_{\eta,\xi}) \begin{pmatrix} 1, & \hat{\rho}_{\zeta,\xi} \\ \hat{\rho}_{\zeta,\xi}, & 1 \end{pmatrix}^{-1} \begin{pmatrix} \hat{\rho}_{\eta,\zeta} \\ \hat{\rho}_{\eta,\xi} \end{pmatrix} = \frac{\hat{\rho}^2_{\eta,\zeta} - 2\hat{\rho}_{\eta,\zeta}\hat{\rho}_{\eta,\xi}\hat{\rho}_{\zeta,\xi} + \hat{\rho}^2_{\eta,\xi}}{1 - \hat{\rho}^2_{\eta,\xi}}$$

(the relations between the operators cov and cor, see p. 71, were used here). As $\overline{y} = 3.857$, $\overline{z} = 3.286$, $\overline{x} = 10.429, \sum_{i=1}^{7} y_i^2 = 119$, $\sum_{i=1}^{7} z_i^2 = 87$ and $\sum_{i=1}^{7} x_1^2 = 855$ we have the following estimates of the correlation coefficients: $r_{\eta,\zeta} = 0.942\,84$, $r_{\eta,\xi} = 0.976\,27$ and $r_{\zeta,\xi} = 0.921\,05$, which means that the square of the empirical coefficient of the multiple correlation is realized by the value $r^2_{\eta,(\zeta,\xi)} = 0.965\,66$. The corresponding realization of the test statistic Z is $z = 56.241$. The critical values of the Fisher - Snedecor random variable with 2 and 4 degrees of freedom at the significance levels $\alpha_1 = 0.01$ and $\alpha_2 = 0.05$ are $F_{2,4}(0.01) = 18$ and $F_{2,4}(0.05) = 6.96$. As $z = 56.241 > 18 = F_{2,4}(0.01)$, we reject the null hypothesis $H_0 : \rho_{\eta,(\zeta,\xi)} = 0$ against the alternative $H_1 : \rho_{\eta,(\zeta,\xi)} \neq 0$ with the risk 0.01 and thus also with the risk 0.05.

For judging the statistical significance of the coefficient of the partial correlation estimated from a random sample of size n from a certain $(p+2)$-dimensional normal probability distribution $(p+2, n)$, i.e., for testing the null hypothesis $H_0 : \rho_{\eta,\zeta\cdot\xi} = 0$ against the alternative $H_1 : \rho_{\eta,\zeta\cdot\xi} \neq 0$ the next assertion is of essential importance.

Let $((\eta_1, \zeta_1, \xi_1')', (\eta_2, \zeta_2, \xi_2')', \ldots, (\eta_n, \zeta_n, \xi_n')')'$ be a random sample from the $p+2$-dimensional normal probability distribution

$$N_{p+2}\left((m_\eta, m_\zeta, m_\xi')', \begin{pmatrix} \sigma^2_\eta, & \rho_{\eta,\zeta}\sigma_\eta\sigma_\zeta, & \Sigma_{\eta\xi} \\ \rho_{\eta,\zeta}\sigma_\eta\sigma_\zeta, & \sigma^2_\zeta, & \Sigma_{\zeta\xi} \\ \Sigma_{\xi\eta}, & \Sigma_{\xi\zeta}, & \Sigma_\xi \end{pmatrix}\right),$$

$\sigma_\eta^2 > 0$, $\sigma_\zeta^2 > 0$ and Σ_ξ is a positive definite matrix. If $\rho_{\eta,\zeta\cdot\xi} = 0$, the random variable

$$T = \frac{\hat{\rho}_{\eta,\zeta\cdot\xi}}{\sqrt{1-\hat{\rho}^2_{\eta,\zeta\cdot\xi}}}\sqrt{n-p-2},$$

where $\hat{\rho}_{\eta,\zeta\cdot\xi}$ is the empirical coefficient of the partial correlation defined by the relation (1.6.14), has the Student's t probability distribution with $n-p-2$ degrees of freedom ([1], p. 128).

Example The problem is to test the null hypothesis $H_0 : \rho_{\eta,\zeta\cdot\xi} = 0$ against the alternative $H_1 : \rho_{\eta,\zeta\cdot\xi} \neq 0$ on the basis of a random sample of the size 7 from a three-dimensional normal probability distribution mentioned in the previous example. In our case of a one-dimensional random variable ξ the empirical correlation coefficient between the random variables η and ζ when the influence of the random variable ξ is eliminated is of the form

$$\hat{\rho}_{\eta,\zeta\cdot\xi} = \frac{\hat{\rho}_{\eta,\zeta} - \hat{\rho}_{\eta,\xi}\hat{\rho}_{\zeta,\xi}}{\sqrt{(1-\hat{\rho}^2_{\eta,\xi})(1-\hat{\rho}^2_{\zeta,\xi})}}$$

(compare this with the relation (1.3.12)). Using the estimates of the correlation coefficients $r_{\eta,\zeta} = 0.949\,82$, $r_{\eta,\xi} = 0.976\,27$ and $r_{\zeta,\xi} = 0.921\,05$ from the preceding example we obtain $r_{\eta,\zeta\cdot\xi} = (r_{\eta,\zeta} - r_{\eta,\xi}r_{\zeta,\xi})/[(1-r^2_{\eta,\xi})(1-r^2_{\zeta,\xi})]^{1/2} = 0.517\,6$ and $t = r_{\eta,\zeta\cdot\xi}(1-r^2_{\eta,\zeta\cdot\xi})^{1/2}(n-p-2)^{1/2} = 1.208$. If we compare this value with the critical values of the Student's t random variable with 4 degrees of freedom, which at the significance levels $\alpha_1 = 0.05$ and $\alpha_2 = 0.01$ are $t_{crit}(4; 0.05) = 2.776$ and $t_{crit}(4; 0.01) = 4.604$, we can state that at both significance levels the test statistic is realized outside the critical region. This is the reason why the null hypothesis $H_0 : \rho_{\eta,\zeta\cdot\xi} = 0$, tested against the alternative $H_1 : \rho_{\eta,\zeta\cdot\xi} \neq 0$, is not rejected. There does not exist a statistically significant relation between the random variables η and ζ when the influence of the random variable ξ is eliminated.

3.4 Tests of goodness of fit

Tests of goodness of fit are nonparametric tests by means of which one judges the degree of accordance of the empirical distribution function $S_n(\cdot)$ (obtained from a random sample of size n) with the distribution function $F(\cdot)$ assumed by the null hypothesis.

3.4.1 The Pearson test of goodness of fit (known parameters of the probability distribution)

The Pearson goodness of fit test is based on the following assertion (the Pearson theorem):

Let the k-dimensional random vector $\nu = (\nu_1, \nu_2, \ldots, \nu_k)'$ possess a multinomial probability distribution[2]. The random variable

$$\zeta_k = \sum_{i=1}^{k} \frac{(\nu_i - np_i)^2}{np_i} \tag{3.4.1}$$

possesses for $n \to \infty$ asymptotically the chi-square probability distribution with $k-1$ degrees of freedom, hence its distribution function is

$$\lim_{n \to \infty} F_n(z) = \begin{cases} \frac{2^{-(k-1)/2}}{\Gamma((k-1)/2)} \int_0^z u^{[(k-1)/2]-1} \exp(-u/2) du, \; z > 0, \\ 0, \; z \leq 0, \end{cases}$$

in other words, the random variable ζ_k defined by the relation (3.4.1) possesses for a sufficiently large n approximately the chi-square probability distribution; [1], p. 194.

The Pearson theorem enables us to test the null hypothesis that the realization $(x_1, x_2, \ldots, x_n)'$ of the random sample $(\xi_1, \xi_2, \ldots, \xi_n)'$ comes from the probability distribution characterized by the known distribution function $F_0(\cdot)$ including the parameter Θ that specifies it; thus the Pearson theorem enables us to test the null hypothesis $H_0 : F(\cdot, \Theta) = F_0(\cdot, \Theta)$ against the alternative $H_1 : F(\cdot, \Theta) \neq F_0(\cdot, \Theta)$, where Θ is known.

---

[2] A k-dimensional random vector $\nu = (\nu_1, \nu_2, \ldots, \nu_k)'$ possesses the multinomial probability distribution if

$$P\{\omega : \nu_1(\omega) = n_1, \nu_2(\omega) = n_2, \ldots, \nu_k(\omega) = n_k\} = \frac{n!}{n_1! n_2! \ldots n_k!} p_1^{n_1} p_2^{n_2} \ldots p_k^{n_k},$$

where $n_1, n_2, \ldots, n_k$ are integers such that $n = n_1 + n_2 + \ldots + n_k$ and $p_1, p_2, \ldots, p_k$ are numbers belonging to the interval (0,1) such that $p_1 + p_2 + \ldots + p_k = 1$.

The following simple example shows how this probability distribution arises: Let us have an urn with marbles of k different colours; the probability p_i of drawing out a marble of the concrete colour is known a priori. If we performe n independent draws (the independence of the draws is ensured by returning the drawn marble to the urn), the probability that the marble of colour 1 is drawn n_1 times, the marble of colour 2 is drawn n_2 times, ... and the marble of colour k is drawn n_k times is determined by the given expression.

The variation region $[a, b)$ of the random variable ξ from which the random sample $(\xi_1, \xi_2, \ldots, \xi_n)'$ comes (as far as the bound points a and b of the variation region are concerned, $F_0(a, \Theta) = P\{\omega : \xi(\omega) < a\} = 0$ and $F_0(b, \Theta) = P\{\omega : \xi(\omega) < b\} = 1$) is divided by the points $a_1, a_2, \ldots, a_k$ into k disjoint intervals $[a_{i-1}, a_i)$, $i = 1, 2, \ldots, k$ in such a way that $a_0 = a$ and $a_k = b$.

The distribution function $F_0(\cdot, \Theta)$ assumed by the null hypothesis determines the theoretical frequencies $np_i(\Theta) = nP\{\omega : \xi(\omega) \in [a_{i-1}, a_i)\} = n[F_0(a_i, \Theta) - F_0(a_{i-1}, \Theta)]$, $i = 1, 2, \ldots, k$, when, self-evidently, $P\{\omega : \xi(\omega) \in [a, b)\} = \sum_{i=1}^{n} P\{\omega : \xi(\omega) \in [a_{i-1}, a_i)\} = 1$.

The realization $(x_1, x_2, \ldots, x_n)'$ of the random sample considered determines uniquely the empirical frequency n_i which corresponds to the theoretical frequency $np_i(\Theta)$; n_i is the number of realizations of the random variable ξ falling inside the interval $[a_{i-1}, a_i)$. It has to be realized that n_i is a realization of a certain random variable ν; for another realization of the same random sample from the same probability distribution the number of the realizations of the random variable falling inside the interval $[a_{i-1}, a_i)$ will differ. The k-dimensional random vector $\nu = (\nu_1, \nu_2, \ldots, \nu_k)'$ is a typical example of a vector having the multidimensional probability distribution, therefore the random variable

$$\zeta_k(\Theta) = \sum_{i=1}^{k} \frac{[\nu_i - np_i(\Theta)]^2}{np_i(\Theta)} \tag{3.4.2}$$

has, due to the Pearson theorem, for a sufficiently large size n of the random sample approximately the chi-square probability distribution with $k - 1$ degrees of freedom; as a consequence of this, the random variable ζ_k represents a suitable test statistic for testing the null hypothesis $H_0 : F(\cdot, \Theta) = F_0(\cdot, \Theta)$ against the alternative $H_1 : F(\cdot, \Theta) \neq F_0(\cdot, \Theta)$ for the known Θ.

Example The problem is to test the null hypothesis that the die by which Meissner carried out his known experiment (1,800 tosses, see p. 20) is homogeneous. The result of the experiment is contained in the two first columns of Table 3.8.

Table 3.8

Number of tosses i	Empirical frequency n_i	Theoretical frequency np_i	$(n_i - np_i)^2/np_i$
1	299	300	0.003 33
2	295	300	0.083 33
3	303	300	0.030 00
4	307	300	0.163 33
5	289	300	0.403 33
6	307	300	0.163 33
$\sum$	1 800	1 800	0.846 65

The toss of a die is a typical example of a discrete random variable that is uniquely described by the probabilities $p_i = P\{\omega : \xi(\omega) = i\}$, $i = 1, 2, \ldots, 6$. Under the condition that the null hypothesis is true, these probabilities are $p_1 = p_2 = \ldots = p_6$. The theoretical frequency calculated on the basis of these probabilities is in the third column. All the data necessary for an application of the Pearson test are at our disposal. The test statistics ζ_6 defined by the relation (3.4.2) was realized by the value $z_6 = 0.846\ 65$. Using the tables of the chi-square probability distribution with 5 degrees of freedom, we obtain $P\{\omega : \zeta_6(\omega) \geq 0.846\ 65\} \doteq 0.97$, or $\chi^2_{crit}(5; 0.05) = 11.070$ and $\chi^2_{crit}(5; 0.01) = 15.086$, which implies that we have no reason for rejecting the null hypothesis on the homogeneity of the die considered at the significance levels we have mentioned.

3.4.2 The Pearson test of goodness of fit (unknown parameters of the probability distribution)

Often the null hypothesis states only the type of the distribution function and the vector parameter specifying it has to be estimated from the realization $(x_1, x_2, \ldots, x_n)'$ of the random sample $(\xi_1, \xi_2, \ldots, \xi_n)'$ by means of which we test the formulated null hypothesis against the alternative. The theoretical distribution function $F(\cdot, \hat{\Theta})$ with an estimated parameter Θ generates the theoretical frequencies $p_i(\hat{\Theta})$ that are also functions of the estimator $\hat{\Theta}$ of the parameter Θ. If the quantity $p_i(\Theta)$ in the relation (3.4.2) is substituted by the quantity $p_i(\hat{\Theta})$, the random variable

$$\zeta_k(\hat{\Theta}) = \sum_{i=1}^{k} \frac{[\nu_i - np_i(\hat{\Theta})]^2}{np_i(\hat{\Theta})}, \tag{3.4.3}$$

which is the function of the estimator of the unknown parameter Θ, appears. Self-evidently, the fact that the theoretical value of the parameter Θ in the definition of the test statistic is substituted by its estimator influences the properties of this modified test statistic (they depend, naturally, also on the concrete method used for estimating the parameter Θ).

If the principle of minimizing the chi-square is applied to estimating the vector parameter Θ, the modified Pearson theorem also has to be used. This theorem forms the basis of the Pearson test of goodness fit if the parameters characterizing the fundamental probability distribution are not known a priori.

The method of the minimum chi-square, which is applied to estimate the unknown parameters, is an analogy of the method of the least squares. The value $\hat{\Theta}$ minimizing for every realization $(n_1, n_2, \ldots, n_k)'$ of the random vector $(\nu_1, \nu_2, \ldots, \nu_k)'$ the expression

$$\sum_{i=1}^{k} \frac{[n_i - np_i(\hat{\Theta})]^2}{np_i(\hat{\Theta})}$$

is considered to be the best estimator of the m-dimensional vector parameter Θ. It means that the optimum (in the sense of the minimum chi-square) estimator $\hat{\Theta}$ of the m-dimensional vector parameter Θ is a solution of the system of m equations

$$\sum_{i=1}^{k} \left[\frac{2[n_i - np_i(\hat{\Theta})]}{p_i(\hat{\Theta})} + \frac{[n_i - np_i(\hat{\Theta})]^2}{np_i^2(\hat{\Theta})} \right] \frac{\partial p_i(\hat{\Theta})}{\partial \Theta_j} = 0,$$
$$j = 1, 2, \ldots, m \tag{3.4.4}$$

(the functions $p_i(\cdot)$, $i = 1, 2, \ldots, k$ are supposed to be sufficiently regulated functions of the parameter Θ, which implies that the partial derivatives in the relation (3.4.4) exist).

The difficulties connected with solving the system of equations in (3.4.4) are caused by the second quadratic term. Its influence on the result of the solution of this system of equations is not very important. Neglecting it, we obtain a simplified system of equations

$$\sum_{i=1}^{k} \frac{n_i}{p_i(\hat{\Theta})} \frac{\partial p_i(\hat{\Theta})}{\partial \Theta_j} = 0, \; j = 1, 2, \ldots, m; \tag{3.4.5}$$

it is necessary to realize that $\sum_{i=1}^{k} \frac{np_i(\hat{\Theta})}{p_i(\hat{\Theta})} \frac{\partial p_i(\hat{\Theta})}{\partial \Theta_j} = n \sum_{i=1}^{k} \frac{\partial p_i(\hat{\Theta})}{\partial \Theta_j} = 0$, as the relation $\sum_{i=1}^{k} p_i(\hat{\Theta}) = 1$ implies that $\sum_{i=1}^{k} \frac{\partial p_i(\hat{\Theta})}{\partial \hat{\Theta}_j} = 0$ for each $j = 1, 2, \ldots, m$.

The estimator, which is a solution of the simplified system of equations (3.4.5) is denoted as the estimator obtained by the modified method of the minimum chi-square.

Now we are prepared to formulate the modified Pearson theorem:

Let $(\nu_1, \nu_2, \ldots, \nu_k)'$ be a k-dimensional random vector with the multinomial probability distribution whose parameters $p_1, p_2, \ldots, p_k$ depend on a priori unknown parameters $\Theta_1, \Theta_2, \ldots, \Theta_m$ $(m < k - 1)$. The random variable

$$\zeta_k(\hat{\Theta}) = \sum_{i=1}^{k} \frac{[\nu_i - np_i(\hat{\Theta})]^2}{np_i(\hat{\Theta})},$$

where $\hat{\Theta}$ is the estimator of the unknown parameter Θ using the modified method of the minimum chi-square, possesses for $n \to \infty$ asymptotically the chi-square probability distribution with $k - m - 1$ degrees of freedom.

3.4.3 The Pearson test of goodness of fit for verifying the normality with unknown m and σ^2

The problem is to test the null hypothesis that the realization $(x_1, x_2, \ldots, x_n)'$ of the random sample $(\xi_1, \xi_2, \ldots, \xi_n)'$ comes from the normal probability distribution whose parameters m and σ^2 are not known a priori.

The solution of the problem using the modified Pearson theorem requires that we estimate the unknown parameters m and σ^2 of the normal probability distribution from the realization $(x_1, x_2, \ldots, x_n)'$ of the random sample $(\xi_1, \xi_2, \ldots, \xi_n)'$ using the modified method of the minimum chi-square.

The variation region of a normal random variable is the interval $(-\infty, \infty)$; by use of the points b_1, $b_2, \ldots,$ b_{k-1} we divide it into k disjoint intervals $(-\infty, b_1)$, $[b_1, b_2), \ldots,$ $[b_{k-2}, b_{k-1})$, $[b_{k-1}, \infty)$. Then

$$p_i(m, \sigma^2) = \frac{1}{\sigma\sqrt{2\pi}} \int_{b_{i-1}}^{b_i} \exp\left(-\frac{(x-m)^2}{2\sigma^2}\right) dx,$$

$$\frac{\partial p(m, \sigma^2)}{\partial m} = \frac{1}{\sigma^3\sqrt{2\pi}} \int_{b_{i-1}}^{b_i} (x - m) \exp\left(-\frac{(x-m)^2}{2\sigma^2}\right) dx,$$

$$\frac{\partial p_i(m,\sigma^2)}{\partial \sigma} = -\frac{p_i(m,\sigma^2)}{\sigma} + \frac{1}{\sigma^4\sqrt{2\pi}} \int_{b_{i-1}}^{b_i} (x-m)^2 \exp\left(-\frac{(x-m)^2}{2\sigma^2}\right) dx,$$

$i = 0, 1, 2, \ldots, k-1, k$.

Inserting the partial derivatives into the relation (3.4.5) we obtain the system of two equations for determining the estimators of m and σ^2 using the modified method of the minimum chi-square:

$$\sum_{i=1}^{k} \frac{n_i}{p_i(\hat{m},\hat{\sigma}^2)} \frac{1}{\hat{\sigma}^3\sqrt{2\pi}} \int_{b_{i-1}}^{b_i} (x-\hat{m}) \exp\left(-\frac{(x-\hat{m})^2}{2\hat{\sigma}^2}\right) dx = 0,$$

$$\sum_{i=1}^{k} \frac{n_i}{p_i(\hat{m},\hat{\sigma}^2)} \left[-\frac{p_i(\hat{m},\hat{\sigma}^2)}{\hat{\sigma}} + \right.$$

$$\left. + \frac{1}{\hat{\sigma}^4\sqrt{2\pi}} \int_{b_{i-1}}^{b_i} (x-\hat{m})^2 \exp\left(-\frac{(x-\hat{m})^2}{2\hat{\sigma}^2}\right) dx \right] = 0;$$

n_i is the empirical frequency, the number of realizations of the random variable belonging to the interval $[b_{i-1}, b_i)$.

In consequence of the fact that

$$\sum_{i=1}^{k} \frac{n_1}{p_i(\hat{m},\hat{\sigma}^2)} \frac{\hat{m}}{\hat{\sigma}^3\sqrt{2\pi}} \int_{b_{i-1}}^{b_i} \exp\left(-\frac{(x-\hat{m})^2}{2\hat{\sigma}^2}\right) dx = n\frac{\hat{m}}{\hat{\sigma}^2},$$

the first equation of the previous system yields

$$\hat{m} = \frac{1}{n} \sum_{i=1}^{k} \frac{n_i}{p_i(\hat{m},\hat{\sigma}^2)} \frac{1}{\hat{\sigma}\sqrt{2\pi}} \int_{b_{i-1}}^{b_i} x \exp\left(-\frac{(x-\hat{m})^2}{2\hat{\sigma}^2}\right) dx \tag{3.4.6}$$

and the second

$$\hat{\sigma}^2 = \frac{1}{n} \sum_{i=1}^{k} \frac{n_i}{p_i(\hat{m},\hat{\sigma}^2)} \frac{1}{\hat{\sigma}\sqrt{2\pi}} \int_{b_{i-1}}^{b_i} (x-\hat{m})^2 \exp\left(-\frac{(x-\hat{m})^2}{2\hat{\sigma}^2}\right) dx. \tag{3.4.7}$$

The system of equations (3.4.6) and (2.4.7) is solved by iteration. The zero approximation is chosen in the form

$$\hat{m}_0 = (1/n) \sum_{i=1}^{k} n_i c_i, \ \hat{\sigma}^2 = (1/n) \sum_{i=1}^{k} n_i c_i^2 - \hat{m}_0^2,$$

where $c_1 = b_1-(b_2-b_1)/2$, $c_2 = (b_1+b_2)/2, \ldots, c_{k-1} = (b_{k-2}+b_{k-1})/2$, $c_k = b_{k-1} + (b_{k-1} - b_{k-2})/2$.

The test statistic

$$\zeta_k(\hat{m}, \hat{\sigma}^2) = \sum_{i=1}^{k} \frac{[\nu_i - np_i(\hat{m}, \hat{\sigma}^2)]^2}{np_i(\hat{m}, \hat{\sigma}^2)},$$

where the theoretical frequencies are calculated using the estimates of the parameters m and σ^2 obtained by solving the equation systems (3.4.6) and (3.4.7), has asymptotically the chi-square probability distribution with $k-3$ degrees of freedom.

With respect to the asymptoticity of the probability distribution of the test statistic, the Pearson test of goodness of fit for verifying the normality with a priori unknown parameters m and σ^2 can be used if the size of the random sample is larger than 100 and simultaneously $np_i \gg 5$, $i = 1, 2, \ldots, k$.

3.4.4 The Kolmogorov test of goodness of fit

The Kolmogorov test of goodness of fit is based on the Kolmogorov theorem. This theorem enables us to test the hypothesis that an n-dimensional random vector is a sample from a probability distribution characterized by a certain concrete continuous distribution function $F_0(\cdot)$ belonging to the class $\mathcal{F}$ of all continuous distribution functions, i.e., to test the null hypothesis $H_0 : F(\cdot) = F_0(\cdot)$ against the alternative $H_1 : F(\cdot) \neq F_0(\cdot)$, $F(\cdot), F_0(\cdot) \in \mathcal{F}$.

Theorem of Kolmogorov Let $S_n(\cdot)$ be the empirical distribution function of the random sample from the probability distribution characterized by a continuous distribution function $F(\cdot)$. The random variable $\sqrt{n}D_n$, where

$$D_n = \sup\{|S_n(x) - F(x)|,\ x \in \mathcal{R}^1\}, \tag{3.4.8}$$

possesses asymptotically the probability distribution characterized by the distribution function

$$Q(y) = \lim_{n\to\infty} Q_n(y) =$$

$$= \begin{cases} 0, \ y \leq 0 \\ \lim_{n\to\infty} P\{\omega : \sqrt{n} D_n(\omega) < y\} = \lim_{n\to\infty} P\left\{\omega : D_n(\omega) < \frac{y}{\sqrt{n}}\right\} = \\ = \sum_{-\infty}^{\infty} (-1)^k \exp(-2k^2 y^2), \ y > 0 \end{cases} \tag{3.4.9}$$

([12], [16], p. 330).

In accordance with this theorem the test statistic

$$D_n = \sup\{|S_n(x) - F_0(x)|, \ x \in \mathcal{R}^1\}$$

serves for testing the null hypothesis $H_0 : F(\cdot) = F_0(\cdot)$ against the alternative $H_1 : F(\cdot) \neq F_0(\cdot)$, $F(\cdot)$, $F_0(\cdot) \in \mathcal{F}$, on the basis of a realization $(x_1, x_2, \ldots, x_n)'$ of a random sample $(\xi_1, \xi_2, \ldots, \xi_n)'$; the empirical distribution function $S_n(\cdot)$ is random, it changes in dependence on the concrete realization of the random sample $(\xi_1, \xi_2, \ldots, \xi_n)'$.

The asymptotical distribution function of the test statistic mentioned is tabulated (see, e.g., [9], p. 241).

Finally, we must make a very important comment. If the continuous distribution function $F_0(\cdot)$ stated by the null hypothesis depends on a priori unknown parameters of the probability distribution, a modification analogous to that carried out in the case of the Pearson test of goodness of fit is not admissible, even if there exists the possibility to estimate the unknown parameters from the realization of the random sample that is at our diposal for testing the null hypothesis.

Example The problem is to test the hypothesis that the realization (0.692, 0.259, 0.463, 0.986, 0.084, 0.466, 0.884, 0.219, 0.992, 0.108, 0.680, 0.792, 0.468, 0.328, 0.565, 0.544)′ of the random sample $(\xi_1, \xi_2, \ldots, \xi_{16})'$ comes from the rectangular probability distribution characterized by the probability density

$$f(x) = \begin{cases} 1, \ x \in (0,1), \\ 0, \ x \notin (0,1). \end{cases}$$

Thus the distribution function assumed by the null hypothesis is

$$F_0(x) = \begin{cases} 0, \ x \leq 0, \\ x, \ 0 < x < 1, \\ 1, \ x \geq 1. \end{cases}$$

The results of calculating the values required by the Kolmogorov test of goodness of fit are in Table 3.9 (the symbols used there accord with the definition of the empirical distribution function, see (1.6.3)).

Table 3.9

i	$x_{(i)}$	$S_{16}(x_{(i)})$	$F_0(x_{(i)})$	$S_{16}(x_{(i)}) - F_0(x_{(i)})$
1	0.084	0.0000	0.084	−0.0840
2	0.108	0.0625	0.108	−0.0455
3	0.219	0.1250	0.219	−0.0940
4	0.259	0.1875	0.259	−0.0715
5	0.328	0.2500	0.328	−0.0780
6	0.463	0.3125	0.463	−0.1505
7	0.466	0.3750	0.466	−0.0910
8	0.468	0.4375	0.468	−0.0305
9	0.554	0.5000	0.544	−0.0440
10	0.565	0.5625	0.565	−0.0025
11	0.680	0.6250	0.680	−0.0550
12	0.692	0.6875	0.692	−0.0045
13	0.792	0.7500	0.792	−0.0420
14	0.884	0.8125	0.884	−0.0715
15	0.986	0.8750	0.986	−0.1110
16	0.992	0.9375	0.992	−0.0545

The test statistic of the form (3.4.9) was realized by the value $d_{16} = \sup\{|S_{16}(x_{(i)}) - F_0(x_{(i)})|,\ x_{(i)} \in [0,1],\ i = 1,2,\ldots,16\} = |S_{16}(x_{(6)}) - F_0(x_{(6)})| = 0.1505$. The critical values of the test statistic D_{16} at the significance levels $\alpha_1 = 0.05$ and $\alpha_2 = 0.01$ are $D_{16crit}(0.05) = 0.327\,33$ and $D_{16crit}(0.01) = 0.392\,01$, which implies that there does not exist any reason for rejecting the null hypothesis at the significance levels considered.

3.4.5 The Rényi test of goodness of fit

The Rényi test of goodness of fit solves the same problem as the Kolmogorov test of goodness of fit. It is based on the following **Rényi theorem** ([31], [6], p. 339):

Let $S_n(\cdot)$ be the empirical distribution function belonging to the random sample of size n from a probability distribution characterized by a continuous distribution function $F(\cdot)$. The random variable

$$\sqrt{n}\sup\left\{\frac{S_n(x) - F(x)}{F(x)},\ x : F(x) \geq a\ \&\ F(a) > 0\right\}$$

possesses asymptotically the probability distribution characterized by the distribution function

$$\lim_{n\to\infty} P\left\{\omega : \sqrt{n}\sup\left\{\frac{S_n(x,\omega)-F(x)}{F(x)}, x : F(x) \geq a \ \&\ F(a) > 0\right\} < y\right\} =$$

$$= \begin{cases} 0, \ y \leq 0, \\ \sqrt{\frac{2}{\pi}} \int_0^{y\sqrt{a/(1-a)}} \exp(-z^2/2)dz, \ y > 0 \end{cases} \tag{3.4.10}$$

and the random variable

$$\sqrt{n}\sup\left\{\left|\frac{S_n(x)-F(x)}{F(x)}\right|, \ x : F(x) \geq a \ \&\ F(a) > 0\right\}$$

possesses asymptotically the probability distribution whose distribution function is of the form

$$\lim_{n\to\infty} P\left\{\omega : \sqrt{n}\sup\left\{\left|\frac{S_n(x,\omega)-F(x)}{F(x)}\right|, x : F(x) \geq a \ \&\ F(a) > 0\right\} < y\right\} =$$

$$= \begin{cases} 0, \ y \leq 0, \\ \frac{4}{\pi}\sum_{j=0}^{\infty}(-1)^j \frac{\exp\{-[\pi(2j+1)^2](1-a)/8ay^2\}}{2j+1}, \ y > 0. \end{cases} \tag{3.4.11}$$

Using the Rényi theorem the null hypothesis $H_0 : F(\cdot) = F_0(\cdot)$ against the alternative $H_1 : F(\cdot) \neq F_0(\cdot)$, where $F(\cdot)$ and $F_0(\cdot)$ are distribution functions belonging to the class of all continuous distribution functions, can be tested. The test statistic is chosen either in the form

$$\sqrt{n}\sup\left\{\frac{S_n(x)-F_0(x)}{F_0(x)}, \ x : F_0(x) \geq a \ \&\ F_0(a) > 0\right\}, \tag{3.4.12}$$

or in the form

$$\sqrt{n}\sup\left\{\left|\frac{S_n(x)-F_0(x)}{F_0(x)}\right|, \ x : F_0(x) \geq a \ \&\ F_0(a) > 0\right\}, \tag{3.4.13}$$

where $F_0(\cdot)$ is the continuous distribution function assumed by the null hypothesis, $S_n(\cdot)$ the empirical distribution function generated by the n-dimensional random sample and the positive constant a is determined by the above mentioned conditions. The Rényi test of goodness of fit in contradistinction to the Kolmogorov test reacts more sensitively to the differences between the distribution function $F_0(\cdot)$ assumed by the null hypothesis and the empirical distribution function $S_n(\cdot)$ for small values of $F_0(\cdot)$.

The asymptotic distribution functions (3.4.10) and (3.4.11) of the test statistics (3.4.12) and (3.4.13) are tabulated, see, e.g., [9], pp. 242 and 243.

3.5 Tests of normality

There exist several tests for testing the hypothesis on the normality of a basic population based on an n-dimensional random sample from the population. The choice of a specific test depends on the actual problem being solved (the null hypothesis states that the probability distribution is normal in the general sense, or it states that it is a normal distribution that has specific parameters which characterises it), based on the a priori information about the basic population whose normality is being tested and on the size of the random sample which serves for testing. Obviously, all tests of goodness of fit from Section 3.4 can be applied for testing normality.

3.5.1 Empirical coefficient of asymmetry and empirical excess as test statistics for verifying normality

The empirical coefficient of asymmetry and empirical excess can be applied as test statistics for testing the null hypothesis that the realization $(x_1, x_2, \ldots, x_n)'$ of the random sample $(\xi_1, \xi_2, \ldots, \xi_n)'$ comes from a population with a normal probability distribution (the parameters m and σ^2 are not specified). As a consequence of this assertion:

Let $(\xi_1, \xi_2, \ldots, \xi_n)'$ be the random sample from a normal probability distribution. The empirical coefficient of asymmetry defined by the relation

$$\hat{\gamma}_1 = \left((1/n)\sum_{i=1}^{n}(\xi_i - \overline{\xi})^3\right) \Big/ \left((1/n)\sum_{i=1}^{n}(\xi_i - \overline{\xi})^2\right)^{3/2},$$

$$\bar{\xi} = (1/n)\sum_{i=1}^{n} \xi_i$$

(see the expression (1.6.8)) has approximately the normal probability distribution $N(0, 6/n)$ and the empirical excess defined by the relation

$$\hat{\gamma}_2 = \left[\left((1/n)\sum_{i=1}^{n}(\xi_i - \bar{\xi})^4\right) \Big/ \left((1/n)\sum_{i=1}^{n}(\xi_i - \bar{\xi})^2\right)^2\right] - 3$$

(see the expression (1.6.9)) approximately the normal probability distribution $N(0, 24/n)$ ([3], p.86).

Example The problem is to test the null hypothesis that the basic population possesses a normal probability if the realization $(0.5, 7.5, 4.5, 2.2, 1.1, 1.7, 0.7, 1.5, 2.8, 3.2)'$ of the random sample $(\xi_1, \xi_2, \ldots, \xi_{10})'$ is available. The test statistics used for this purpose are the empirical coefficient of asymmetry and the empirical excess, as the random variables $\hat{\gamma}_1$ and $\hat{\gamma}_2$, subject to the condition that the null hypothesis is true, have approximately the normal probability distribution, namely $\hat{\gamma}_1^{approx} \sim N(0, 0.6)$ and $\hat{\gamma}_2^{approx} \sim N(0, 2.4)$. Hence, the null hypothesis $H_0 : E[\hat{\gamma}_1] = 0$ is tested against the alternative $H_1 : E[\hat{\gamma}_1] \neq 0$ and the null hypothesis $H_0 : E[\hat{\gamma}_2] = 0$ against the alternative $H_1 : E[\hat{\gamma}_2] \neq 0$.

The test statistic realized by the value

$$g_1 = \left((1/10)\sum_{i=1}^{10}(x_i - \bar{x})^3\right) \Big/ \left((1/10)\sum_{i=1}^{10}(x_i - \bar{x})^2\right)^{3/2} =$$

$$= 10.675\ 476/8.138\ 698 = 1.311\ 693.$$

Because of

$$P\{\omega : -g_1 < \hat{\gamma}_1(\omega) < g_1\} =$$

$$= P\left\{\omega : -\frac{g_1}{\sqrt{D[\hat{\gamma}_1]}} < \frac{\hat{\gamma}_1(\omega)}{\sqrt{D[\hat{\gamma}_1]}} < \frac{g_1}{\sqrt{D[\hat{\gamma}_1]}}\right\} =$$

$$= P\{\omega : -1.6934 < \xi_{tab}(\omega) < 1.6934\} = 0.910,$$

$P\{\omega : \hat{\gamma}_1(\omega) > g_1\} + P\{\omega : \hat{\gamma}_1(\omega) < -g_1\} = 0.09$ and, therefore, there exists no reason for rejecting the hypothesis on the normality of the population, neither at the significance level $\alpha_1 = 0.01$ nor at $\alpha_2 = 0.05$. Nevertheless, it is necessary to state that the risk of rejecting the true hypothesis is relatively small, the realized value 1.311 693 of the test statistic comes near to the critical value 1.518 at the significance level $\alpha_2 = 0.05$. The critical region of the applied $\hat{\gamma}_1$ test of normality at the significance levels $\alpha_1 = 0.01$ and $\alpha_2 = 0.05$ are $(-\infty, -1.995] \cup [1.995, \infty)$ and $(-\infty, -1.518] \cup [1.518, \infty)$, respectively.

The test statistic $\hat{\gamma}_2$ realized by the value

$$g_2 = \left[\left(\sum_{i=1}^{1} 0 (x_i - \overline{x})^4 / 10 \right) \Big/ \left(\sum_{i=1}^{10} (x_i - \overline{x})^2 / 10 \right)^2 \right] - 3 =$$

$$= 64.192398 / 16.370925 - 3 = 0.92112.$$

Since

$$P\{\omega : -g_2 < \hat{\gamma}_2(\omega) < g_2\} =$$

$$= P\left\{ \omega : -\frac{g_2}{\sqrt{D[\hat{\gamma}]}} < \frac{\hat{\gamma}_2(\omega)}{\sqrt{D[\hat{\gamma}_2]}} < \frac{g_2}{\sqrt{D[\hat{\gamma}_2]}} \right\} =$$

$$= P\{\omega : -0.5946 < \xi_{tab}(\omega) < 0.5946\} = 0.724 - 0.276 = 0.448,$$

$P\{\omega : \hat{\gamma}_2(\omega) > g_2\} + P\{\omega : \hat{\gamma}_2(\omega) < -g_2\} = 0.552$. Hence the realization $g_2 = 0.921$ of the test statistics $\hat{\gamma}_2$ does not contradict the null hypothesis on normality of the basic population. The critical region at the significance level $\alpha_1 = 0.01$ is $(-\infty, -3.990) \cup [3.990, \infty)$ and at the significance level $\alpha_2 = 0.05$ $(-\infty, -3.036] \cup [3.036, \infty)$.The realized value 0.92112 of the test statistic lies definitely outside the critical regions considered.

3.5.2 The Shapiro — Wilk test of normality

The problem is to test, on the basis of a sample of size n $(7 \leq n \leq 30)$ the null hypothesis that the general population possesses the normal probability distribution.

The test statistic is of the form

$$W = \left(\sum_{i=1}^{m} a_i^{(n)} (\xi_{(n-i+1)} - \xi_{(i)}) \right)^2 \Big/ \sum_{i=1}^{n} (\xi_i - \overline{\xi})^2,$$

where the random variable $\xi_{(i)}$ is the ith order statistic; it is a random variable that is realized in the random sample of size n by the value $x_{(i)}$, i.e., by the value being in the ith place in the set of measured data arranged according to magnitude from the smallest to the largest; if n is even, $m = n/2$, if n is odd, then $m = (n-1)/2$; for the coefficients a_i see Table I of Appendix.

While applying the Shapiro — Wilk test we arrange the components of the realization $(x_1, x_2, \ldots, x_n)'$ of the random sample $(\xi_1, \xi_2, \ldots, \xi_n)'$ according to their magnitude into the sequence $x_{(1)}, x_{(2)}, \ldots, x_{(n)}$; by means of this arrangement we calculate the realization w of the test statistic W and compare it with its critical values that, at the significance levels $\alpha_1 = 0.01$ and $\alpha_2 = 0.05$, are tabulated in Table II of Appendix. If the realized value w of the test statistic W at the significance level α satisfies the inequality $w_{crit}(\alpha) < w$, we are not in contradiction with the assertion stated by the null hypothesis; the general population can be considered to be normal [32].

Example Using the Shapiro — Wilk test let us test the null hypothesis that the general population is normal; the realization $(0.5, 7.5, 4.5, 2.2, 1.1, 1.7, 0.7, 1.5, 2.8, 3.2)'$ of the random sample $(\xi_1, \xi_2, \ldots, \xi_n)'$ is at our disposal (see the previous example).

In our case $\overline{x} = (1/10)\sum_{i=1}^{10} x_i = 2.57, m = n/2 = 5, x_{(1)} = 0.5, x_{(2)} = 0.7, x_{(3)} = 1.1, x_{(4)} = 1.5, x_{(5)} = 1.7, x_{(6)} = 2.2, x_{(7)} = 2.8, x_{(9)} = 4.5, x_{(10)} = 7.5$ and $\sum_{i=1}^{10}(x_{(i)} - \overline{x})^2 = \sum_{i=1}^{10}(x_i - \overline{x})^2 = 40.4610$. For the following calculation see Table 3.10.

Table 3.10

i	$a_i^{(10)}$	$x_{(10-i+1)}$	$x_{(i)}$	$x_{(10-i+1)} - x_{(i)}$	$a_i^{(10)}(x_{(10-i+1)} - x_{(i)})$
1	0.5739	7.5	0.5	7.0	4.01730
2	0.3291	4.5	0.7	3.8	1.25058
3	0.2141	3.3	1.1	2.1	0.44961
4	0.1224	2.8	1.5	1.3	0.15912
5	0.0399	2.2	1.7	0.5	0.01995

The test statistic realized by the value $w = 34.769\,42/40.461 = 0.859\,33$. Since $w_{crit}(0.01) = 0.781 < 0.859\ 33$ and similarly $w_{crit}(0.05) = 0.842 < 0.859\ 33$, the realization is not in contradiction with the null hypothesis; however, we note that it is relatively close to the critical value at the significance level $\alpha = 0.05$.

3.5.3 The D'Agostini test of normality

The problem is to test the hypothesis that the general population is normal on the basis of a realization of a random sample of size n, $30 < n \leq 100$.

The test statistic of the form

$$\eta = \sqrt{n}(D - 0.2820948)/0.0299860$$

is defined by means of the random variable

$$D = \sum_{i=1}^{m}\left(\frac{n+1}{2} - i\right)\left(\xi_{(n-i+1)} - \xi_{(i)}\right) \Big/ \sqrt{n^3 \sum_{i=1}^{n}(\xi_i - \bar{\xi})^2},$$

where $\xi_{(i)}$ is the ith order statistic (see the Shapiro — Wilk test), $m = n/2$ if n is even and $m = (n-1)/2$ if n is odd.

If the test statistic η is realized within the interval $(y_{crit}(\alpha/2), y_{crit}(1-\alpha/2))$, the realized random sample is not in contradiction with the null hypothesis on the normality of the general population at the significance level α.

The critical values of the random variable η at the significance levels $\alpha_1 = 0.01$ and $\alpha_2 = 0.05$ are in Table III (see Appendix).

If the size of the random sample used to test the null hypothesis on normality of some population is larger than 100, the Pearson test of goodness of fit is applied.

3.6 Tests of homogeneity

In this section, we shall deal with three groups of tests of homogeneity, namely the tests verifying the same probability distribution of k general populations, the tests verifying that k general normal populations are characterized by the same dispersion and the tests for verifying that two normal general populations are characterized by the same mean value.

3.6.1 Tests for verifying the same probability distribution

3.6.1.1 The Smirnov test

The problem is to test the null hypothesis that two independent random samples $(\xi_{11}, \xi_{21}, \ldots, \xi_{n_1 1})'$ and $(\xi_{12}, \xi_{22}, \ldots, \xi_{n_2 2})'$ come from the same general population characterized by the continuous distribution function, thus $H_0 : F_1(\cdot) = F_2(\cdot)$, against the alternative $H_1 : F_1(\cdot) \neq F_2(\cdot)$.

The test is based on the **Smirnov theorem** ([33], [6], p. 331):

Let $S_{1n_1}(\cdot)$ and $S_{2n_2}(\cdot)$ be sample distribution functions of two independent random samples with the sizes n_1 and n_2, respectively, from a general population characterized by the continuous distribution function $F(\cdot)$. The random variable

$$D_{n_1 n_2} = \max\{|S_{1n_1}(x) - S_{2n_2}(x)|,\ x \in \mathcal{R}^1\} \tag{3.6.1}$$

possesses asymptotically the probability distribution that is characterized by the distribution function

$$K(y) = \lim_{\substack{n_1\to\infty \\ n_2\to\infty}} P\{\omega : D_{n_1 n_2}(\omega) < y/\sqrt{n_1 n_2/(n_1+n_2)}\} =$$

$$= \begin{cases} 0, & y \le 0, \\ \sum_{-\infty}^{\infty}(-1)^j \exp(-2j^2y^2), & y > 0; \end{cases} ; \tag{3.6.2}$$

the random variable

$$D^+_{n_1 n_2} = \max\{S_{1n_1}(x) - S_{2n_2}(x),\ x \in \mathcal{R}^1\} \tag{3.6.3}$$

possesses asymptotically the probability distribution characterized by the distribution function

$$K^+(y) = \lim_{\substack{n_1\to\infty \\ n_2\to\infty}} P\{\omega : D^+_{n_1 n_2}(\omega) < y/\sqrt{n_1 n_2/(n_1+n_2)}\} = 1 - \exp(-2y^2). \tag{3.6.4}$$

The values of the distribution functions of the forms (3.6.2) and (3.6.4) are tabulated, e.g., in [9], where also the values of the distribution function of the exact probability distribution are given; they are important mainly if the sizes of the samples are small.

Example The problem is to test the hypothesis that the following two random samples $(1.220, 1.270, 1.235, 1.224, 1.261, 1.298)'$ and $(1.212, 1.246,$

1.225, 1.274, 1.257, 1.207, 1.248, 1.289)′, where $n_1 = 6$ and $n_2 = 8$ are samples from the same probability distribution which is characterized by a continuous distribution function. The test statistic $D_{n_1 n_2}$ defined by the relation (3.6.1) will be applied. The calculation is contained in Table 3.11.

Table 3.11

i	$x_{(i)}$	$S_{1,6}(x)$	$S_{2,8}(x)$	$S_{1,6}(x) - S_{2,8}(x)$
1	1.207	0	0	0
2	1.212	0	3/24	3/24
3	1.220	0	6/24	6/24
4	1.224	4/24	6/24	2/24
5	1.225	8/24	6/24	2/24
6	1.235	8/24	9/24	1/24
7	1.246	12/24	9/24	3/24
8	1.248	12/24	12/24	0
9	1.257	12/24	15/24	3/24
10	1.261	12/24	18/24	6/24
11	1.270	16/24	18/24	2/24
12	1.274	20/24	18/24	2/24
13	1.289	20/24	21/24	1/24
14	1.298	20/24	24/24	4/24

We see that $d_{6,8} = \max\{|S_{1,6}(x) - S_{2,8}(x)|,\ x \in \mathcal{R}^1\} = 6/24$. The tables for the exact probability distribution of the random variable $D_{6,8}$ yield

$$\begin{aligned} & P\{\omega : \max\{|S_{1,6}(x,\omega) - S_{2,8}(x,\omega)|,\ x \in \mathcal{R}^1\} \geq 6/24\} = \\ = \ & 1 - P\{\omega : \max\{|S_{1,6}(x,\omega) - S_{2,8}(x,\omega)|,\ x \in \mathcal{R}^1\} \leq 5/24\} = \\ = \ & 1 - 0.04795 = 0.95205; \end{aligned}$$

the tables of the asymptotic probability distribution of the test statistic $D_{6,8}$ give $K(0.4629) = 0.017$, hence

$$P\{\omega : \max\{|S_{1,6}(x,\omega) - S_{2,8}(x,\omega)|,\ x \in \mathcal{R}^1\} \geq 6/24\} \doteq 0.983.$$

The result of testing shows a very good agreement, the null hypothesis is not rejected.

The Smirnov test is relatively sensitive to the mean value as one of the main parameters of the probability distribution. It is possible to demonstrate this using the example of the random samples (−1.37, −1.20, −0.80, −0.42, 0.11, 0.33, 0.50, 1.00, 1.35)′ and (0.13, 0.30, 0.70, 1.08, 1.30, 1.61, 1.83,

2.00, 2.50, 2.85)'. We proceed as in the previous case. The partial results are in Table 3.12.

Table 3.12

i	$x_{(i)}$	$S_{1,10}(x)$	$S_{2,10}(x)$	$S_{1,10}(x) - S_{2,10}(x)$
1	−1.37	0	0	0
2	−1.20	1/10	0	1/10
3	−0.80	2/10	0	2/10
4	−0.42	3/10	0	3/10
5	−0.20	4/10	0	4/10
6	0.11	5/10	0	5/10
7	0.13	6/10	0	6/10
8	0.30	6/10	1/10	5/10
9	0.33	6/10	2/10	4/10
10	0.50	7/10	2/10	5/10
11	0.70	8/10	2/10	6/10
12	1.00	8/10	3/10	5/10
13	1.08	9/10	3/10	6/10
14	1.30	9/10	4/10	5/10
15	1.35	9/10	5/10	4/10
16	1.61	10/10	5/10	5/10
17	1.83	10/10	6/10	4/10
18	2.00	10/10	7/10	3/10
10	2.50	10/10	8/10	2/10
20	2.58	10/10	9/10	1/10

The test statistic $D_{10,10}$ realized in the value $d_{10,10} = 6/10$. The tables of the exact probability distribution of the test statistic $D_{10,10}$ yield

$$\begin{aligned} & P\{\omega : \max\{|S_{1,10}(x,\omega) - S_{2,10}(x,\omega)|\} \leq 6/10\} = \\ = \; & 1 - P\{\omega : \max\{|S_{1,10}(x,\omega) - S_{2,10}(x,\omega)|\} \geq 5/10\} = \\ = \; & 1 - 0.94755 = 0.05245 \end{aligned}$$

(the risk of rejecting the null hypothesis); using the tables for the asymptotic probability distribution of the statistic $D_{10,10}$ we obtain

$$P\{\omega : \max\{|S_{1,10}(x,\omega) - S_{2,10}(x,\omega)|\} \leq 6/10\} = 0.05513.$$

3.6.1.2 The Wang test

The Wang test, similar to the Smirnov test, is used for testing the null hypothesis that two independent random samples $(\xi_{11}, \xi_{21}, \ldots, \xi_{n_1 1})'$ and $(\xi_{12}, \xi_{22}, \ldots, \xi_{n_2 2})'$ come from the same general population characterized by the continuous distribution function, thus $H_0 : F_1(\cdot) = F_2(\cdot)$, against the alternative $H_1 : F_1(\cdot) \neq F_2(\cdot)$.

S.J.Wang, in order to improve the test's sensibility to the differences between the empirical distributions when the values of the distribution function $S_{2n_2}(\cdot)$ are small, has chosen the test statistic of the form

$$\left(\frac{n_1 n_2}{n_1 + n_2}\right)^{1/2} \sup\left\{\frac{S_{1n_1}(x) - S_{2n_2}(x)}{S_{2n_2}(x)},\ x :\ S_{2n_2}(x) \geq a > 0\right\};$$

here $S_{1n_1}(\cdot)$ is the empirical distribution function belonging to the first random sample of size n_1 and analogously S_{2n_2} is the empirical distribution function belonging to the second random sample of size n_2; the order of the random samples depends on the user, evidently.

If $(\xi_{11}, \xi_{21}, \ldots, \xi_{n_1 1})'$ and $(\xi_{12}, \xi_{22}, \ldots, \xi_{n_2 2})'$ are independent random samples from the same probability distribution with a continuous distribution function, the Wang test statistic defined by the relation (3.6.5) possesses asymptotically the probability distribution characterized by the distribution function

$$\lim_{\substack{n_1 \to \infty \\ n_2 \to \infty}} P\left\{\omega : \left(\frac{n_1 n_2}{n_1 + n_2}\right)^{1/2} \sup\left\{\frac{S_{1n_1}(x,\omega) - S_{2n_2}(x,\omega)}{S_{2n_2}(x,\omega)},\ x :\right.\right.$$

$$\left.\left. S_{2n_2}(x) \geq a > 0\right\} < \lambda\right\} = \begin{cases} 0, \ \lambda \leq 0, \\ \sqrt{\frac{2}{\pi}} \int_0^{\lambda\sqrt{a(1-a)}} \exp(-t^2/2)dt, \ \lambda > 0, \end{cases}$$

under the condition that $(n_1/n_2) \to d \geq 1$ for $n_1 \to \infty$ and $n_2 \to \infty$ ([34], [6], p. 340).

The Wang test arose from the Smirnov test by a modification, similar as to how Rényi test of goodness of fit originated from the Kolmogorov test; the Rényi modification of the Kolmogorov test was published in 1953, the Wang modification of the Smirnov test in 1955.

3.6.1.3 The Wilcoxon test

Let $(\xi_{11}, \xi_{21}, \ldots, \xi_{n_1 1})'$ and $(\xi_{12}, \xi_{22}, \ldots, \xi_{n_2 2})'$ be two independent random samples. The problem is to test the null hypothesis that they come from the same probability distribution characterized by a continuous distribution function.

The Wilcoxon test is based on the following assertion: Let the sizes n_1 and n_2 be sufficiently large and the independent random samples $(\xi_{11}, \xi_{21}, \ldots, \xi_{n_1 1})'$ and $(\xi_{12}, \xi_{22}, \ldots, \xi_{n_2 2})'$ come from the same probability distribution characterized by a continuous distribution function. Then the random variable

$$U_0 = \frac{U_1 - \frac{n_1 n_2}{2}}{\sqrt{\frac{n_1 n_2}{12}(n_1 + n_2 + 1)}},$$

where

$$U_1 = n_1 n_2 + \frac{n_1(n_1+1)}{2} - T_1$$

and T_1 is the sum of the ranks of the values $x_{11}, x_{21}, \ldots, x_{n_1 1}$ realized in the joint random sample $(\xi_{11}, \xi_{21}, \ldots, \xi_{n_1 1}, \xi_{12}, \xi_{22} \ldots, \xi_{n_2 2})'$ has for $n_1 \to \infty$ and $n_2 \to \infty$ asymptotically the normal probability distribution $N(0,1)$ (see [1], p.230).

If it is not possible to apply the asymptotic probability distribution, the following assertion forms the basis of the Wilcoxon test:

Let the independent random samples $(\xi_{11}, \xi_{21}, \ldots, \xi_{n_1 1})'$ and $(\xi_{12}, \xi_{22}, \ldots, \xi_{n_2 2})'$ come from the same probability distribution characterized by a continuous distribution function. Let U_1 and U_2 be the random variables defined by the relations

$$U_1 = n_1 n_2 + \frac{n_1(n_1+1)}{2} - T_1, \; U_2 = n_1 n_2 + \frac{n_2(n_2+1)}{2} - T_2,$$

where T_1 is the sum of the ranks of the values $x_{11}, x_{21}, \ldots, x_{n_1 1}$ realized in the joint random sample $(\xi_{11}, \xi_{21}, \ldots, \xi_{n_1 1}, \xi_{12}, \xi_{22}, \ldots, \xi_{n_2 2})'$ and T_2 is the sum of the ranks of the values $x_{12}, x_{22}, \ldots, x_{n_2 2}$ realized in the same joint random sample (the expression $t_1 + t_2 = (n_1 + n_2)(n_1 + n_2 + 1)/2$ can serve for checking the correctness of the calculus).

Then the test statistic

$$W = \min\{U_1, U_2\}$$

has the probability distribution whose critical values at the significance levels $\alpha_1 = 0.01$ and $\alpha_2 = 0.05$ are given in Tables IV and V of Appendix (see [1], p. 229).

If the test statistic W is realized as a value w, smaller or equal to the critical value, we reject the null hypothesis at the chosen significance level α.

Example Use the Wilcoxon test for testing the homogeneity of the two sets from the example concerning the Smirnov test. The problem is to test the null hypothesis that the random samples $(1.220, 1.270, 1.235, 1.224, 1.261, 1.298)'$ and $(1.212, 1.246, 1.225, 1.274, 1.257, 1.207, 1.248, 1.289)'$ correspond to the same probability distribution that is characterized by a continuous distribution function. The calculus of the order of the values realized in the first and the second sample are in Table 3.13 (the numbers in parentheses following the data in the first column denotes the random sample in which it was realized).

Table 3.13

Realization of the joint sample	Rank of the values realized in the 1st sample	Rank of the values realized in the 2nd sample
1.207(2)		1
1.212(2)		2
1.220(1)	3	
1.244(1)	4	
1.225(2)		5
1.235(1)	6	
1.246(2)		7
1.248(2)		8
1.257(2)		9
1.261(1)	10	
1.270(1)	11	
1.274(2)		12
1.289(2)		13
1.298(1)	14	
$\sum$	$t_1 = 48$	$t_2 = 57$

The realizations of the test statistics U_1 and U_2 are

$$u_1 = n_1 n_2 + \frac{n_1(n_1+1)}{2} - t_1 = 48 + 21 - 48 = 21,$$

$$u_0 = \frac{n_1 - n_1 n_2/2}{\frac{n_1 n_2}{12}(n_1 + n_2 + 1)} = \frac{21 - 24}{48(6+8+1)/12} = -0.387.$$

The critical region of the normal probability distribution $N(0,1)$ at the significance level $\alpha_1 = 0.05$ is given by the union $(-\infty, 1.960] \cup [1.960, \infty)$ and at the significance level $\alpha_2 = 0.01$ by the union $(-\infty, -2.576] \cup [2.576, \infty)$. The realized value $u_0 = -0.378$ lies outside both critical regions, hence there is no reason for rejecting the null hypothesis $H_0 : F_1(\cdot) = F_2(\cdot)$ tested against the alternative $H_1 : F_1(\cdot) \neq F_2(\cdot)$.

The realizations of the test statistics U_1, U_2 and W are:

$$u_1 = 21, \;\; u_2 = n_1 n_2 + \frac{n_2(n_2+1)}{2} - t_2 = 48 + 36 - 57 = 27,$$

$$w = \min\{u_1, u_2\} = 21,$$

$$w_{krit}(6, 8; 0.05) = 8, \;\; w_{krit}(6, 8; 0.01) = 4.$$

The test statistic W was realized as the value $w = 21$ which is essentially larger than the critical values at both chosen significance levels; thus there exists no reason to reject the null hypothesis.

The sensitivity of the Wilcoxon test of homogeneity to a shift in the mean value will be demonstrated on the same example as in the case of the Smirnov test.

Example The problem is to test the null hypothesis that the random samples $(-1.37, -1.20, -0.80, -0.42, -0.20, 0.11, 0.33, 0.50, 1.00, 1.35)'$ and $(0.13, \; 0.30, 0.70, 1.08, 1.30, 1.61, 1.83, 2.00, 2.50, 2.85)'$ are from the same probability distribution. The realizations of the ranks are in Table 3.14.

The test statistics U_1 and U_2 were realized by the values $u_1 = 88$ and $u_0 = -2.8723$. The latter value lies inside the critical region $(-\infty, -2.576] \cup [2.576, \infty)$ at the significance level 0.01 (more at the significance level 0.05), therefore the null hypothesis H_0 is rejected, the realized random samples are not from the same probability distribution. The same conclusion has to be drawn after applying the test statistic W. In this case $u_1 = 88$, $u_2 = 12$, $w = \min\{u_1, u_2\} = 12$, $w_{crit}(10, 10, 0.05) = 23$, $w_{crit}(10, 10, 0.01) = 16$.

Table 3.14

Realization of the joint sample	Rank of the values realized in the 1st sample	Rank of the values realized in the 2nd sample
−1.37(1)	1	
−1.20(1)	2	
−0.80(1)	3	
−0.42(1)	4	
−0.20(1)	5	
0.11(1)	6	
0.13(2)		7
0.30(2)		8
0.33(1)	9	
0.50(1)	10	
0.70(2)		11
1.00(1)	12	
1.08(2)		13
1.30(2)		14
1.35(1)	15	
1.61(2)		16
1.83(2)		17
2.00(2)		18
2.50(2)		19
2.58(2)		20
$\sum$	$t_1 = 67$	$t_2 = 143$

3.6.1.4 The Kruskal — Wallis test of homogeneity

This test represents a generalization of the Wilcoxon test of homogeneity for k $(k > 2)$ independent random samples.

The problem is to test the hypothesis that k $(k > 2)$ independent random samples $(\xi_{11}, \xi_{21}, \ldots, \xi_{n_1 1})'$, $(\xi_{12}, \xi_{22}, \ldots, \xi_{n_2 2})', \ldots, (\xi_{1k}, \xi_{2k}, \ldots, \xi_{n_k k})'$ are from the same probability distribution with the same continuous distribution function.

The Kruskal — Wallis test is based on the following assertion: The random variable

$$Q = \frac{12}{n(n+1)} \sum_{i=1}^{k} \frac{T_i^2}{n_i} - 3(n+1),$$

where $n = \sum_{i=1}^{k} n_i$, T_i is the sum of ranks of the values $x_{1i}, x_{2i}, \ldots, x_{n_i i}$, $i = 1, 2, \ldots, k$, realized in the joint random sample $(\xi_{11}, \xi_{21}, \ldots, \xi_{n_1 1}, \xi_{12}, \xi_{22}, \ldots, \xi_{n_2 2}, \ldots, \xi_{1k}, \xi_{2k}, \ldots, \xi_{n_k k})'$ under the assumption that $(\xi_{11}, \xi_{21}, \ldots, \xi_{n_1 1})'$, $(\xi_{12}, \xi_{22}, \ldots, \xi_{n_2 2})'$, $\ldots$, $(\xi_{1k}, \xi_{2k}, \ldots, \xi_{n_k k})'$ are independent random samples from the same probability distribution with a continuous distribution function, has for $n_i \to \infty$ $(i = 1, 2, \ldots, k)$ asymptotically the chi-square probability distribution with $k-1$ degrees of freedom; (the correctness of the calculation of the orders can be checked by the relation $\sum_{i=1}^{k} t_i = n(n+1)/2$, [1], p. 231).

3.6.2 Tests verifying the identity of dispersions

3.6.2.1 The Fisher test

The problem is to test the null hypothesis that two independent random samples come from normal probability distributions with the same dispersions, i.e., the null hypothesis $H_0 : \sigma_1^2 = \sigma_2^2$ against the alternative $H_1 : \sigma_1^2 \neq \sigma_2^2$ is tested.

The test is based on the assertion:

Let $(\xi_{11}, \xi_{21}, \ldots, \xi_{n_1 1})'$ be a random sample from the normal probability distribution $N(m_1, \sigma^2)$ and $(\xi_{12}, \xi_{22}, \ldots, \xi_{n_2 2})'$ be a random sample from the normal probability distribution $N(m_2, \sigma^2)$, $n_1 \geq 2$, $n_2 \geq 2$, $\sigma^2 > 0$, and let the random samples be independent. Then the test statistic

$$F_{n_1-1,n_2-1} = \frac{[1/(n_1-1)]\sum_{i=1}^{n_1}(\xi_{i1} - \bar{\xi}_1)^2}{[1/(n_2-1)]\sum_{i=1}^{n_2}(\xi_{i2} - \bar{\xi}_2)^2}, \quad \bar{\xi}_j = (1/n_j)\sum_{i=1}^{n_j} \xi_{ij}, \; j = 1, 2,$$

possesses the Fisher — Snedecor probability distribution with $n_1 - 1$ and $n_2 - 1$ degrees of freedom. Its probability density is given by the relation (1.5.11).

This assertion enables us, on the basis of the realizations $(x_{11}, x_{21}, \ldots, x_{n_1 1})'$ and $(x_{12}, x_{22}, \ldots, x_{n_2 2})'$ of two random samples, where $(\xi_{11}, \xi_{21}, \ldots, \xi_{n_1 1})'$ is a random sample from the normal probability distribution $N(m_1, \sigma_1^2)$ and $(\xi_{12}, \xi_{22}, \ldots, \xi_{n_2 2})'$ is a random sample from the probability distribution $N(m_2, \sigma_2^2)$, to test the null hypothesis $H_0 : \sigma_1^2 = \sigma_2^2$ against the alternative $H_1 : \sigma_1^2 \neq \sigma_2^2$ (without taking into account the relation between the mean values m_1 and m_2, since neither their values nor their estimators appear within the testing procedure).

The expressions in the numerator and denominator of the test statistic F_{n_1-1,n_2-1} are unbiased estimators of the dispersions of the first and the second general populations obtained on the basis of the samples of sizes n_1 and n_2, respectively. If both samples come from the same general population, the corresponding estimates do not differ essentially, hence the realized test statistic will be close to 1. Those values of the estimated dispersions indicate departure from the null hypothesis whose ratio differs substantially from 1.

Example The problem is to test the null hypothesis $H_0 : \sigma_1^2 = \sigma_2^2$ against the alternative $H_1 : \sigma_1^2 \neq \sigma_2^2$, if the realizations $(62, 54, 55, 60, 53, 58)'$ and $(142, 146, 139, 140, 141)'$ of two independent random samples $(\xi_{11}, \xi_{12}, \ldots, \xi_{16})'$ and $(\xi_{21}, \xi_{22}, \ldots, \xi_{25})'$ are available.

The Fisher test assumes the normality of the general populations, thus the first step in solving the problem consists in testing the normality. In both cases, the empirical coefficient of asymmetry and the empirical excess have to be calculated. We obtain $g_1 = 0.2583$ and $D[\hat{\gamma}_1] = 1$, therefore $P\{\omega : \hat{\gamma}_1(\omega) \leq 0.2538\} = 0.602$ and $P\{\omega : \hat{\gamma}_1(\omega) > 0.2583\} = 0.408$. Analogously, $g_2 = -1.4473$ and $D[\hat{\gamma}_2] = 4$, therefore $P\{\omega : \hat{\gamma}_2(\omega) \leq -1.4473\} = P\{\omega : \gamma_2(\omega)/2 \leq -0,7236\} = 0.235$ and $P\{\omega : \hat{\gamma}_2(\omega) > -1.4473\} = 0.765$, whereby we have convinced ourselves that the first sample does not contradict the assumption of the normality of the general population. A similar conclusion may be drawn for the second sample. In this case, $g_1 = 0.898$ and $D[\hat{\gamma}_1] = 1.2$, therefore $P\{\omega : \hat{\gamma}_1(\omega) \leq 0.898\} = 0.794$ and $P\{\omega : \hat{\gamma}_1(\omega) > 0.898\} = 0.206$. Further, $g_2 = -0.4947$ and $D[\hat{\gamma}_2] = 4.8$, therefore, $P\{\omega : \hat{\gamma}_2(\omega) \leq -0.4947\} = 0.4106$ and $P\{\omega : \hat{\gamma}_2(\omega) > -0.4947\} = 0.589$.

Now we may apply the Fisher test for testing the null hypothesis $H_0 : \sigma_1^2 = \sigma_2^2$ against the alternative $H_1 : \sigma_1^2 \neq \sigma_2^2$. Since $(1/5)\sum_{i=1}^{6}(x_{i1} - \overline{x}_1)^2 = 12.8$ and $(1/4)\sum_{i=1}^{5}(x_{i2} - \overline{x}_2)^2 = 7.3$ the test statistic $F_{5,4}$ is realized as the value $f_{5,4} = 1.753$. The tables of the values of the distribution function of the Fisher — Snedecor random variable with 5 and 4 degrees of freedom yield $P\{\omega : F_{5,4}(\omega) < 0.528\} = 0.25$, $P\{\omega : F_{5,4}(\omega) \geq 2.072\} = 0.25$, $P\{\omega : F_{5,4}(\omega) < 0.284\} = 0.10$, $P\{\omega : F_{5,4}(\omega) \geq 4.051\} = 0.10$, $P\{\omega : F_{5,4}(\omega) < 0.193\} = 0.05$, $P\{\omega : F_{5,4}(\omega) \geq 6.256\} = 0.05$, $P\{\omega : F_{5,4}(\omega) < 0.135\} = 0.025$, $P\{\omega : F_{5,4}(\omega) \geq 9.365\} = 0.025$, $P\{\omega : F_{5,4}(\omega) < 0.088\} = 0.01$, $P\{\omega : F_{5,4}(\omega) \geq 15.522\} = 0.01$, $P\{\omega : F_{5,4}(\omega) < 0.064\} = 0.005$, $P\{\omega : F_{5,4}(\omega) \geq 22.456\} = 0.005$. These values enable us to construct the following table (Table 3.15) for the critical regions of the test of the null hypothesis $H_0 : \sigma_1^2 = \sigma_2^2$ against the alternative $H_1 : \sigma_1^2 \neq \sigma_2^2$ at different significance levels. The value $f_{5,4}$ realized for the test statistic $F_{5,4}$ lies outside the critical region at the significance levels mentioned.

Table 3.15

Significance level	Critical region
0.50	$(0, 0.528] \cup [2.072, \infty)$
0.20	$(0, 0.284] \cup [4.051, \infty)$
0.10	$(0, 0.193] \cup [6.256, \infty)$
0.05	$(0, 0.135] \cup [9.365, \infty)$
0.02	$(0, 0.088] \cup [15.522, \infty)$
0.01	$(0, 0.064] \cup [22.456, \infty)$

3.6.2.2 The Bartlett test

The Bartlett test enables us, on the basis of the realizations $(x_{11}, x_{21}, \ldots, x_{n_1 1})'$, $(x_{12}, x_{22}, \ldots, x_{n_2 2})'$, $\ldots$, $(x_{1k}, x_{2k}, \ldots, x_{n_k k})'$ of k $(k > 2)$ independent random samples such that $(\xi_{11}, \xi_{21}, \ldots, \xi_{n_1 1})'$ is from the normal probability distribution $N(m_1, \sigma_1^2)$, $(\xi_{12}, \xi_{22}, \ldots, \xi_{n_2 2})'$ from the normal probability distribution $N(m_2, \sigma_2^2)$, $\ldots, (\xi_{1k}, \xi_{2k}, \ldots, \xi_{n_k k})'$ from the normal probability distribution $N(m_k, \sigma_k^2)$, to test the null hypothesis H_0 : $\sigma_1^2 = \sigma_2^2 = \ldots = \sigma_k^2$ against the alternative that all the dispersions are not equal.

The Bartlett test is based on the assertion:

Consider k $(k > 2)$ independent random samples such that $(\xi_{11}, \xi_{21}, \ldots, \xi_{n_1 1})'$ is from the normal probability distribution $N(m_1, \sigma^2)$, $(\xi_{12}, \xi_{22}, \ldots, \xi_{n_2 2})'$ from the normal probability distribution $N(m_2, \sigma^2)$, $\ldots, (\xi_{1k}, \xi_{2k}, \ldots, \xi_{n_k k})'$ from the normal probability distribution $N(m_k, \sigma^2)$. Then the test statistic

$$B = \frac{2.3026}{C}\left[\sum_{i=1}^{k}(n_i - 1)\log_{10}\hat{\sigma}^2 - \sum_{i=1}^{k}(n_i - 1)\log_{10}\hat{\sigma}_i^2\right],$$

where

$$C = 1 + \frac{1}{3(k-1)}\left(\sum_{i=1}^{k}\frac{1}{n_i - 1} - 1 \Big/ \sum_{i=1}^{k}(n_i - 1)\right),$$

$$\hat{\sigma}_i^2 = [1/(n_i - 1)]\sum_{j=1}^{k_i}(\xi_{ji} - \overline{\xi}_i)^2,\ \overline{\xi}_i = (1/n_i)\sum_{j=1}^{n_i}\xi_{ji},\ i = 1, 2, \ldots, k,$$

$$\hat{\sigma}^2 = \sum_{i=1}^{k}(n_i - 1)\hat{\sigma}_i^2 \Big/ \sum_{i=1}^{k}(n_i - 1),$$

possesses asymptotically the chi-square probability distribution with $k - 1$ degrees of freedom (in consequence of the asymptoticity, the requirement $n_i > 6$, $i = 1, 2, \ldots, k$, has to be fulfilled).

Example The unknown first order parameter Θ was estimated from 4 independent random samples of the sizes $n_1 = 8$, $n_2 = 12$, $n_3 = 10$ and $n_4 = 10$. The result of the processing of each of the realized samples is the estimator of the parameter Θ characterized by an unbiased estimate of the dispersion. The problem is to find out whether all the estimators of the first order parameter can be considered to be of the same accuracy.

The measurement errors can be assumed to possess a normal probability distribution and the estimators of the dispersions are unbiased, therefore the problem can be solved using the Bartlett test with the null hypothesis $H_0 : \sigma_1^2 = \sigma_2^2 = \sigma_3^2 = \sigma_4^2$ (see Table 3.16).

Table 3.16

i	n_i	$\hat{\sigma}_i^2$	$(n_1 - 1)\hat{\sigma}_i^2$	$\log_{10} \hat{\sigma}_i^2$	$(n_i - 1) \log_{10} \hat{\sigma}_i^2$
1	8	$5.0275\ 10^{-5}$	$35.19250\ 10^{-5}$	-4.29865	-30.09054
2	12	$5.6847\ 10^{-5}$	$62.53170\ 10^{-5}$	-4.24529	-46.69822
3	10	$5.0348\ 10^{-6}$	$4.53132\ 10^{-5}$	-5.29802	-47.68216
4	10	$8.5620\ 10^{-6}$	$7.70580\ 10^{-5}$	-5.06742	-45.60682

$$C = 1.04758,\ b = 2.3026 - 162.5423 + 170.0777/1.0476 = 16.56.$$

The critical region at the significance level $\alpha_1 = 0.05$ is of the form $(0, 0.22) \cup (9.35, \infty)$ and at the significance level $\alpha_2 = 0.01$ is defined by the union $(0, 0.07] \cup [12.84, \infty)$, which means that the null hypothesis on the identity of the dispersions has to be rejected.

3.6.2.3 The Cochran test

The Cochran test enables us, on the basis of the realizations $(x_{11}, x_{21}, \ldots, x_{n1})'$, $(x_{12}, x_{22} \ldots, x_{n2})', \ldots$, $(x_{1k}, x_{2k}, \ldots, x_{nk})'$ of k $(k \geq 2)$ random samples of the same size n such that $(\xi_{11}, \xi_{21}, \ldots, \xi_{n1})'$ comes from the normal probability distribution $N(m_1, \sigma_1^2)$, $(\xi_{12}, \xi_{22}, \ldots, \xi_{n2})'$ from the probability distribution $N(m_2, \sigma_k^2)$ $\ldots$, $(\xi_{1k}, \xi_{2k}, \ldots, \xi_{nk})'$ is from the normal probability distribution $N(m_k, \sigma_k^2)$, to test the null hypothesis $H_0 : \sigma_1^2 = \sigma_2^2 = \ldots = \sigma_k^2$. The Cochran test is based on the following assertion:

Have k $(k \geq 2)$ independent random samples of the same size n, where $(\xi_{11}, \xi_{21}, \ldots, \xi_{n1})'$ is from the normal probability distribution $N(m_1, \sigma^2)$, $(\xi_{12}, \xi_{22}, \ldots, \xi_{n2})'$ from $N(m_2, \sigma^2)$, $\ldots$, $(\xi_{1k}, \xi_{2k}, \ldots, \xi_{nk})'$ is the random sample from the normal probability distribution $N(m_k, \sigma^2)$. Then the test statistic

$$G(k, n) = \hat{\sigma}_{max}^2 / (\hat{\sigma}_1^2 + \hat{\sigma}_2^2 + \ldots + \hat{\sigma}_k^2),$$

where

$$\hat{\sigma}_i^2 = [1/(n-1)]\sum_{j=1}^{n}(\xi_{ji} - \overline{\xi}_i)^2, \ \overline{\xi}_i = \sum_{j=1}^{n}\xi_{ji}, \ i = 1, 2, \ldots, k,$$

and $\hat{\sigma}_{max}^2 = \max\{\hat{\sigma}_1^2, \hat{\sigma}_2^2, \ldots, \hat{\sigma}_k^2\}$, whose approximate critical values are tabulated (see [9]) is used for testing the null hypothesis $H_0 : \sigma_1^2 = \sigma_2^2 = \ldots = \sigma_k^2$. If the realized value of the test statistic $G(k, n)$ exceeds the critical value at the given significance level, the null hypothesis stating the identity of the dispersions of individual probability distributions is rejected.

Example Each of the five sets containing the results of measuring the same quantity is characterized by the unbiased estimate of the dispersion; it acquires sequentially the values: 239, 653, 156, 120 and 111. The problem is to decide whether the sets characterized by these empirical dispersions can be considered to have the same accuracy under the conditions that $n = 5$ and $n = 41$, respectively.

The null hypothesis $H_0 : \sigma_1^2 = \sigma_2^2 = \ldots = \sigma_5^2$ will be tested using the Cochran test and the results will be compared with those of the Bartlett test. Since the results of measurement are taken into account here, the assumption of normality may be considered to be fulfilled.

Let $k = 5$ and $n = 5$. The test statistic $G(5, 5)$ is realized as the value

$$g(5, 5) = 653/(239 + 653 + 156 + 120 + 111) = 0.5106;$$

comparing it with the critical values $g_{crit}(5, 5; 0.05) = 0.5063$ and $g_{crit}(5, 5; 0.010) = 0.5875$ we state that the null hypothesis at the significance level 0.05 has to be rejected; there does not exist a reason for rejecting it at the significance level 0.01, however the realized value is relatively close to the critical value.

Let $k = 5$ and $n = 41$. The test statistic $G(5, 41)$ is realized by the value 0.5106. The critical values at the significance levels $\alpha_1 = 0.05$ and $\alpha_2 = 0.01$ are $g_{crit}(5, 41; 0.05) = 0.30$ and $g_{crit}(5, 41; 0.01) = 0.33$, respectively, therefore the null hypothesis has to be rejected at both significance levels mentioned.

The Bartlett test statistic for $n_1 = n_2 = \ldots = n_k = n$ acquires the form

$$B = \frac{2.3026(n-1)}{C}\left[k\log_{10}\left((1/k)\sum_{i=1}^{k}\hat{\sigma}_i^2\right) - \sum_{i=1}^{k}\log_{10}\hat{\sigma}_i^2\right],$$

$$C = 1 + \frac{1}{3(k-1)}\left(\frac{k}{n-1} - \frac{1}{k(n-1)}\right).$$

In our case

i	$\hat{\sigma}_i^2$	$\log_{10}\hat{\sigma}_i^2$
1	239	2.37840
2	653	2.81491
3	156	2.19312
4	120	2.07918
5	111	2.04532

$$\sum_{i=1}^{5}\hat{\sigma}_i^2 = 1\ 279 \text{ and } \sum_{i=1}^{5}\log_{10}\hat{\sigma}_i^2 = 11.510\ 94.$$

If $n = 5$, the Bartlett statistic B is realized as the value $b = 4.426$ ($C = 1.1$, $b = 4(2.3026/1.1)(12.03950 - 11.51094)$); the critical regions at the significance levels $\alpha_1 = 0.01$ and $\alpha_2 = 0.05$ are $(0, 0.21] \cup [14.86.\infty)$ and $(0, 0.48] \cup [11.14, \infty)$, respectively, the null hypothesis stating the identity of dispersions is not rejected.

If $n = 41$, the Bartlett statistic is realized as the value $b = 48.203$ ($C = 1.01$, $b = 40(2.3026/1.1)(12.03960 - 11.51094)$); the critical regions are the same as in the previous case, thus the null hypothesis on the identity of the dispersions has to be rejected.

3.6.3 Tests verifying the identity of mean values

3.6.3.1 The Student's t test

The Student's t test serves for testing, on the basis of the realizations $(x_{11}, x_{21}, \ldots, x_{n_1 1})'$ and $(x_{12}, x_{22} \ldots, x_{n_2 2})'$ of independent random samples $(\xi_{11}, \xi_{21}, \ldots, \xi_{n_1 1})'$ and $(\xi_{12}, \xi_{22}, \ldots, \xi_{n_2 2})'$ from normal probability distributions, the null hypothesis that the difference of their mean values equals a certain value (the dispersion need not be known), i.e., the null hypothesis is of the form $H_0 : m_1 - m_2 = \Delta$, and the alternative is $H_1 : m_1 - m_2 \neq \Delta$.

If $\Delta = 0$, the Student's t test tests, on the basis of the realization of two independent random samples, the null hypothesis that the general two populations possess the same mean value.

The Student's t test is based on the following assertion:

Let $(\xi_{11}, \xi_{21}, \ldots, , \xi_{n_1 1})'$ be a random sample from the probability distribution $N(m_1, \sigma_1^2)$ and $(\xi_{12}, \xi_{22}, \ldots, \xi_{n_2 2})'$ a random sample from the probability distribution $N(m_2, \sigma_2^2)$, where $n_1 \geq 2$, $n_2 \geq 2$, $\sigma_1^2 = \sigma_2^2 = \sigma^2 > 0$;

the dispersions need not be known a priori. Then the random variable

$$T = \frac{\bar{\xi}_1 - \bar{\xi}_2 - (m_1 - m_2)}{[(n_1-1)\hat{\sigma}_1^2 + (n_2-1)\hat{\sigma}_2^2]^{1/2}} \left(\frac{n_1 n_2 (n_1 + n_2 - 2)}{n_1 + n_2} \right)^{1/2},$$

where

$$\bar{\xi}_i = (1/n_i) \sum_{j=1}^{n_i} \xi_{ji}, \; \hat{\sigma}_i^2 = [1/(n_i - 1)] \sum_{j=1}^{n_i} (\xi_{ji} - \bar{\xi}_i)^2, \; i = 1, 2,$$

has the Student's t probability distribution with $n_1 + n_2 - 2$ degrees of freedom ([1], p. 91).

The Student's t test can be obviously applied only after verifying that the dispersions of both general populations are the same.

Example The problem is to judge if two different methods, e.g., of working a material have the same effect if the results of their influence on two different material samples are available; the effects of the method A are given by the values 62, 54, 55, 60, 53, 58 and of the method B by 52, 56, 49, 50, 51.

It was verified that both random samples are not in contradiction to the assumption of the normality of the corresponding general populations and, simultaneously, that both these samples do not contradict the assumption of their identical dispersions. Hence the Student's t test testing the null hypothesis $H_0 : m_1 = m_1$ on the same effect of the working methods considered against the alternative $H_1 : m_1 \neq m_2$ may be used.

In our case $\bar{x}_1 = 57.0$, $\bar{x}_2 = 51.6$, $(n_1 - 1)\hat{\sigma}_1^2 = 64.0$, $(n_2 - 1)\hat{\sigma}_2^2 = 29.2$ and $t = (57.0 - 51.6)(64.0 + 29.2)^{-1/2}(270/11)^{1/2} = 2.7712$. As the critical values of the test statistic are $t_{crit}(9; 0.05) = 2.262$ and $t_{crit}(9; 0.01) = 3.250$, the null hypothesis on the equivalent effect of both methods considered at the significance level 0.05 is rejected, at the significance level 0.01 we are not yet in contradiction with the formulated null hypothesis.

3.6.3.2 The Fisher — Behrens problem

Let $(\xi_{11}, \xi_{21}, \ldots, \xi_{n_1 1})'$ be a random sample from the normal probability distribution $N(m_1, \sigma_1^2)$ and $(\xi_{12}, \xi_{22}, \ldots, \xi_{n_2 2})'$ a random sample from the normal probability distribution $N(m_2, \sigma_2^2)$, where $\sigma_1^2 > 0$, $\sigma_2^2 > 0$ and $\sigma_1^2 \neq \sigma_2^2$. The problem is to test the null hypothesis $H_0 : m_1 = m_2$ against the alternative $H_1 : m_1 \neq m_2$.

If the size of the independent random samples mentioned is the same, $n_1 = n_2 = n$, the following assertion may be applied:

Let $(\xi_{11}, \xi_{21}, \ldots, \xi_{n1})'$ be a random sample from the normal probability distribution $N(m_1, \sigma_1^2)$ and $(\xi_{12}, \xi_{22}, \ldots, \xi_{n2})'$ be a random sample from the normal probability distribution $N(m_2, \sigma_2^2)$ and let these random samples be independent; the random variable

$$T = \frac{\overline{\eta}}{\left(\frac{1}{n(n-1)} \sum_{i=1}^{n} (\eta_i - \overline{\eta})^2\right)^{1/2}},$$

where

$$\overline{\eta} = (1/n) \sum_{i=1}^{n} \eta_i, \ \eta_i = \xi_{i1} - \xi_{i2}, \ i = 1, 2, \ldots, n,$$

has the Student's t probability distribution with $n - 1$ degrees of freedom.

If the sizes of the random samples differ, the following assertion has to be applied:

Let $(\xi_{11}, \xi_{21}, \ldots, \xi_{n_1 1})'$ be a random sample from the normal probability distribution $N(m_1, \sigma_1^2)$ and $(\xi_{12}, \xi_{22}, \ldots, \xi_{n_2 2})'$ be a random sample from the normal probability distribution $N(m_2, \sigma_2^2)$ such that $n_1 < n_2$, and let these random samples be independent; the random variable

$$T = \frac{\overline{\eta}}{\left(\frac{1}{n_1(n_1-1)} \sum_{i=1}^{n_1} (\eta_i - \overline{\eta})^2\right)^{1/2}},$$

where

$$\overline{\eta} = (1/n_1) \sum_{i=1}^{n_1} \eta_i, \ \eta_i =$$

$$= \xi_{i1} - \sqrt{\frac{n_1}{n_2}} \xi_{i2} + \frac{1}{\sqrt{n_1 n_2}} \sum_{j=1}^{n_1} \xi_{j2} - (1/n_2) \sum_{j=1}^{n_2} \xi_{j2}, \ i = 1, 2, \ldots, n_1,$$

possesses the Student's t probability distribution with $n_1 - 1$ degrees of freedom ([2], p. 165).

If the test statistic T is realized by the value t such that $|t| \geq t_{crit}(n_1 - 1; \alpha)$, the null hypothesis $H_0 : m_1 = m_2$ against the alternative $H_1 : m_1 \neq m_2$ is rejected at the significance level α.

Example The problem is to test the hypothesis on the same effectiveness of two different methods based of the measured effects on two sets of material samples. For the effects obtained by the two different methods mentioned see Table 3.17 (the index 1 belongs to the method A and 2 to the method B).

Table 3.17

i	x_{i1}	x_{i2}	$\sqrt{(n_1/n_2)}x_{i2}$	$x_{i1} - \sqrt{(n_1/n_2)}x_{i2}$	y_i	$y_i - \overline{y}$
1	55	62	56.598	−1.598	−6.747	−6.347
2	39	54	49.295	−10.295	−15.444	−15,044
3	67	55	50.208	16.792	11.643	12.043
4	63	60	54.772	8.228	3.079	3.497
5	59	53	48.382	10.618	5.469	5.869
6		58				

The test statistic T is realized as the value $t = -0.4/4.78638 = -0.083\,57$. The critical regions of the Student's t probability distribution with 4 degrees of freedom are $t_{crit}(4; 0.01) = 4.604$ and $t_{crit}(4; 0.05) = 2.773$, therefore there does not exist any reason for rejecting the null hypothesis; both methods may be considered to have the same effect.

For the sake of complexity, let us convince ourselves that the dispersions of the samples considered significantly. Using the Fisher test, we obtain $f(n_1 - 1, n_2 - 1) = f(4, 5) = 116.8/12.8 = 9.125$; the critical region at the significance level $\alpha = 0.05$ is $(0, 0.11) \cup (7.39, \infty)$, thus at the significance level 0.05 the hypothesis on the identical dispersions of the populations considered is rejected.

3.7 Test for judging the significance of extremely deviated data

The main characteristic that enables us to judge the statistical significance of the deviation of an experimental result is the distribution function of the corresponding general population.

Let $F(\cdot)$, $x \in \mathcal{R}^1$, be the distribution function of the general population from which the set of measured data $x_1, x_2, \ldots, x_n$ comes. In accordance with the p. 95, the elements of this set ordered against their magnitude will be denoted by $x_{min} = x_{(1s)} < x_{(2s)}, \ldots < x_{(ns)} = x_{max}$.

The test of the hypothesis that the value x_{min} or x_{max} cannot be considered to be extremely deviated data is based on the assertions concerning the distribution functions of the random variables ξ_{min} and ξ_{max} realizations of which the values x_{min} and x_{max}, respectively, are:

Let $\xi_1, \xi_2, \ldots, \xi_n$ be a random sample of the size n taken from the probability distribution characterized by the distribution function $F(\cdot)$, $x \in \mathcal{R}^1$. The distribution function of the random variable ξ_{max} (the notation $F_{max}(\cdot)$, $x \in \mathcal{R}^1$ is used for it) is

$$F_{max}(x) = F^n(x),\ x \in \mathcal{R}^1,$$

and the distribution function of the random variable ξ_{min} (it is denoted by the symbol $F_{min}(\cdot)$, $x \in \mathcal{R}^1$) reads

$$F_{min}(x) = \sum_{m=1}^{n} \binom{n}{m} F^m(x)[1 - F(x)]^{n-m} = 1 - [1 - F(x)]^n.$$

If the probability density $f(x) = dF(x)/dx$ of the random variable ξ exists, the probability densities of the random variables ξ_{max} and ξ_{min} are

$$f_{max} = nF^{n-1}(x)f(x)$$

and

$$f_{min}(x) = n[1 - F(x)]^{n-1} f(x),$$

respectively (see [6], p. 312).

The value realized within the random sample of the size n by the value x_{max} is considered to be significantly deviated at the significance level α if

$$P\{\omega : \xi_{max}(\omega) \geq x_{max}\} \leq \alpha,$$

i.e., if

$$F_{max}(x_{max}) = P\{\omega : \xi_{max}(\omega) < x_{max}\} \geq 1 - \alpha;$$

analogously, the value x_{min} within the random sample of the size n is considered to be significantly deviated at the significance level α if

$$F_{min}(x_{min}) = P\{\omega : \xi_{min}(\omega) \leq x_{min}\} \leq \alpha.$$

In the cases just mentioned the distribution function of the general population is assumed to be known, inclusive of the parameters determining it.

Next, we shall deal with the assertion necessary for judging the statistical significance of extremly deviated data under the condition that the general population is normally distributed. Under this condition the following four alternatives occur:

1. Let $\xi_1, \xi_2, \ldots, \xi_n$ be a random sample from the normal probability distribution $N(m, \sigma^2)$, where m and σ^2 are known a priori. Then the random variable $(\xi_{max} - m)/\sigma)$ satisfies the relation

$$P\left\{\omega : \frac{\xi_{max}(\omega) - m}{\sigma} \geq x\right\} = 1 - \left(\frac{1}{\sqrt{2\pi}} \int_{-\infty}^{x} \exp(-u^2/2)du\right)^n. \tag{3.7.1}$$

2. Let $\xi_1, \xi_2, \ldots, \xi_n$ be a random sample from the normal probability distribution $N(m, \sigma^2)$, where m is known a priori and σ^2 is unknown a priori, it has to be estimated from the realization of the random sample mentioned above using the relation

$$\hat{\sigma}^2 = (1/n)\sum_{i=1}^{n}(\xi_i - m)^2$$

(it is an unbiased and efficient estimator). Then the random variable

$$G(1,n) = \frac{(\xi_{max} - m)^2}{\sum_{i=1}^{n}(\xi_i - m)^2} \tag{3.7.2}$$

represents the Cochran test statistic for $k = 1$ and n which equals the size of the random sample. Its critical values are tabulated, e.g., in [9]. This test statistic is used for testing the null hypothesis H_0 : "x_{max} is not a significantly deviated datum" against the alternative H_1 : "x_{max} significantly deviates from the other data of the set of measurements".

3. Let $\xi_1, \xi_2, \ldots, \xi_n$ be a random sample from the normal probability distribution $N(m, \sigma^2)$, where m is unknown a priori and σ^2 is known a priori; the unbiased and efficient estimator of the unknown parameter of the mean value m is $\hat{m} = \overline{\xi} = (1/n)\sum_{i=1}^{n}\xi_i$. Then the probability distribution of the random variable $(\xi_{max} - \overline{\xi})/\sigma$ satisfies the relation

$$n\left[1 - \Phi\left(z\sqrt{\frac{n}{n-1}}\right)\right]\left\{1 - \frac{n}{2}\left[1 - \Phi\left(z\sqrt{\frac{n}{n-1}}\right)\right]\right\} \leq$$
$$\leq P\left\{\omega : \frac{\xi_{max}(\omega) - \overline{\xi}(\omega)}{\sigma} \geq z\right\} \leq n\left[1 - \Phi\left(z\sqrt{\frac{n}{n-1}}\right)\right], \; z \in \mathcal{R}^1, \tag{3.7.3}$$

where $\Phi(x) = \frac{1}{\sqrt{2\pi}}\int_{-\infty}^{x}\exp(-u^2/2)du$ is the distribution function of the normal probability distribution $N(0,1)$ and n is the size of the random sample considered.

4. Let $\xi_1, \xi_2, \ldots, \xi_n$ be a random sample from the normal probability distribution $N(m, \sigma^2)$, where m and σ^2 are unknown a priori. The unbiased and efficient estimators of the mean value and the dispersion are $\hat{m} = \overline{\xi} =$

$(1/n)\sum_{i=1}^{n}\xi_i$ and $\hat{\sigma}^2 = [(1/(m-1)]\sum_{i=1}^{n}(\xi_i - \overline{\xi})^2$, respectively. Then the probability distribution of the test statistic of the form $(\xi_{max} - \overline{\xi})/\hat{\sigma}$ satisfies the inequalities

$$n\left[1 - H_{n-2}\left(z\sqrt{\frac{n-2}{n-1-z^2}}\right)\right]\cdot$$

$$\cdot\left\{1 - \frac{n}{2}\left[1 - H_{n-2}\left(z\sqrt{\frac{n-2}{n-1-z^2}}\right)\right]\right\} \leq$$

$$\leq P\left\{\omega : \frac{\xi_{max}(\omega) - \overline{\xi}(\omega)}{\hat{\sigma}(\omega)} \geq z\right\} \leq n\left[1 - H_{n-2}\left(z\sqrt{\frac{n-2}{n-1-z^2}}\right)\right],$$

$$z \in \mathcal{R}^1, \tag{3.7.4}$$

where

$$H_{n-2}(t) = \frac{\Gamma[(n-1)/2]}{\sqrt{\pi(n-2)}\Gamma[(n-2)/2]}\int_{-\infty}^{t}\left(1 + \frac{x^2}{n-2}\right)^{-(n-1)/2} dx$$

is the distribution function of the Student's t random variable with $n-2$ degrees of freedom.

Example The set of 21 measured data (see Table 3.18, where they are arranged according to their magnitude from the smallest to the largest) that is a realization of the random sample from the probability distribution $N(m, \sigma^2)$, will be analysed under the assumption that m and σ^2 are known a priori, that σ^2 is known a priori and the unknown m is estimated, that m is known a priori and σ^2 estimated and, finally, that both the parameters m and σ^2 are a priori unknown and therefore they have to be estimated.

Table 3.18

i	x_i	i	x_i	i	x_i	i	x_i	i	x_i
1	−1.37	5	−0.80	9	−0.20	13	0.33	17	1.00
2	−1.35	6	−0.50	10	−0.11	14	0.42	18	1.20
3	−1.20	7	−0.42	11	0.11	15	0.50	19	1.35
4	−1.00	8	−0.33	12	0.20	16	0.80	20	1.37
								21	3.00

1st alternative: m and σ^2 are a priori known. In this case if m=0 and $\sigma^2 = 1$, the test statistic $(\xi_{max} - m)/\sigma$ is realized by the value $(x_{max} - m)/\sigma = x_{max} = 3$. Since

$$P\{\omega : \xi_{max}(\omega) \geq 3\} = 1 - P\{\omega : \xi_{max}(\omega) < 3\} = 1 - \Phi^{21}(3) =$$

$$= 1 - (0.99865)^{21} = 0.02797,$$

the null hypothesis that $x_{max} = 3$ is not a significantly deviated datum is rejected with the risk 0.028.

2nd alternative: $\sigma^2 = 1$, the unbiased and efficient estimator of the unknown mean value m is realized as the value $\hat{m} = \overline{\xi} = (1/21)\sum_{i=1}^{21} x_i = 0.1429$. The test statistic $(\xi_{max} - \overline{\xi})/\sigma$ realized as the value 2.8571; in accordance with the relation (3.7.3) we obtain

$$-220.5\left[1 - \Phi\left(2.8571\sqrt{\frac{21}{20}}\right)\right]^2 + 21\left[1 - \Phi\left(2.8571\sqrt{\frac{21}{20}}\right)\right] =$$

$$= -0.00065 + 0.03593 = 0.03528$$

and

$$21\left[1 - \Phi\left(2.8571\sqrt{\frac{21}{20}}\right)\right] = 0.03593,$$

hence

$$P\{\omega : \xi_{max}(\omega) - \overline{\xi}(\omega) \geq 2.8571\} \in [0.03528,\ 0.03593],$$

which means that under the assumption just mentioned, the null hypothesis that x_{max}=3 is not significantly deviated from the set of the measured data, is rejected at the significance level 0.036.

3rd alternative: $m = 0$, the unbiased and efficient estimator of the unknown parameter σ^2 assumes the value $\hat{\sigma}^2 = (1/21)\sum_{i=1}^{21} x_i^2 = 1.103\,02$. The test statistic $G(1, n) = \frac{(\xi_{max} - m)^2}{\sum_{i=1}^{n}(\xi_i - m)^2}$ is realized in our case as the value $g(1,\ 21) = \frac{x_{max}^2}{\sum_{i=1}^{21} x_i^2} = 9/23.7336 = 0.379\,21$. The critical values of the Cochran test statistic $G(1, 21)$ at the significance levels $\alpha_1 = 0.05$ and $\alpha_2 = 0.01$ are $g_{crit}(1, 21; 0.05) = 0.3779$ and $g_{crit}(1, 21; 0.01) = 0.4662$, respectively, which means that the null hypothesis that $x_{max} = 3$ under the given circumstances is not a significantly deviated datum is rejected with the risk $\alpha = 0.05$; however, we note that at the significance level the realized value is close to the critical value.

4th alternative: The unbiased and efficient estimator of the mean value is realized by the value $\hat{m} = \overline{x} = 0.1429$, the unbiased and efficient estimator of the dispersion is realized by the value $\hat{\sigma}^2 = (1/20)\sum_{i-1}^{21}(x_i - 0.1429)^2 = 1.1653$. The test statistic $(\xi_{max} - \overline{\xi})/\hat{\sigma}^2$ is realized by the value $(x_{max} - \overline{x})^2/\hat{\sigma} = 2.6467$ and simultaneously, due to the relation (3.7.3),

$$21\left[1 - H_{19}\left(2.6467\sqrt{\frac{19}{12.995}}\right)\right] = 0.050,$$

$$220.5\left[1 - H_{19}\left(2.6467\sqrt{\frac{19}{12.995}}\right)\right]^2 = 0.00123,$$

which means that

$$P\left\{\omega : \frac{\xi_{max}(\omega) - \overline{\xi}(\omega)}{\hat{\sigma}(\omega)} \geq 2.6467\right\} \in [0.0483,\ 0.0496],$$

thus the value $x_{max} = 3$, as a value significantly deviated from the set of measured data, is eliminated with a risk smaller than 0.05.

Let us consider the following law which is illustrated by our numerical example. The smallest risk of rejecting the null hypothesis that x_{max} is not a significantly deviated datum, in other words the risk of eliminating x_{max} as a significantly deviated value from the set of measured data considered occurs in the first alternative (m and σ^2 are supposed to be known a priori); this risk increases if one of the determining characteristics m and σ^2 of the normal probability distribution or both of them are unknown a priori and have to be estimated from the set of measured data.

The four alternatives mentioned in the analysis of a set of measured data, which is a realization of the random sample from a normal probability distribution $N(m, \sigma^2)$, are often in practice substituted by the "three-sigma" rule. Due to it, that datum is considered to be significantly deviated within a set of measured data whose value differs from the mean value of the set more than by the threefold of the standard deviation which characterizes the set of measured data. This rule has an asymptotic character; this is demonstrated in Table 3.19, which yields, depending on the size n of the sample available the probabilities that the value differs from the mean value of the set of measured data by more than 3σ; the probabilities given there correspond to the assumption that the mean value m and the dispersion σ^2 are known a priori. This fact, together with our knowledge concerning the increasing risk if the estimated values of m and σ^2 are at our disposal, has only to be taken into account to interpret the results of an application of the "three-sigma" rule.

Table 3.19

The size of the random sample	$P\left\{\omega : \frac{\xi_{max}(\omega)-m}{\sigma} \geq 3\right\} = 1 - \Phi^n(3)$
5	$1 - 0.99327 = 0.00673$
10	$1 - 0.98658 = 0.01342$
50	$1 - 0.93469 = 0.06531$
100	$1 - 0.87364 = 0.12636$
150	$1 - 0.81659 = 0.18341$
200	$1 - 0.76326 = 0.23674$
250	$1 - 0.71341 = 0.28659$
300	$1 - 0.66681 = 0.33319$
350	$1 - 0.62326 = 0.37674$
400	$1 - 0.58256 = 0.41744$
450	$1 - 0.54451 = 0.45549$
500	$1 - 0.50895 = 0.49105$
1000	$1 - 0.25900 = 0.74100$
5000	$1 - 0.00117 = 0.99883$

The table implies the following practical conclusion: The greater the size of the measured data set considered, the greater is the probability of a realization of an element which differs from the mean value of the set by more than threefold the standard deviation.

TABLES

Table I

Coefficients $a_i^{(n)}$ of the Shapiro – Wilk test

i \ n	7	8	9	10	11	12	13	14
1	0.6233	0.6052	0.5888	0.5739	0.5601	0.5475	0.5359	0.5251
2	0.3031	0.3164	0.3244	0.3291	0.3315	0.3325	0.3325	0.3318
3	0.1401	0.1743	0.1976	0.2141	0.2260	0.2347	0.2412	0.2460
4	0	0.0561	0.0947	0.1224	0.1429	0.1586	0.1707	0.1802
5	0	0	0	0.0399	0.0695	0.0922	0.1099	0.1240
6	0	0	0	0	0	0.0303	0.0539	0.0727
7	0	0	0	0	0	0	0	0.0240

i \ n	15	16	17	18	19	20	21	22
1	0.5150	0.5056	0.4968	0.4886	0.4808	0.4734	0.4643	0.4590
2	0.3306	0.3290	0.3273	0.3253	0.3232	0.3211	0.3185	0.3156
3	0.2495	0.2521	0.2540	0.2553	0.2561	0.2565	0.2578	0.2571
4	0.1878	0.1939	0.1988	0.2027	0.2059	0.2085	0.2119	0.2131
5	0.1353	0.1447	0.1524	0.1587	0.1641	0.1686	0.1736	0.1764
6	0.0880	0.1005	0.1109	0.1197	0.1271	0.1334	0.1399	0.1443
7	0.0433	0.0593	0.0725	0.0837	0.0932	0.1013	0.1092	0.1150
8	0	0.0196	0.0359	0.0496	0.0612	0.0711	0.0804	0.0878
9	0	0	0	0.0163	0.0303	0.0422	0.0530	0.0618
10	0	0	0	0	0	0.0140	0.0263	0.0368
11	0	0	0	0	0	0	0	0.0122

i \ n	23	24	25	26	27	28	29	30
1	0.4542	0.4493	0.4450	0.4407	0.4366	0.4328	0.4291	0.4254
2	0.3126	0.3098	0.3069	0.3043	0.3018	0.2992	0.2968	0.2944
3	0.2563	0.2554	0.2543	0.2533	0.2522	0.2510	0.2499	0.2487
4	0.2139	0.2145	0.2148	0.2151	0.2152	0.2151	0.2150	0.2148
5	0.1787	0.1807	0.1822	0.1836	0.1848	0.1857	0.1864	0.1870
6	0.1480	0.1512	0.1539	0.1563	0.1584	0.1601	0.1616	0.1630
7	0.1201	0.1245	0.1283	0.1316	0.1346	0.1372	0.1395	0.1415
8	0.0941	0.0997	0.1046	0.1089	0.1128	0.1162	0.1192	0.1219
9	0.0696	0.0764	0.0823	0.0876	0.0923	0.0965	0.1002	0.1036
10	0.0459	0.0539	0.0610	0.0672	0.0728	0.0778	0.0822	0.0862
11	0.0228	0.0321	0.0403	0.0476	0.0540	0.0598	0.0650	0.0697
12	0	0.0107	0.0200	0.0284	0.0358	0.0424	0.0483	0.0537
13	0	0	0	0.0094	0.0178	0.0253	0.0320	0.0381
14	0	0	0	0	0	0.0084	0.0159	0.0227
15	0	0	0	0	0	0	0	0.0076

Table II

α-quantiles of the Shapiro – Wilk random variable W

n \ α	0.01	0.05		n \ α	0.01	0.05
7	0.730	0.803		19	0.863	0.901
8	0.749	0.818		20	0.868	0.905
9	0.764	0.829		21	0.873	0.908
10	0.781	0.842		22	0.878	0.911
11	0.792	0.850		23	0.881	0.914
12	0.805	0.859		24	0.884	0.916
13	0.814	0.866		25	0.888	0.918
14	0.825	0.874		26	0.891	0.920
15	0.835	0.881		27	0.894	0.923
16	0.844	0.887		28	0.896	0.924
17	0.851	0.892		29	0.898	0.926
18	0.858	0.897		30	0.900	0.927

Table III

α-quantiles of the D'Agostin random variable D

n \ α	0.005	0.025	0.975	0.995
30	−4.19	−2.91	0.830	0.941
32	−4.16	−2.88	0.862	0.983
34	−4.12	−2.86	0.891	1.02
36	−4.09	−2.85	0.917	1.05
38	−4.06	−2.83	0.941	1.08
40	−4.03	−2.81	0.964	1.11
42	−4.00	−2.80	0.986	1.14
44	−3.98	−2.78	1.01	1.17
46	−3.95	−2.77	1.02	1.19
48	−3.93	−2.75	1.04	1.22
50	−3.91	−2.74	1.06	1.24
60	−3.81	−2.68	1.13	1.34
70	−3.73	−2.64	1.19	1.42
80	−3.67	−2.60	1.24	1.48
90	−3.61	−2.57	1.28	1.54
100	−3.57	−2.54	1.31	1.59

Table IV

Critical values $w_{crit}(n_1, n_2; 0.01)$ for the Wilcoxon test

n_1	n_2																			
	1	2	3	4	5	6	7	8	9	10	11	12	13	14	15	16	17	18	19	20
1	-																			
2	-	-																		
3	-	-	-																	
4	-	-	-	-																
5	-	-	-	-	0															
6	-	-	-	0	1	2														
7	-	-	-	0	1	3	4													
8	-	-	-	1	2	4	6	7												
9	-	-	0	1	3	5	7	9	11											
10	-	-	0	2	4	6	9	11	13	16										
11	-	-	0	2	5	7	10	13	16	18	21									
12	-	-	1	3	6	9	12	15	18	21	24	27								
13	-	-	1	3	7	10	13	17	20	24	27	31	34							
14	-	-	1	4	7	11	15	18	22	26	30	34	38	42						
15	-	-	2	5	8	12	16	20	24	29	33	37	42	46	51					

n_1	n_2																			
	1	2	3	4	5	6	7	8	9	10	11	12	13	14	15	16	17	18	19	20
16	-	-	2	5	9	13	18	22	27	31	36	41	45	50	55	60				
17	-	-	2	6	10	15	19	24	29	34	39	44	49	54	60	65	70			
18	-	-	2	6	11	16	21	26	31	37	42	47	53	58	64	70	75	81		
19	-	0	3	7	12	17	22	28	33	39	45	51	57	63	69	74	81	87	93	
20	-	0	3	8	13	18	24	30	36	42	48	54	60	67	73	79	86	92	99	105
21	-	0	3	8	14	19	25	32	38	44	51	58	64	71	78	84	91	98	105	112
22	-	0	4	9	14	21	27	34	40	47	54	61	68	75	82	89	96	104	111	118
23	-	0	4	9	15	22	29	35	43	50	57	64	72	79	87	94	102	109	117	125
24	-	0	4	10	16	23	30	37	45	52	60	68	75	83	91	99	107	115	123	131
25	-	0	5	10	17	24	32	39	47	55	63	71	79	87	96	104	112	121	129	138
26	-	0	5	11	18	25	33	41	49	58	66	74	83	92	100	109	118	127	135	144
27	-	1	5	12	19	27	35	43	52	60	69	78	87	96	105	114	123	132	142	151
28	-	1	5	12	20	28	36	45	54	63	72	81	91	100	109	119	128	138	148	157
29	-	1	6	13	21	29	38	47	56	66	75	85	94	104	114	124	134	144	154	164
30	-	1	6	13	22	30	40	49	58	68	78	88	98	108	119	129	139	150	160	170

Table V

Critical values $w_{crit}(n_1, n_2; 0.05)$ for the Wilcoxon test

n_1	n_2 = 1	2	3	4	5	6	7	8	9	10	11	12	13	14	15	16	17	18	19	20
1	-																			
2	-	-																		
3	-	-	-																	
4	-	-	-	0																
5	-	-	0	1	2															
6	-	-	1	2	3	5														
7	-	-	1	3	5	6	8													
8	-	0	2	4	6	8	10	13												
9	-	0	2	4	7	10	12	15	17											
10	-	0	3	5	8	11	14	17	20	23										
11	-	0	3	6	9	13	16	19	23	26	30									
12	-	1	4	7	11	14	18	22	26	29	33	37								
13	-	1	4	8	12	16	20	24	28	33	37	41	45							
14	-	1	5	9	13	17	22	26	31	36	40	45	50	55						
15	-	1	5	10	14	19	24	29	34	39	44	49	54	59	64					

n_1	n_2																			
	1	2	3	4	5	6	7	8	9	10	11	12	13	14	15	16	17	18	19	20
16	-	1	6	11	15	21	26	31	37	42	47	53	59	64	70	75				
17	-	2	6	11	17	22	28	34	39	45	51	57	63	69	75	81	87			
18	-	2	7	12	18	24	30	36	42	48	55	61	67	74	80	86	93	99		
19	-	2	7	13	19	25	32	38	45	52	58	65	72	78	85	92	99	106	113	
20	-	2	8	14	20	27	34	41	48	55	62	69	76	83	90	98	105	112	119	127
21	-	3	8	15	22	29	36	43	50	58	65	73	80	88	96	103	111	119	126	134
22	-	3	9	16	23	30	38	45	53	61	69	77	85	93	101	109	117	125	133	141
23	-	3	9	17	24	32	40	48	56	64	73	81	89	98	106	115	123	132	140	149
24	-	3	10	17	25	33	42	50	59	67	76	85	94	102	111	120	129	138	147	156
25	-	3	10	18	27	35	44	53	62	71	80	89	98	107	117	126	135	145	154	163
26	-	4	11	19	28	37	46	55	64	74	83	93	102	112	122	132	141	151	161	171
27	-	4	11	20	29	38	48	57	67	77	87	97	107	117	127	137	147	158	168	178
28	-	4	12	21	30	40	50	60	70	80	90	101	111	122	132	143	154	164	175	186
29	-	4	13	22	32	42	52	62	73	83	94	105	116	127	138	149	160	171	182	193
30	-	5	13	23	33	43	54	65	76	87	98	109	120	131	143	154	166	177	189	200

CONCLUSION

At present, measurement represents a tool for solving a whole series of problems that arise in various areas of human activities, beginning with disciplines concerned with the technical, economic, and social affairs of mankind, and which range through the natural sciences including biology, chemistry, medicine, agriculture, and archeology, and also embrace the social sciences. But most important of all, the sets of measured data are sources of new knowledge about reality, at which point measurement becomes a tool of scientific research.

The information, necessary for solving practical or scientific problems, which is contained in sets of measured data, sometimes obtained in very exacting ways, must be extracted from the raw in an explicit form and with minimum loss.

Let us demonstrate the unique value of statistical procedures, in support of which this publication is dedicated, by citing two typical examples of how statistics are applied to solve concrete problems, often using various different approaches.

One such general problem is the processing of long series of observations of a certain phenomenon carried out in equidistant or nonequidistant steps when the phenomenon in question can be modelled by superposition of a trend component in the form of a polynomial of a certain degree and several periodic functions with various frequencies, amplitudes, and phase displacements. The problem is to determine on the basis of the measured data the degree of the polynomial which models the trend component, the number of superposing periodic functions, and their unknown parameters, together with the corresponding characteristics of accuracy. This analysis then provides a basis for determining the course of the mean value of the phenomenon together with the confidence region and to predict the mean value at points outside the set of observation points, and to characterize these predicted values by their confidence region.

The graph of the measured results gives us preliminary information on the degree of the polynomial modelling the trend (if it does not follow from the theory of the investigated, measured object). The number of superposing components and their approximate frequencies (if they are again not known a priori) can be determined by means of a periodogram. By this means, all the input data necessary for an application of a simple linear Model 2 (an indirect measurement of a vector parameter) are at our disposal. After linearizing the function which models the phenomenon being studied (linearization concerns the periods of the individual periodic components) the relations which determine the LS-estimators of the unknown vector parameters and its covariance matrix can be applied. The statistical significance of the numerically small coefficients of the trend component or the amplitudes of the superposing periodic components can be judged by using the relevant parametric tests. By doing this, the degree of the polynomial and the number of the superposing periodic components of the theoretical model are determined. If some of the tested parameters are statistically insignificant, the unknown parameters in the changed model must be determined once more.

If the unit dispersion of the observation vector is not known a priori, it must be estimated, using the relations from Section 2.7. For determining the predicted values, the relations from Section 2.5 have to be applied.

The model described was elaborated and used in internal medicine. It was aimed at determining the time course of the blood pressure of a patient. In its unchanged form, it was applied to analysing long series of meteorological data and in geophysics for analysing the tidal residues.

It is evident that the results of rigorous processing of measured data may only serve as a background for expert interpretation within the framework of a given branch of science and for further applications of the results obtained in practice. The consequences of errors caused by unreasonable and inadequate procedures applied when processing observed data can be, to put it mildly, unconscionable (think, for example, of its possible impact in medicine).

The second example, in addition, shows the economic contributions that are made by rigorously choosing appropriate procedures for processing measured data.

The problem is to construct a gravimetric map of a given territory and to determine its accuracy. The usual way of solving this problem consists of performing a measurement within a network of k points suitably placed on the territory. The gravitymeters enable us to measure the differences in the acceleration of gravity between two points only, i.e., each of the gravitymeters used works according to its own relative scale, which means that there are two possibilities how to process the measurement results: either by means of the gravity differences or according to the scale of one chosen gravitymeter, into which all the other results have to be transformed. It is evident that the second way means an essential increase in the unknown parameters being determined from the results of measurement. Moreover, the gravitymeter is typical equipment with a moving zero point which causes the drift, which in turn affects the measurement results. Within a one-day measurement this drift can be modelled by a polynomial of a certain degree, whose unknown coefficients have to be determined from the results of measurement replicated in a suitable way. Choosing the procedure of processing measured data by means of the gravity differences, the value of the rate of each gravitymeter used at the moment of the first measurement can be considered to be zero; if the latter method is applied, another unknown parameter for each gravitymeter must be determined, transforming the measurement into a unique relative system of calculating.

The reader has surely noticed the difference between these two approaches, which consists in varying numbers of basic equations having to be solved in order to determine the estimates of the unknown parameters.

The measurements performed by gravitymeters, which are affected by drift, contain full information about drift in the measured results, and thus provide a typical example of measurements that are influenced by a systematic effect. In order to obtain estimates of the useful parameters free of systematic influence (estimates in which the systematic influence is suppressed in an optimum way), it is necessary to apply the procedures elaborated in detail in Section 2.6. The theorems of equivalent estimators offer us several procedures, equivalent from the point of view of the quality of the resulting estimators, which differ essentially in their requirements for computer technology and, hence, on the resources and costs needed to implement them.

It was shown that from the point of view of economical effectiveness, the most efficient procedure consists in separate processing of individual measurements carried out by single gravitymeters, followed by interlinking the stations and the subordinate results into a common network, simultaneously joining this network to the absolute gravity point. The advantage of this procedure in comparison with the procedure of totally independent processing of all the individual results consists in the fact that the former approach to determine the numerical value of the LS-estimator of the useful parameters requires that one solve a system of normal equations whose number is essentially smaller than that which would have to be solved if the latter approach were to be applied. A significant advantage of the latter approach is the fact that the intermediary results obtained in the separate processing represent an important check on whether the prescribed technology of measurement has been respected.

The stage in measurement which we have described and the result of its processing is described by a k-tuple of the optimum estimates of the gravity accelerations at the points in the network characterized by the corresponding covariance matrix.

This first-stage network can be condensed by further points. The gravity accelerations are determined by means of connection measurements, based at first on the acceleration of gravity determined during the first stage, but also using procedures for constructing the multistage network.

REFERENCES

[1] ANDĚL, J.: Mathematical Statistics. SNTL, Prague 1987 (in Czech) - (Matematická statistika).

[2] ANDERSON, T. W.: An Introduction to Multivariate Statistical Analysis. J.Wiley, New York 1958.

[3] BOL'SHEV, L. N., SMIRNOV, N. V.: Tables of Mathematical Statistics. Nauka, Moscow 1965 (in Russian) - (Tablicy matematicheskoj statistiky).

[4] BROWNLEE, K. A.: Statistical Theory and Methodology in Science and Engineering. J. Wiley, New York, London, Sydney 1965.

[5] COCHRAN, W. G.: The distribution of the largest of a set of estimated variances as a fraction of their total. Annals of eugenics. Vol. 11, 1941.

[6] FISZ, M.: Wahrscheinlichkeitsrechnung and mathematische Statistik. Deutscher Verlag der Wissenschaften, Berlin 1962.

[7] GAUSS, C. F.: Theoria motus corporum coelestium. Perthes und Besser, Hamburg 1809.

[8] GRADSHTEYN, I. S., RYZHIK, I. M.: Tables of Integrals, Series and Products. Academic press, New York 1965.

[9] JANKO, J.: Statistical Tables, NČSAV, Prague 1958 (in Czech) - (Statistické tabulky).

[10] KOCH, K. R.: Parammeterschätzung und Hypothesentests in linearen Modellen. Dűmmler, Bonn 1980.

[11] KOLMOGOROV, A. N.: Grundbegriffe der Wahrscheilichkeitsrechnung. Egreb. Math., Bd. 2, Nr. 3, Berlin 1933.

[12] KOLMOGOROV, A. N.: Sulla determinazione empirica di una legge di distribuzione. Giorn. 1st Ital Atturari, 4, 1933, 83–91.

[13] KUBÁČEK, L., WIMMER, G., VOLAUFOVÁ, J.: Estimations in finite-dimensional spaces. Research report, Institute of measurement and measuring equipments SAS 1977 (not published) (in Slovak) - (Odhady v konečnorozmerných priestoroch).

[14] KUBÁČEK, L., KUBÁČKOVÁ, L.: The present approach to the study of the least-squares method. Studia geoph. et geod., 22, 1978, 140–147.

[15] KUBÁČEK, L., PÁZMAN, A.: Statistical Methods in Measurement. Veda, Bratislava 1979 (in Slovak) — Štatistické metódy v meraní).

[16] KUBÁČEK, L.,: About paralel computation algorithms. In: Present problems of gravimetry. Geophysical Institute of CSAV Praha and Geophysics Brno, 1979, 223–232 (in Slovak) - (O paralelizme výpočtových algoritmov. In: Problémy současné gravimetrie).

[17] KUBÁČEK, L.: Foundations of Estimation Theory. Veda Bratislava 1983 (in Slovak) - (Základy teórie odhadu).

[18] KUBÁČEK, L., KUBÁČKOVÁ, L.: The effect of stochastic relations on the statistical properties of an estimator. Contr. Geophys. Inst. Slov Acad. Sci., 17, 1987, 31–42.

[19] KUBÁČEK, L.: Foundations of Estimation Theory. Elsevier, Amsterdam, Oxford, New York, Tokyo 1988.

[20] KUBÁČKOVÁ, L., KUBÁCEK, L., KUKUČA, J.: Probability end Statistics in Geodesy and Geophysics. Veda, Bratislava 1982 (in Slovak) - (Pravdepodobnosť a štatistika v geodézii a geofyzike).

[21] KUBÁČKOVÁ, L., KUBÁCEK, L., KUKUČA, J.: Probability and Statistics in Geodesy and Geophysics. Elsevier, Amsterdam, Oxford, New York, Tokyo 1987.

[22] KUBÁČKOVÁ, L.:The design matrix in statistical collocation problems. Contr. Geophys. Inst. Slov. Acad. Sci., 18, 1988, 20–25.

[23] LEHMANN, E. L.: Testing Statistical Hypotheses. J. Wiley, New York 1959.

[24] LIKEŠ, J., LAGA, J.: Fundamental Statistical Tables. SNTL, Prague 1978 (in Czech) - (Základní statistické tabulky).

[25] PÁZMAN, A.: Foundations of Optimum Experimental Design. Veda, Bratislava 1980 (in Slovak) - (Základy optimalizácie experimentu).

[26] PÁZMAN, A.: Foundations of Optimum Experimental Design. D. Reidel, Dordrecht, Boston, Lancaster, Tokyo 1986.

[27] PÁZMAN, A., MIKULECKÁ, J., RAFFAJ, V., TOKOŠOVÁ, M.: Solved Problems from Designing Experiments. Alfa, Bratislava 1986, (in Slovak) - (Riešené situácie z navrhovania experimentu).

[28] RAO, C. R.: Unified theory of linear estimation. Sankhya, 33, 1971, 371–394.

[29] RAO, C. R., MITRA, S. K.: Generalized Inverse of Matrices and Its Applications. J. Wiley, New York 1971.

[30] RAO, C. R.: Linear Statistical Inference and Its Applications. J. Wiley, New York 1973.

[31] RÉNYI, A.: On the theory of order statistics. Acta Math. Acad. Sci. Hung., 4, 1953, 191–231.

[32] SHAPIRO, S. S., WILK, M. B., CHEN, H. J.(Mrs): A comparative study of various tests for normality. Journ. Amer. Statist. Assoc., 63, 1968, 1343–1372.

[33] SMIRNOV, N. V.: An estimator of the difference between empirical distributions of two independent samples. Byull. MGU, ser. A, vyp. 2, 1939, 3–14, (in Russian) - (Ocenka raskhozdeniya mezhdu empiricheskimi krivymi raspredeleniya v dvukh nezavisimykh vyborkakh).

[34] WANG, S. J.: On the limiting distribution of the ratio of two empirical distributions. Act. Math. Sinica, 5, 1955, 253.

[35] WIMMER, G.: On the equivalence problem in linear regression models. Part I. BLUE of the mean value. Aplikace matematiky, 25, 1980, 417–422.

[36] WIMMER, G.: On the equivalence problem in linear regression models. Part II. Unbiased estimation of the covariance matrix scalar factor. Aplikace matematiky, 25, 1980, 423–431.

SUBJECT INDEX

acceptance region 276
additivity of the probability measure 21
σ-additivity of the probability measure 21
σ-algebra 19
alternative hypothesis 275
analysis of a set of measured data 274
asymmetry 44
asymptotic probability distribution 104
axiomatic definition of probability 17

Bartlett test 338
basic linear models of measurement 129
beta function 83

Cartesian product 92
central chi-square probability distribution 81
centred random variable 44
characteristic function
  of a one-dimensional random variable 73
    its properties 73
  of a random vector 75
    its properties 75
Chebyshev inequality 114
chi-square probability density 81
σ^2 chi-square probability density 83
Cochran test 339
collocation model 146
complementary event 20
composite hypothesis 275
conditional
  mean value 56
  probability 23
    density 41
    distribution 38
confidence
  ellipsoid 220
  interval 213
  level 214
  regions of the first and second order parameters in the case of a normal observation vector 213
continuous random
  variable 29
  vector 35
correctness check in computing
  the covariance matrix of the estimator of the vector function 268
  the dispersion in the estimator of the linear function 268
  the estimates of the first order parameters 246
  the unit dispersion estimate 259
correlation
  coefficient 54, 70
    multiple 70
    partial 72
  matrix
covariance 52

matrix 53
its properties 53
critical region 276

D'Agostoni test 326
degrees of freedom 81
design matrix 115
discrete random
variable 28
vector 34
disjoint random events 20
dispersion 43
of the unit dispersion estimator 193, 198
distribution function 25, 30
its properties 27, 32

elementary event 18, 19, 28
empirical coefficient
of asymmetry 100
as test for verifying normality 323
of multiple correlation 101
of partial correlation 102
empirical
correlation coefficient 100
covariance matrix 101
dispersion 100
distribution function 96
excess 100
as test for verifying normality 323
mean value 99
moments 97
error
of the first kind 276
of the second kind 276
estimated covariance matrix of a normal observation vector in a replicated model 226
its influence on the first order parameter estimators 227-234
estimation
of an unknown covariance matrix in a replicated linear model of measurement 199
theory 111
estimator
consistent 111
jointly efficientt 111
linear 115
of the covariance matrix 205
unbiased 205
of the parameters of a covariance matrix 199
of the unit dispersion
from data affected by a systematic influence 199
in mixed linear models 195
in simple models of measurement 190
invariant 192
unbiased 193
its dispersion 194
in universal model 245
of unkown parameters
in basic mixed linear models 148-162
in basic simple linear models 129-145
in universal model 224
its dispersion 244
unbiased 111
excess 44
existence of an unbiased estimator 238

Fisher test 336
Fisher — Behrens problem 342
Fisher — Snedecor probability distribution 88
fundamental
assertion on the elimination of systematic affects 184
properties of moments 43
theorem of mathematical statistics 96, 97

gamma function 81
Gauss (normal) probability distribution 79
generalized inverse of a matrix 183, 240

Hadamard multiplication of matrices 204
hypothesis testing 274

idempotent matrix 207
impossible event 20
independent random
  events 24
  variables 41
invariant estimator
  of the second order parameters 201
  of the unit dispersion 192
inverse of a block matrix 150

Kolmogorov test of goodness of fit 319
kriging 146
Kronecker multiplication of matrices 194
Kruskal — Wallis test 335

LS-estimator 129
least squares method 128
Lebesgue — Riemann integral 99
linear
  estimator 114
  measurement models 114
  model for processing data from
    direct measurement of a scalar parameter 117
    direct measurement of a vector parameter with a system of conditions 118
    incomplete direct measurement of a vector parameter with a system of conditions 118
    incomplete indirect measurement of a vector parameter with a system of conditions 119
    indirect measurement of a vector parameter 117
    indirect measurement of a vector parameter with a system of conditions 119
  regression in estimation theory 163
  unbiased and jointly efficient estimator 129, 148
Lindeberg — Lévy central limit theorem 77, 80, 105

marginal probability distribution 35
mean value 43
median 44
method of least squares 128
minimum N-seminorm generalized inverse of a matrix 183, 241
mixed linear model 147
mode (modal value) 45
Model 1 (a direct measurement of a scalar parameter) 117
Model 2 (an indirect measurement of a vector parameter) 117
Model 3 (a direct measurement of a vector parameter with a system of conditions) 118
Model 4 (an incomplete indirect measurement of a vector parameter with a system of conditions) 118
Model 5 (an indirect measurement of a vector parameter with a system of conditions) 119
Model 6 (an incomplete indirect measurement of a vector parameter with a system of conditions) 119
modified Pearson theorem 317
moments
  of a random variable 42
    central 43
    initial 43
  of a random vector 49, 50
    central 51
    initial 51
most powerful level-α test 288
multiple correlation coefficient 70

Neyman — Pearson lemma 285
noise 147
  part of the observation vector 176

non-central
  chi-square probability density 84
  Fisher — Snedecor probability density 89
  Student's t probability density 87
  σ^2 chi-square probability density 85
norm of a vector 241
normal (Gauss)
  probability distribution 46
  random variable
    characteristic function 77
    dispersion 48
    mean value 47
    median 48
    quantiles 48
  random vector 79
    characteristic function 79
    mean value 79
    variance matrix 79
normality of the observation vector 206
  confidence region of the estimated
    first order parameter 213–225
    standard deviation 225–226
    unit dispersion 225
  properties of estimators of the first and second order parameter 206–208
normed random variable 44
null hypothesis 274

observation vector 3
operational characteristic 281
optimum
  approximation of a random variable 63, 64
  critical region 287
  elimination of systematic influences 174
    in basic linear models of measurement 176–178
      equivalent formulas 185–189
  experimental design 108
  prediction of useful components in basic linear models of measurement 169–172

Pandora's box matrix 249
parameters of the first order 190
parameters of the second order 190
partial correlation coefficient 72
Pearson
  lemma 84
  test of goodness of fit
    (known parameters) 313
    (unknown parameters) 315
    for verifying the normality 317
  theorem 313
power function of a test 281
probability
  density 29
    its properties 29
  of a transformed random vector 86
  distribution od sample variables 102
  field 19
  function 29
  measure 21
  space 22
properties of estimators of the first and second order parameters in the case of a normal observation vector 206

quantile 44
  function 44

random
  error 109
  event 18
  sample 91, 92
  variable 25
    continuous 29
    discrete 28
  vector 29
    continuous 35
    discrete 34
regression 62
  of the first kind 64
  of the second kind 67

replicated model of measurement 204, 226
Rényi
test of goodness of fit 321
theorem 321
residuals 190
risk 276

$\mathcal{S}$-measurable mapping 25
sample
characteristics 91
random variable 97
space 110
seminorm of a vector 241
set of measured data 94
Shapiro — Wilk test 235
significance level 276
simple hypothesis 275
Smirnov
test 327
theorem 327
standard deviation 44
statistical
hypothesis 274
alternative 275
composite 275
nonparametric 274
null 274
parametric 274
simple 275
model 110
test 275
stochastic model 108, 110
Student's
probability density 85
test 341
sure event 20
systematic influence 174
deterministic 174
stochastic 174

test
for judging the significance of extremaly deviated data 344
for verifying the same probability distribution 327
of goodness of fit 312
of homogeneity 327
of normality 323
on stochastic relations 302–312
statistic 275
verifying the identity
of dispersions 336
of mean values 341
verifying the linear hypothesis on the first order parameters 292
theoretical (deterministic) model 107
theorem
of Kolmogorov 319
of the equivalence of estimators 129
of the probability density
of a transformed random variable 84
of a transformed random vector 86
of the sum of two random variables 82
three-sigma rule 349
trace of a matrix 91, 193

uniformly most powerful test 291
unit dispersion 190
universal model 235
useful
component 147
part of the parameter 175

vector of residuals 190
in basic simple linear models of measurement 191–192
in mixed linear models of measurement 196–197

Wang test 331
Wilcoxon test 332
Wishart probability distribution 90